Conference Proceedings of the Society for Experimental Mechanics Series

Series Editor

Kristin B. Zimmerman, Ph.D.
Society for Experimental Mechanics, Inc.,
Bethel, CT, USA

More information about this series at http://www.springer.com/series/8922

Jamie Kimberley • Leslie Lamberson • Steven Mates
Editors

Dynamic Behavior of Materials, Volume 1

Proceedings of the 2017 Annual Conference on Experimental and Applied Mechanics

Springer

Editors
Jamie Kimberley
New Mexico Institute of Mining and Technology
Socorro, NM, USA

Leslie Lamberson
Drexel University
Philadelphia, PA, USA

Steven Mates
National Institute of Standards and Technology
Gaithersburg, MD, USA

ISSN 2191-5644 ISSN 2191-5652 (electronic)
Conference Proceedings of the Society for Experimental Mechanics Series
ISBN 978-3-319-62955-1 ISBN 978-3-319-62956-8 (eBook)
DOI 10.1007/978-3-319-62956-8

Library of Congress Control Number: 2017956714

Printed on acid-free paper

This Springer imprint is published by Springer Nature
The registered company is Springer International Publishing AG
The registered company address is: Gewerbestrasse 11, 6330 Cham, Switzerland

Preface

Dynamic Behavior of Materials represents one of nine volumes of technical papers presented at the 2017 SEM Annual Conference and Exposition on Experimental and Applied Mechanics organized by the Society for Experimental Mechanics and held in Indianapolis, IN, June 12–15, 2017. The complete proceedings also includes volumes on: *Challenges in Mechanics of Time-Dependent Materials; Advancement of Optical Methods in Experimental Mechanics; Mechanics of Biological Systems, Materials and Other Topics in Experimental and Applied Mechanics; Micro- and Nanomechanics; Mechanics of Composite, Hybrid and Multifunctional Materials; Fracture, Fatigue, Failure and Damage Evolution; Residual Stress, Thermomechanics and Infrared Imaging, Hybrid Techniques and Inverse Problems; and Mechanics of Additive and Advanced Manufacturing.*

Each collection presents early findings from experimental and computational investigations on an important area within experimental mechanics. Dynamic behavior of materials is one of these areas.

The Dynamic Behavior of Materials track was initiated in 2005 and reflects our efforts to bring together researchers interested in the dynamic behavior of materials and structures, and provide a forum to facilitate technical interaction and exchange. In the past years, this track has represented an ever growing area of broad interest to the SEM community, as evidenced by the increased number of papers and attendance.

The contributed papers span numerous technical divisions within SEM, which may be of interest not only to the dynamic behavior of materials community but also to the traditional mechanics of materials community.

The track organizers thank the authors, presenters, organizers, and session chairs for their participation, support, and contribution to this track. We are grateful to the SEM TD chairs for co-sponsoring and/or co-organizing the sessions in this track. They would also like to acknowledge the SEM support staff for their devoted efforts in accommodating the large number of paper submissions this year, making the 2017 Dynamic Behavior of Materials Track successful.

New Mexico Institute of Mining and Technology, Socorro, NM, USA — Jamie Kimberley
Drexel University, Philadelphia, PA, USA — Leslie Lamberson
National Institute of Standards and Technology, Gaithersburg, MD, USA — Steven Mates

Contents

1 **Effects of Adiabatic Heating Estimated from Tensile Tests with Continuous Heating** 1
N. Vazquez Fernandez, M. Isakov, M. Hokka, and V.-T. Kuokkala

2 **Effect of Pre-strain, Processing Conditions, and Impact Velocity on Energy Dissipation in Silicone Foams and Rubber** 9
Brett Sanborn and Bo Song

3 **Strain Rate Sensitivity of Richtmyer-Meshkov Instability Experiments for Metal Strength** 13
Michael B. Prime

4 **Impact Response of Density Graded Cellular Polymers** 17
Behrad Koohbor, Suraj Ravindran, and Addis Kidane

5 **Dynamic Mixed-Mode Crack Initiation and Growth in PMMA and Polycarbonate** 25
Balamurugan M. Sundaram and Hareesh V. Tippur

6 **Dynamic Response of Alumina Ceramics Under Brazilian Disc Test Conditions** 31
Ping Hong Lin, Liren Tsai, and N.S. Liou

7 **Digital Gradient Sensing Method to Visualize and Quantify Crack-Tip Deformations in Soda-Lime Glass Under Static and Dynamic Loading** 39
Balamurugan M. Sundaram and Hareesh V. Tippur

8 **Construction of Phase Diagrams of Mg-Zn with Selected Rare Earth (R.E) Elements** 45
C.G. Fountzoulas

9 **High Strain Rate Transverse Compression Response of Ballistic Single Fibers** 51
Subramani Sockalingam, Daniel T. Casem, Tusit Weerasooriya, and John W. Gillespie Jr.

10 **WIAMan ATD Polymeric Material Characterization for Under-Body Blast Environment Simulation** 57
Mostafiz R. Chowdhury and Dawn M. Crawford

11 **Investigation of Transmission of a Shock Wave Through Thin Films** 63
Hongjoo Jeon and Veronica Eliasson

12 **Experimental Testing and Computational Analysis of Viscoelastic Wave Propagation in Polymeric Split Hopkinson Pressure Bar** 67
M. Bustamante, D.S. Cronin, and D. Singh

13 **Mode I Rigid Double Cantilever Beam Test and Analysis Applied to Structural Adhesives** 73
C.-H. Liao, B. Watson, M.J. Worswick, and D.S. Cronin

14 **Brute Force Ceramic Constitutive Model Parameterization** 83
Brady Aydelotte and Brian E. Schuster

15 **High-Strain Rate Compressive Behavior of a "Natural Soil" Under Uniaxial Strain State** 87
Huiyang Luo, Zhenxing Hu, Tingge Xu, Lei Guo, and Hongbing Lu

16 **Latest Results for Elasto-Plastic Identification at High Rates Using Inertial Impact** ... 93
Sarah Dreuilhe, Frances Davis, Clive Sivour, and Fabrice Pierron

17 **Mesoscale Modeling of Porous Materials Using New Methodology for Fracture and Frictional Contact in the Material Point Method** ... 97
M.A. Homel and E.B. Herbold

18 **Underwater Blast Response of Weathered Carbon Composite Plates** ... 103
Helio Matos, Carlos Javier, James LeBlanc, and Arun Shukla

19 **Characterization of a Visco-Hyperelastic Synthetic Gel for Ballistic Impacts Assessment** ... 109
A. Bracq, G. Haugou, B. Bourel, R. Delille, C. Maréchal, F. Lauro, S. Roth, and O. Mauzac

20 **On the Microstructural Aspects of Shock Induced Failure in Magnesium Alloys** ... 115
C.L. Williams, J. Ligda, L. Farbaniec, and N. Krywopusk

21 **High Speed Imaging Techniques to Study Effects of Pressure Waves from Detonating Explosive Charges on Biological Materials** ... 123
Thuvan Piehler, Nicole Zander, Rohan Banton, Richard Benjamin, Ray Sparks, Kimberly Byrnes, Josh Duckworth, and Ben A. Bahr

22 **Computational Study on the Driver Section Design of an Explosively Driven Conical Shock Tube** ... 127
Joel B. Stewart

23 **Unidirectional Carbon-Epoxy Composite Plates Subjected to Extreme Marine Environment** ... 137
C. Javier, J. LeBlanc, and A. Shukla

24 **Hybrid Computational and Experimental Approach to Identify the Dynamic Initiation Fracture Toughness at High Loading Rate** ... 141
Ali Fahad Fahem and Addis Kidane

25 **High-Rate Mechanical Response of Aluminum Using Miniature Kolsky Bar Techniques** ... 147
Daniel T. Casem, Jonathan P. Ligda, Brian E. Schuster, and Shane Mims

26 **Application of 3D Digital Image Correlation In Ballistic Testing** ... 155
Phillip Jannotti and Brian E. Schuster

27 **Compaction Wave Propagation Characteristics in Polymer Bonded Explosives at Macro-Meso Scale** ... 163
Suraj Ravindran, Peter Malchow, Addis Tessema, and Addis Kidane

28 **Direct Compression Loading Using the Pre-stretched Bar Technique: Application to High Strains Under Moderate Strain Rates** ... 169
G. Haugou, H. Morvan, and N. Leconte

29 **Compaction Wave Characteristics of Polymeric Foams Under Dynamic Loading** ... 175
Suraj Ravindran, Behrad Koohbor, Peter Malchow, and Addis Kidane

30 **Comparison of Numerical Simulations with Experiments of Blast-Induced Pressure Wave Impact on a Surrogate Head Model** ... 181
Rohan Banton, Thuvan Piehler, Nicole Zander, Richard Benjamin, and Josh Duckworth

31 **Pressure Sensing in Clay: A New Metric for Characterizing the Ballistic Backface Deformation Response of Personnel Protection Equipment** ... 189
M.H. Merrill, C.J. Kindle, and J.P. Thomas

32 **Evaluation of Concrete Penetration Resistance Using Small Caliber Bullets** ... 193
Stephan Bless, Michael McAleer, and Rey Guzman

33 **Dynamic Analysis of a Bi-stable Buckled Structure for Vibration Energy Harvester** ... 199
Masoud Derakhshani, Thomas Berfield, and Kevin D. Murphy

34 **Effects of Strain Rate on Mechanical Properties and Fracture Mechanisms in a Dual Phase Steel** ... 209
Sukanya M. Sharma, Kishlay Mishra, Omar Rodriguez, Wilburn R. Whittington, Paul Allison, Shrikant P. Bhat, Arun M. Gokhale, and Naresh N. Thadhani

35 Nonlinear and Inertant Acoustic Metamaterials and Their Device Implications 217
Prateek P. Kulkarni and James M. Manimala

36 Evaluation of Stress Equilibrium in Dynamic Tests on Agglomerated Cork 235
Marco Sasso, Fabrizio Sarasini, Gianluca Chiappini, Edoardo Mancini, and Jacopo Tirillò

37 High Strain Rate Induced Phase Transition of Polymer 243
G. Montella and C.M. Roland

Chapter 1
Effects of Adiabatic Heating Estimated from Tensile Tests with Continuous Heating

N. Vazquez Fernandez, M. Isakov, M. Hokka, and V.-T. Kuokkala

Abstract The mechanical behavior of metastable austenitic stainless steels is strongly influenced by the strain induced phase transformation of austenite into martensite. The phase transformation rate is significantly affected by the strain rate and by the adiabatic heating at higher strain rates. Uncoupling of the effects of strain rate and adiabatic heating can lead to a better understanding of the strain-induced martensitic transformation and allow more accurate material modeling. This paper presents a preliminary analysis of the effects of adiabatic heating during a tensile test. The adiabatic heating as a function of strain was calculated from the stress-strain curves obtained in adiabatic conditions. Then the tensile tests were carried out at a lower strain rate while continuously heating the specimen at the same rate as obtained in the adiabatic conditions. With this method, the thermal conditions of the adiabatic tests were reproduced in the low rate conditions, which would normally be isothermal without the external heating. The martensite fraction was evaluated using the magnetic balance method. In this paper, we present a detailed description of the experimental procedure and discuss the observed changes in the mechanical behavior and microstructure of the studied steel.

Keywords Stainless steel • Adiabatic heating • Martensite transformation • Metastable austenite • Magnetic balance method

1.1 Introduction

The successful product development in the steel industry is strongly linked with a proper understanding of the microplastic deformation mechanisms of the material. Many applications require materials with optimal mechanical properties, such as high strength, good weldability, good corrosion resistance, and good formability. To obtain the desired properties, many modern steels have multiphase microstructures so that a suitable mixture of the mechanical properties of the different phases in the steel fulfills the overall requirements. However, some microstructures can undergo phase transformations due to the elastic and plastic deformations. These microstructures are often very complex and their behavior usually changes significantly during the deformation as the phase fractions evolve.

In recent years, many researchers have focused their attention on the phase transformations in austenite containing steels. Xu et al. [1] and Huang et al. [2] studied the effects of heat treatments and cold rolling reductions on the mechanical behavior of different steels. These studies focused on the effects of the initial microstructures and processing parameters on certain properties. However, their work did not provide many details on the microplastic deformation mechanisms and the microstructural changes during the deformation. Therefore, further investigations are needed to understand the microplasticity and its mechanisms, which depend on many factors including local chemical composition, temperature, amount of deformation, and strain rate. Regarding the strain rate, it has been shown by Hecker et al. [3] that higher strain rates result in less martensite formation. Moreover, lower deformation temperatures enhance the martensitic phase transformations. This can be explained by the thermodynamics of the transformation, as the driving force for the phase transformation is higher at lower temperatures. Significant adiabatic heating occurs usually at strain rates above $\sim$0.5 s^{-1} [4], when the heat induced by the plastic deformation cannot dissipate fast enough from the material into the surroundings and, therefore, the temperature of the material increases. However, many scientists [3–7] have suggested that also the strain rate could have a direct effect on the martensitic transformation rate. Hokka [8] demonstrated that the phase transformation rates in the quasi-static conditions are different from those in the dynamic conditions even at small plastic strains. Furthermore, Isakov [9] discovered that the strain hardening rate after a strain rate jump from quasi-static to dynamic conditions decreases instantaneously following the

N. Vazquez Fernandez (✉) • M. Isakov • M. Hokka • V.-T. Kuokkala
Tampere University of Technology, Laboratory of Materials Science, POB 589, FI-33101, Tampere, Finland
e-mail: naiara.vazquez@tut.fi

J. Kimberley et al. (eds.), *Dynamic Behavior of Materials, Volume 1*, Conference Proceedings of the Society for Experimental Mechanics Series, DOI 10.1007/978-3-319-62956-8_1

jump. These two studies indicate that the phase transformation rate changes already at very low strains where the adiabatic heating has not yet increased the temperature of the material notably.

The temperature change can be estimated from the stress-strain curves using the Taylor-Quinney coefficient. This coefficient describes the fraction of the energy applied to the material that is converted into heat, as well as the amount of energy that is permanently stored in the (defect) structure of the material. However, the formation of martensite is an exothermal process. Thus, the temperature increases also due to the phase transformation itself, not only because of the plastic deformation. Consequently, the temperature increase produced by the phase transformation is complex and difficult to estimate precisely during high strain rate deformation. Moreover, Rusinek et al. [10] investigated the effect of adiabatic heating and concluded that the Taylor-Quiney coefficient varies with plastic deformation between 0.82 and 1.00, implying that the fraction of mechanical energy converted into heat changes during the deformation process of the material.

The aforementioned studies were qualitative and demonstrated that the martensitic phase transformation rate is influenced by both adiabatic heating and the strain rate. However, more work is needed to understand the true effects of strain rate and adiabatic heating on the micromechanisms of plastic deformation and martensitic transformation in austenitic stainless steels. Therefore, this work presents an experimental method for uncoupling the effects of adiabatic heating from the effects of strain rate on the mechanical behavior of the metastable austenitic steel EN 1.4318 using a tensile test with continuous heating. The preliminary results presented in this paper give good examples of how the described experiments can help to deepen the understanding of the effects of deformation induced phase transformations in this material.

1.2 Experimental Procedure

The test material is a metastable austenitic stainless steel EN 1.4318 (AISI 301LN) manufactured and supplied by Outokumpu Stainless. Table 1.1 shows the chemical composition of the material. The samples were laser cut from a cold rolled (2H/C850) sheet with a thickness of 2 mm. Figure 1.1 describes the geometry of the specimens used in the mechanical testing, where RD is the rolling direction.

The material was tested at room temperature at the strain rates of 2×10^{-4} s^{-1}, 10^{-2} s^{-1}, and 1 s^{-1}. The tests at the lowest strain rate are essentially isothermal, the tests at 10^{-2} s^{-1} are quasi-isothermal and the temperature change is modest, while the tests at the highest strain rate are more or less adiabatic. The increase of the specimen temperature during the high strain rate test was estimated using Eq. (1.1):

$$\Delta T\ (\varepsilon) = \frac{1}{\rho \cdot c} \int \beta_T\ \sigma\ d\varepsilon_p \tag{1.1}$$

where $\rho = 7.9$ Kg/m^3 and $c = 500$ J/(Kg K) are the density and the heat capacity of the material, respectively, ΔT is the temperature increase, and β is the Taylor-Quiney coefficient, for which a value of 0.95 was used. Finally, the data obtained from Eq. (1.1) (temperature increase vs. strain) was used to estimate a target heating rate for a low strain rate test with

Table 1.1 Chemical composition of EN 1.4318 in wt-%

C	Cr	Ni	Mn	Si	Cu	N	Co	Mo	P	S	Fe
0.017	17.8	6.44	1.27	0.5	0.18	0.117	0.08	0.06	0.027	0.001	Bal

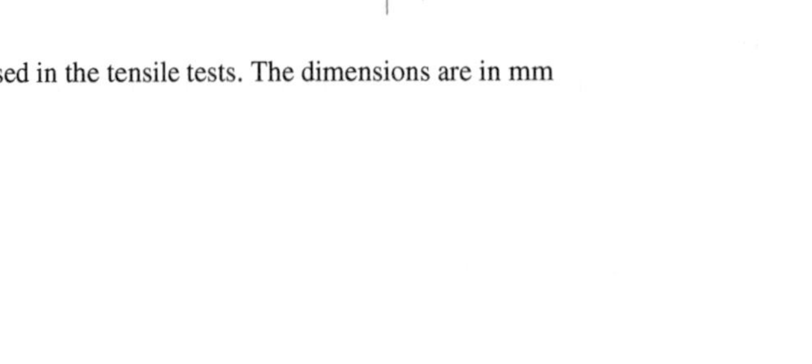

Fig. 1.1 Geometry of the sample used in the tensile tests. The dimensions are in mm

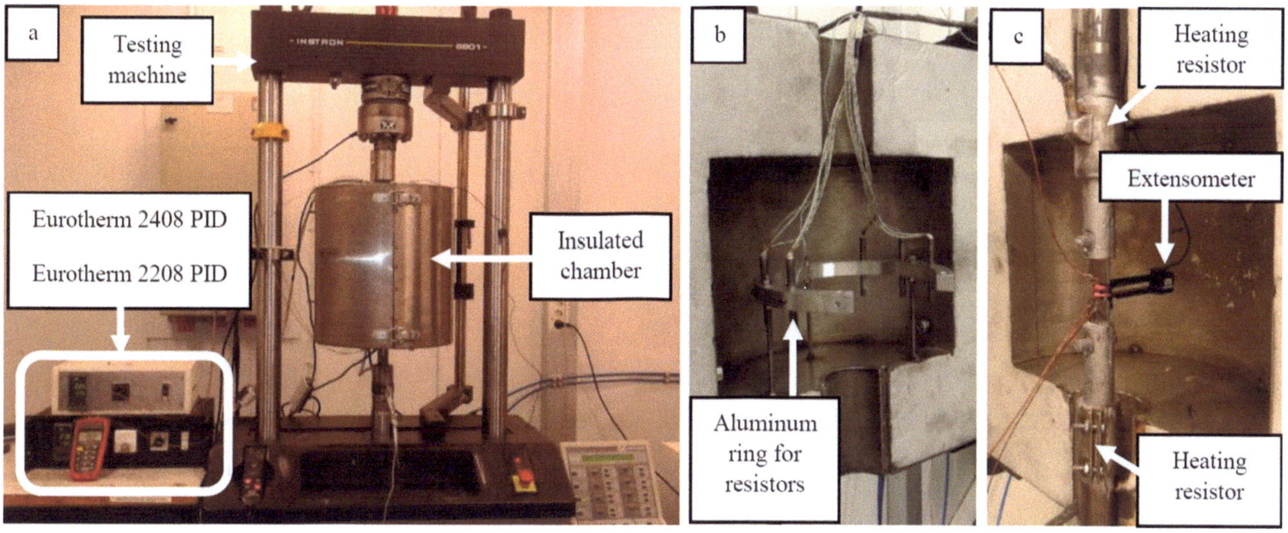

Fig. 1.2 Equipment used in the tensile tests: (**a**) servohydraulic testing machine Instron 8800 with an insulated heating chamber and PID controllers, (**b**) aluminum ring with the heating resistors, (**c**) heating resistors mounted on the pull rods with the sample and the extensometer ready for a test

continuous heating. With this heating rate, the test at the low strain rate had macroscopically similar thermal conditions as the high strain rate (adiabatic) test. The heating was synchronized with the tensile test so that the sample temperature starts to increase immediately after the elastic part. Kendall et al. [11] have previously used a similar approach for testing polymer composites, demonstrating that it is possible to simulate the performance of the material at high strain rate adiabatic conditions by conducting low strain rate experiments with continuous heating and proper control of the temperature.

The low strain rate tests were carried out with and without the continuous heating to various final strains. Figure 1.2 shows the experimental equipment used in this study. The tests were performed using an Instron 8800 servohydraulic machine with a load cell of 100 kN (Fig. 1.2a). The hydraulic grips of the Instron were replaced with pull rods where the sample could be mounted with pins (Fig. 1.2c). An extensometer with a gage length of 6 mm was used to measure the deformation.

The tests with the continuous heating were carried out inside an insulated chamber (Fig. 1.2a). In the chamber, the sample was heated by two sets of heating resistors, which were independently controlled and heated to match the calculated heating rate of the specimen. The first set of resistors (Fig. 1.2b) was fixed to an aluminum ring holding six individual resistors at a fixed distance around the specimen. The temperature of the resistors was adjusted by a Eurotherm 2208 PID controller. The annular resistors (Fig. 1.2c) were placed on the pull-rods and controlled by a Eurotherm 2408 PID controller. The temperature of the sample was measured with 0.1 mm thick K-type thermocouples spot-welded on the gauge section of the specimen. The temperature was monitored at three points: the center and the upper and lower parts of the gage section. The data was recorded with a multilogger at a frequency of 1 Hz. Preliminary measurements were done to evaluate the temperature gradients between the three thermocouple locations. However, no significant differences were observed.

A magnetic balance method was used to evaluate the martensitic phase fractions after the deformation. This method relies on measuring the force required to detach a permanent magnet from the sample [12]. In this study, the permanent magnet was a needle with a tip diameter of 1.5 mm. The martensite fractions were measured using an Instron 5967 machine with a load cell of 5 N. The setup is shown in Fig. 1.3. The sample was placed on a non-magnetic sample holder, and two x-y manipulators (Fig. 1.3a) were used to position the sample so that the magnetic needle was pointing exactly to the area under examination. The load cell measured the force required to detach the magnetic needle from the sample surface, when the actuator frame was moving upwards away from the sample (Fig. 1.3b). The force was measured at three locations along the gage section to check whether the martensite phase fractions were homogeneous. The magnetic scale method requires a calibration to convert the force values into martensite phase fractions. In this work, the calibration was obtained with a Ferritoscope and Satmagan measurements carried out at Outokumpu Inc. The calibration values are presented in Fig. 1.4. A linear calibration between the martensite fraction and the magnetic force can be used up to 50% martensite content, while above this a nonlinear calibration is required.

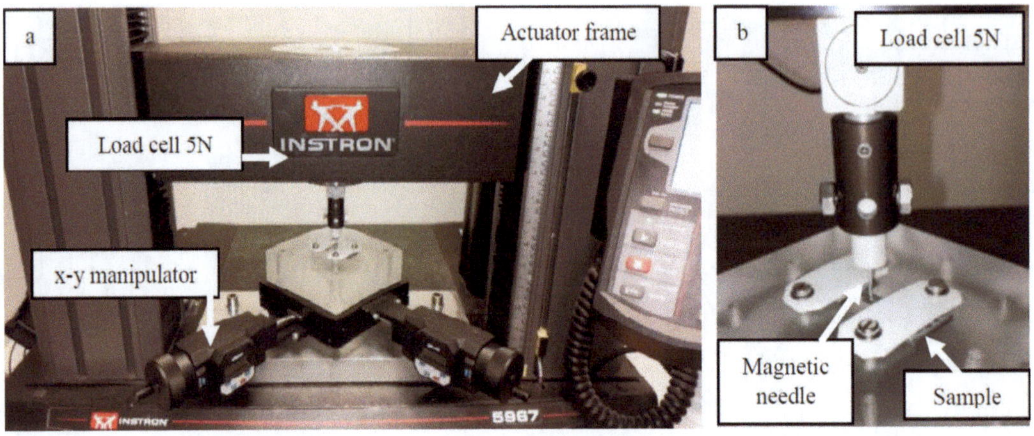

Fig. 1.3 Experimental setup for the magnetic balance method: (**a**) sample holder and the x-y manipulators, and (**b**) a close-up of the magnetic needle

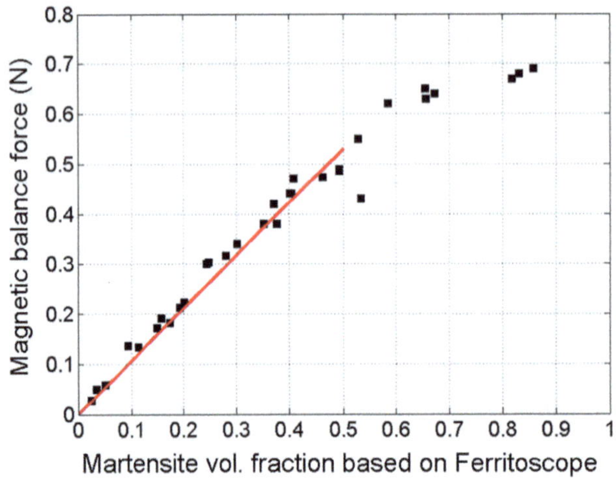

Fig. 1.4 Calibration used for the magnetic balance method

1.3 Results and Discussion

Figure 1.5 shows the true stress - true strain curves for the austenitic stainless steel EN 1.4318 at three different strain rates of $2 \times 10^{-4}\ s^{-1}$, $10^{-2}\ s^{-1}$, and $1\ s^{-1}$, as well as the corresponding strain hardening rates. At the lowest strain rate (isothermal), the material has a yield strength of 610 MPa. As the strain rate is increased to $1 \times 10^{-2}\ s^{-1}$ and $1\ s^{-1}$, the yield strength reaches 640 MPa and 720 MPa, respectively. In contrast, the ultimate tensile strength decreases as the strain rate increases (Fig. 1.5a). A very interesting feature in these plots is the shape of the curve, which differs from a standard ductile material curve at the beginning of the plastic regime. The rapid hardening that starts after about 8% of plastic strain is due to the phase transformation occurring in the material. The phase transformation induces a change in the mechanical properties of the material, especially in the strain hardening rate and consequently the strain rate sensitivity at higher plastic strains. According to previous studies [7], the phase transformation rate from austenite to martensite decreases with increasing strain rate due to the adiabatic heating and, therefore, the strength of the material decreases as well. This behavior can be more clearly analyzed from the strain hardening rates (Fig. 1.5b), where the strain hardening rate decreases with increasing strain rate because the adiabatic heating slows down the phase transformation rate.

Figure 1.6 shows the true stress – true strain curve for the sample tested at the strain rate of $2 \times 10^{-4}\ s^{-1}$ with the external heating, the measured temperature of the sample, and the target temperature, which was calculated using Eq. (1.1) based on the stress-strain data obtained at the strain rate of $1\ s^{-1}$ (presented in Fig. 1.5). As can be seen in Fig. 1.6, the material

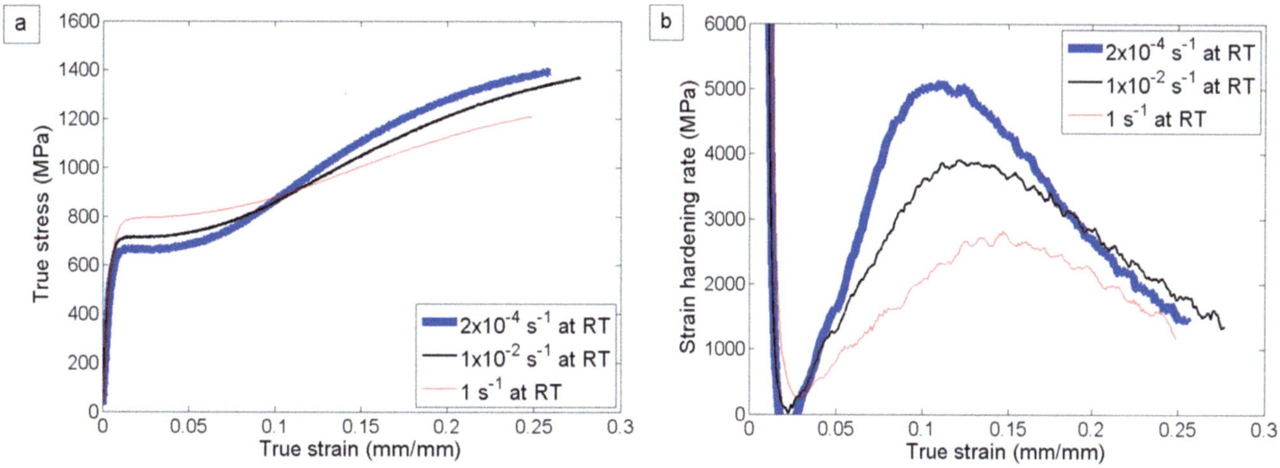

Fig. 1.5 Results of the tensile tests at three different strain rates at room temperature: (**a**) stress-strain curves, and (**b**) strain hardening rates

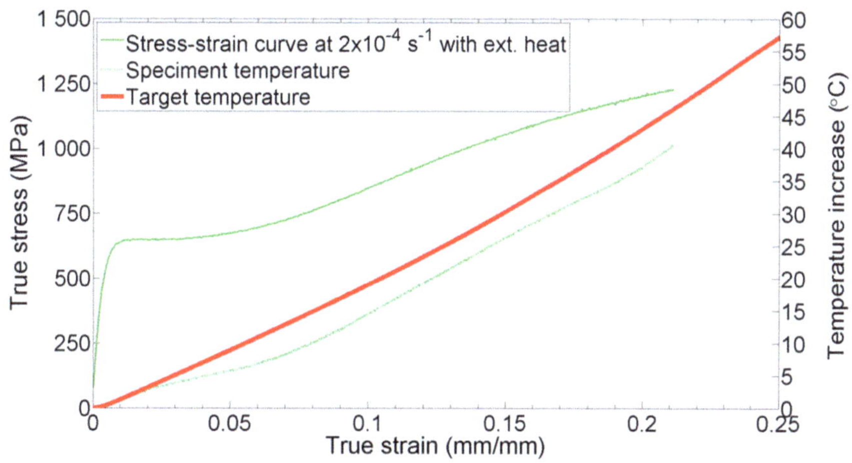

Fig. 1.6 Stress-strain curve obtained at the strain rate of 2×10^{-4} s^{-1} with external heating, and the target and measured specimen temperatures

temperature resulting from the external heating is very close to the target temperature, having a maximum difference of approximately five degrees. For all strains, the actual measured temperature of the specimen is either the same or lower than the target temperature.

Figure 1.7a shows the stress – strain curve with the continuous heating at the strain rate of 2×10^{-4} s^{-1} and the stress - strain curves at room temperature at the strain rates of 2×10^{-4} s^{-1}, 1×10^{-2} s^{-1}, and 1 s^{-1}. The corresponding strain hardening rate curves are presented in Fig. 1.7b. It can be clearly seen that up to 10% of true strain the sample with the continuous heating has a similar stress-strain response as the sample tested at the same strain rate at room temperature, which is expected since the difference in the material temperature between the specimens is only modest at this stage. It is also noteworthy that after 10% strain the strain hardening rate of the externally heated material is clearly higher than that of the material deforming in the adiabatic conditions (strain rate 1 s^{-1}). In fact, the external heating in the low rate conditions seems to have only relatively small influence on the strain hardening rate. The overall strain-hardening behavior of the sample with the continuous heating is relatively close to that observed at the strain rate of 10^{-1} s^{-1}, but notably different from that of the strain rate 1 s^{-1}, which was the basis of the heating calculations. The thermal softening effect through increased thermal activation of dislocation motion cannot explain all the observed differences in the mechanical response. The temperature of the specimen at 10% of plastic strain has increased only by about 15–20 °C, which is very little compared to the melting temperature of the steel. Furthermore, at higher strain rates the dynamic recovery of dislocations is typically reduced leading to a higher hardening rate. This is also not evident in the stress – strain curves. Therefore, the most likely explanation to the observed changes are the changes in the phase transformation kinetics during the deformation. In fact, the M_d temperature of the material is only slightly above room temperature, and therefore even a modest increase in the specimen temperature can have a strong effect on the phase transformation rate. However, as shown in Fig. 1.7b,

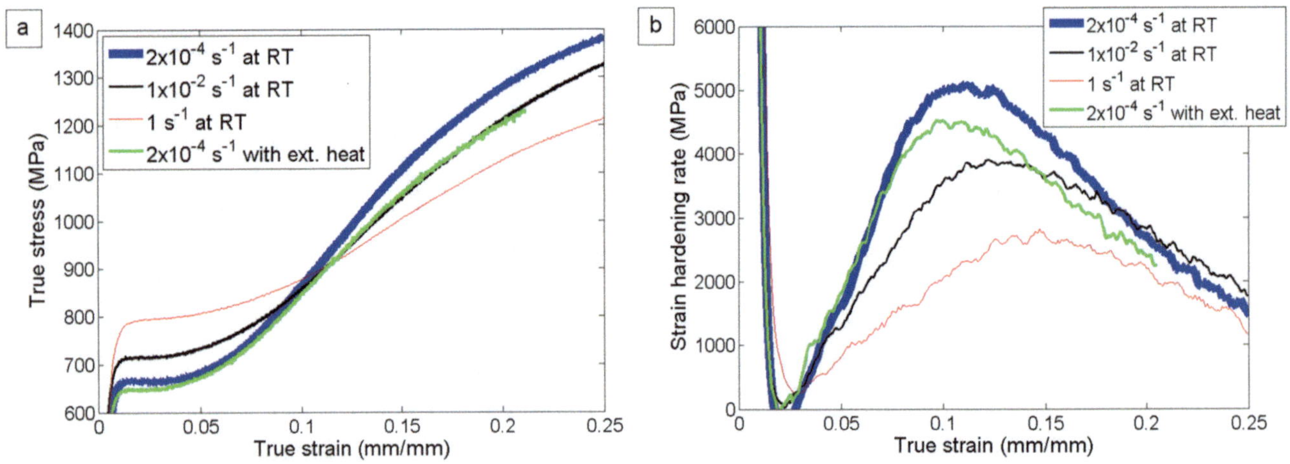

Fig. 1.7 Results of the tensile tests with and without continuous heating (**a**) the true stress – true strain curves, and (**b**) the strain hardening rates

Table 1.2 Results of the magnetic balance measurements

Strain rate	True strain	Thermal conditions	Magnetic force (N)	Martensite fraction
–	(as-received 0%)	–	0.0484	< 5%
2×10^{-4} (s^{-1})	10%	Adiabatic heating simulated from the test at 1 s^{-1}	0.455	~45%
2×10^{-4} (s^{-1})	10%	Room temperature	0.514	~51%
2×10^{-4} (s^{-1})	16%	Adiabatic heating simulated from the test at 1 s^{-1}	0.585	~60%
2×10^{-4} (s^{-1})	21%	Adiabatic heating simulated from the test at 1 s^{-1}	0.604	~65%
2×10^{-4} (s^{-1})	26%	Room temperature	0.723	~100%

the considerable external heating subjected to the material does not lead to similar strain-hardening behavior as observed in the adiabatic deformation conditions. This indicates that there might be also other strain-rate dependent effects present in addition to adiabatic heating. A cautioning remark should, however, be made; the tests presented here are based on macroscopic material heating, i.e., the possible effects of adiabatic material heating in the grain level cannot be reproduced with the current tests.

Table 1.2 shows the results of the magnetic balance measurements. The as-received material contains some martensite that had formed already during cold rolling, and therefore the magnetic force obtained for the as-received sample is not zero. The phase fractions measured from the sample deformed at the strain rate of 2×10^{-4} s^{-1} with the continuous heating were compared with the phase fractions measured from the sample deformed at the same strain rate at room temperature. The sample tested with the continuous heating was deformed up to 21% of true strain. Unfortunately, reaching exactly the same amount of true deformation in a test was challenging and the samples had somewhat different amounts of deformation. However, the measured phase fractions are considerably different, which clearly indicates that the phase transformation rate in the room temperature test is much higher. Moreover, two tests with external heating were carried out at the strain rate of 2×10^{-4} s^{-1} to the final deformations of 10% and 16%, corresponding to the maximum strain hardening rate and the downward slope following the peak value. At 10% of deformation, the martensite phase fractions are 45% and 51% for the samples with and without the heating, respectively. This is in agreement with the stress-strain behavior shown in Fig. 1.7. The phase fractions evolve during the next 6% of deformation, and the martensite fraction increases to 60% in the sample deformed with the continuous heating. However, further deformation from 16% to 21% of strain increases the martensite fraction only to about 65%. The phase transformation rate is clearly decreasing, which is also evident in the hardening rate that decreases rapidly after 10% of deformation. However, the hardening rate has a fairly constant downward slope between the maximum and the end of the test at 21%. Based on the magnetic balance measurements, the phase transformation rate decreases similarly to the strain hardening rate. The hardening rate is influenced by other factors as well, such as increased thermal activation and dynamic recovery, both of which decrease the hardening rate, and therefore the decrease in the hardening rate does not correspond exactly to the phase transformation rate. In fact, if one estimates the slope of the hardening rate vs. strain plot after the maximum value, the slopes are quite similar for all stress strain curves, whereas the upward slopes are significantly different for all cases. Therefore, it seems that the microstructural changes that occur before approximately 15% have the strongest effect on the macroscopic behavior of the material.

1.4 Conclusions

This work presents an experimental method for evaluating the effects of adiabatic heating on the mechanical behavior and microstructural changes of a metastable austenitic stainless steel. The mechanical behavior of the studied material is strongly affected by the strain rates that involve adiabatic heating, which affects the phase transformation rate. The method used in this work allows experimental simulation of the adiabatic heating produced in a high strain rate test by applying the same heating rate in a low strain rate test. The results show that the low strain rate hardening behavior of the studied steel in a tensile test with external heating is between the observed isothermal and adiabatic mechanical behavior of the material. The characterization of the phase fractions with the magnetic scale method shows that the external heating in a low strain rate test somewhat decreases the martensitic phase transformation rate in the microstructure. It is noteworthy that based on the preliminary results presented in this paper, adiabatic heating alone cannot fully explain the effect of strain rate on the hardening behavior of the test material. There exists, therefore, clearly a need for future research on the topic.

References

1. Xu, D., Li, J., Meng, Q., Liu, Y., Li, P.: Effect of heating rate on microstructure and mechanical properties of TRIP-aided multiphase steel. J. Alloys Compd. **614**, 94–101 (2014)
2. Huang, J.-x., Ye, X.-n., Xu, Z.: Effect of cold rolling on microstructure and mechanical properties of AISI 301LN metastable austenitic stainless steels. J. Iron Steel Res. Int. **19**(10), 59–63 (2012)
3. Hecker, S., Stout, M., Staudhammer, K., Smith, J.: Effects of strain state and strain rate on deformation-induced transformation in 304 stainless steel: part I. Magnetic measurements and mechanical behavior. Metall. Trans. A. **13**(4), 619–626 (1982)
4. Talyan, V., Wagoner, R., Lee, J.: Formability of stainless steel. Metall. Mater. Trans. A. **29**(8), 2161–2172 (1998)
5. Talonen, J.: Doctoral Thesis, Helsinki University of Technology, Espoo, Finland (2007)
6. Lichtenfeld, J.A., Van Tyne, C.J., Mataya, M.C.: Effect of strain rate on stress-strain behavior of alloy 309 and 304L austenitic stainless steel. Metall. Mater. Trans. A. **37**(1), 147–161 (2006)
7. Andrade-Campos, A., Teixeira-Dias, F., Krupp, U., Barlat, F., Rauch, E., Grácio, J.: Effect of strain rate, adiabatic heating and phase transformation phenomena on the mechanical behaviour of stainless steel. Strain. **46**(3), 283–297 (2010)
8. Hokka, M.: Doctoral Thesis, Tampere University of Technology (2008)
9. Isakov, M., Hiermaier, S., Kuokkala, V.-T.: Effect of strain rate on the martensitic transformation during plastic deformation of an austenitic stainless steel. Metall. Mater. Trans. A. **46**(6), 2352–2355 (2015)
10. Rusinek, A., Klepaczko, J.: Experiments on heat generated during plastic deformation and stored energy for TRIP steels. Mater. Des. **30**(1), 35–48 (2009)
11. Kendall, M., Siviour, C.: Experimentally simulating adiabatic conditions: approximating high rate polymer behavior using low rate experiments with temperature profiles. Polymer. **54**(18), 5058–5063 (2013)
12. Livitsanos, C., Thomson, P.: Rapid determination of the deformation-induced martensite in metastable stainless steels. J. Mater. Sci. **12**(11), 2209–2213 (1977)

N. Vazquez Fernandez holds Bachelor degree in Chemistry at University of Barcelona – Spain; Bachelor degree in Materials Engineering at Polytechnical University of Catalonia – Spain; MSc- 2015 in Materials Engineering at Technical University of Madrid – Spain; Now PhD student at Tampere University of Technology – Finland.

Chapter 2
Effect of Pre-strain, Processing Conditions, and Impact Velocity on Energy Dissipation in Silicone Foams and Rubber

Brett Sanborn and Bo Song

Abstract Silicone foams and rubber are used in a variety of applications to protect internal components from external shock impact. Understanding how these materials mitigate impact energy is a crucial step in designing more effective shock isolation systems for components. In this study, a Kolsky bar with pre-compression and passive radial confinement capabilities was used to investigate the response of silicone foams and rubber subjected to impact loading at different speeds. Using the preload capability, silicone foam samples were subjected to increasing levels of pre-strain. Frequency-based analyses were carried out on results from silicone foams and rubber to study the effect of both pre-strain and material processing conditions on the mechanism of energy dissipation in the frequency domain. Additionally, effects of impact speed on energy dissipation through silicone foams and rubber were investigated.

Keywords Silicone foam • Silicone rubber • Kolsky bar • Material processing • Frequency response

2.1 Introduction

Mitigation of high frequency components that are potentially damaging to internal electronics is a critical design concept in applications ranging from portable electronics to defense applications [1, 2]. Selection of shock mitigation materials is typically dictated by the ability of different materials to dissipate energy over a range of frequencies for a specific application. Improved design in vibration control and shock-resistant packaging is achieved through better understanding of how the constituent materials behave in those environments. Hyperelastic polymer foams and rubbers are used in applications for vibration reduction and shock mitigation. Many factors such as stress-state, strain rate, and temperature may affect the shock mitigation behavior of foams and rubbers in these applications. Silicone foams may also be subjected to pre-strain during assembly or manufacture of components. Understanding the response of silicone foams in the frequency domain is of paramount importance for applications where shock or vibration energies at specific frequencies may damage internal components. In this study, the frequency-domain energy dissipation behavior of silicone foams and rubbers was investigated. The silicone foam was subjected to varying levels of pre-strain under passive confinement to determine the effect of stress-state on energy dissipation. Energy dissipation behavior of silicone foam and rubber is compared to try to discover the mechanism by which silicone foam dissipates energy in the frequency domain.

2.2 Materials and Specimens

Silicone foam and rubber were evaluated in this study for energy dissipation capability. Silicone foam and rubber were processed using the same chemistry; however, small pellets were introduced into the silicone foam mixture which produced a porosity of approximately 50%, whereas the silicone rubber had essentially 0% porosity. This is clearly shown when the densities of the foam and rubber were measured. The silicone foam had a density of 608 ± 21.85 kg/m^3 while the silicone rubber had a density of 1153 kg/m^3. To investigate the effect of stress-state, the silicone foam was pre-compressed to strains of 0, 13, 23.3 and 33.5% prior to dynamic loading while the silicone rubber was not held under pre-strain. Both specimen types were confined in a steel tube in the experimental setup to provide passive lateral confinement. The silicone foam and rubber were subjected to multiple impact speeds to study the energy dissipation behavior in the frequency domain.

B. Sanborn (✉) • B. Song
Sandia National Laboratories, Albuquerque, NM, USA
e-mail: bsanbor@sandia.gov

J. Kimberley et al. (eds.), *Dynamic Behavior of Materials, Volume 1*, Conference Proceedings of the Society for Experimental Mechanics Series, DOI 10.1007/978-3-319-62956-8_2

2.3 Experiments and Results

A Kolsky bar with pre-load capability was used to conduct experiments on silicone foam and rubber, as shown in Fig. 2.1. A tapered tungsten striker was used to extend the frequency content of the incident pulse. The specimens were confined in a steel tube surrounding the gage section of the Kolsky bar. Such passive confinement replicates the stress-state of shock mitigation materials in real applications. Prior to dynamic loading, the silicone foam was confined in the steel tube and static pre-strain was increased to different levels measured using a linear laser and photodetector system. After the pre-strain was applied, the striker bar was fired to initiate dynamic loading. The incident, reflected, and transmitted signals were recorded and subsequently converted to the frequency domain using FFT.

The energy dissipation behavior of silicone foam as a function of pre-strain for a single impact speed is shown in Fig. 2.2a. As Fig. 2.2a shows, the energy dissipation capability of the silicone foam decreases with increasing pre-strain. For 0 and 13% pre-strain, the silicone foam dissipated nearly 100% of the impact energy over the entire frequency range. However, at the highest level (33.5%) of pre-strain, the silicone foam only dissipated approximately 80% of the impact energy. Silicone rubber, as shown in Fig. 2.2b, had a lower overall energy dissipation ratio (25%) compared to silicone foam. This suggests that the silicone foam dissipates more energy through cellular structures than the matrix material. Furthermore, no trend can be seen between impact speed and energy dissipation ratio for the silicone rubber.

In an effort to better understand the energy dissipation with respect to densification of the foam and rubber, the overall energy dissipation ratio for all foam and rubber experiments is plotted against the final density of the specimen at the conclusion of dynamic loading is shown in Fig. 2.3. The final density can be calculated by using the total engineering strain that is reached as a result of quasi-static pre-strain and dynamic loading. The outside diameter is constant due to the confining

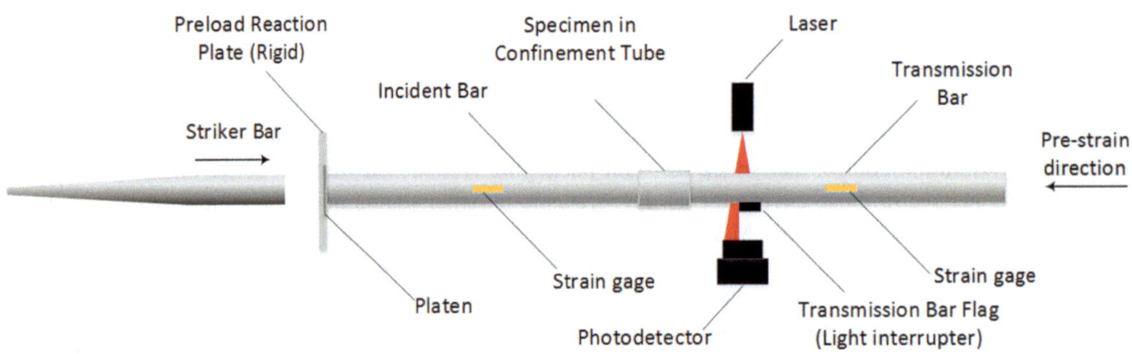

Fig. 2.1 Kolsky compression bar for passive confinement frequency-based experiments

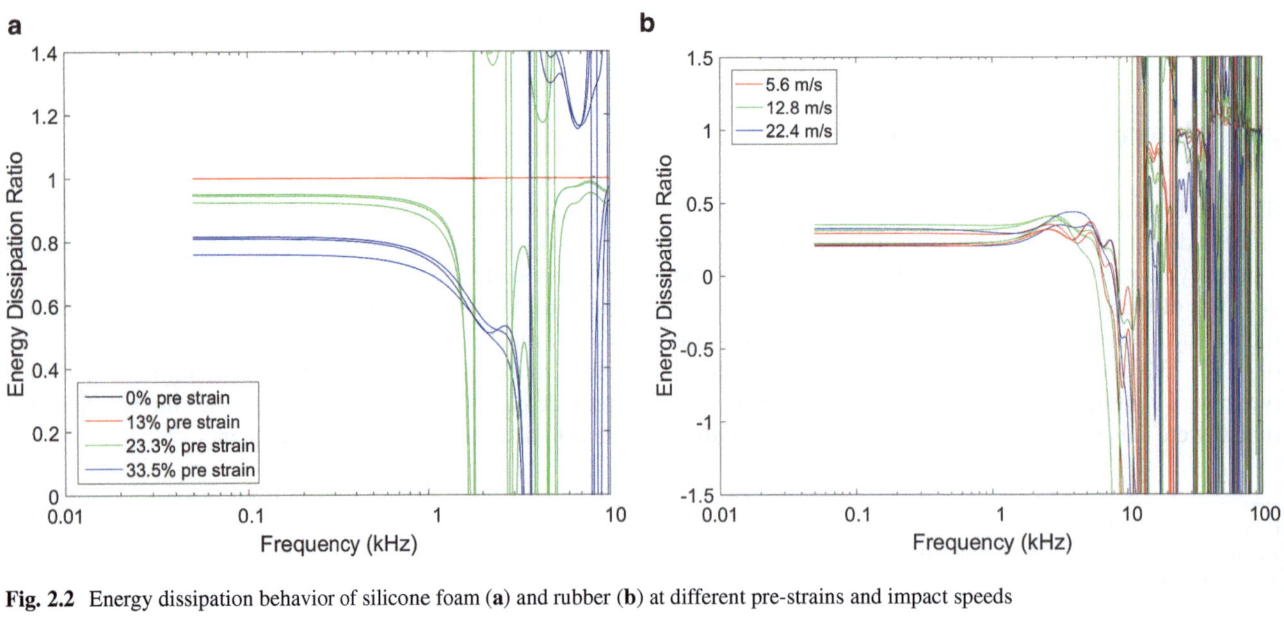

Fig. 2.2 Energy dissipation behavior of silicone foam (**a**) and rubber (**b**) at different pre-strains and impact speeds

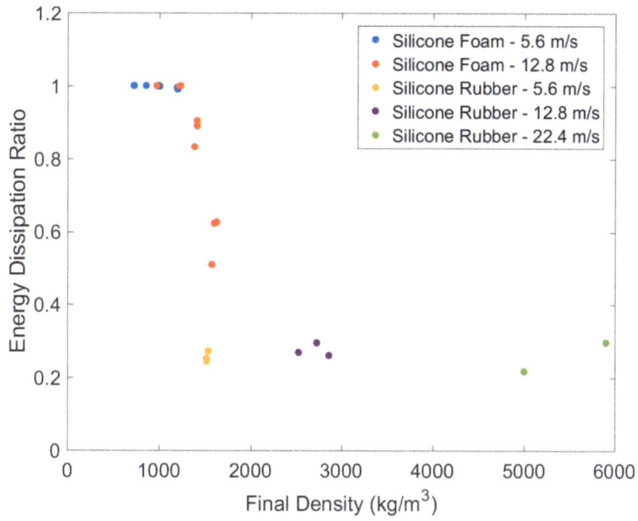

Fig. 2.3 Energy dissipation ratio as a function of final density for silicone foams and rubbers

tube; with the initial mass known, the final volume and hence density are calculated. Based on the behavior shown in Fig. 2.3, there appears to be a critical density between 1200 and 1400 kg/m^3 at which the energy dissipation ratio begins to drop off to a saturated value of 0.25. Hence, the state of densification of the material may be the driving factor determining the energy dissipation capability of silicone foams and rubbers.

2.4 Summary

In this study, a Kolsky compression bar with a pre-load capability was used to subject passively silicone foams and rubbers to different levels of pre-strain and dynamic loading to simulate the stress-state seen inside components. The silicone foams and rubbers were evaluated for their energy dissipation capability in the frequency domain. The energy dissipation ratio of silicone foam decreased when the foam was subjected to increasing pre-strain at the same impact velocity. At low levels of pre-strain, the foam dissipated nearly 100% of the impact energy. The energy dissipation ratio of silicone foam was found to be independent of impact speed. The overall state of densification of the material may be the driving factor for energy dissipation. A critical density between 1200 and 1400 kg/m^3 was identified at which the energy dissipation ratio dropped to a saturated value of 0.25.

Acknowledgment Sandia National Laboratories is a multi-mission laboratory managed and operated by Sandia Corporation, a wholly owned subsidiary of Lockheed Martin Corporation, for the U.S. Department of Energy's National Nuclear Security Administration under contract DE-AC04-94AL85000.

References

1. Zhou, C.Y., Yu, T.X.: Analytical models for shock isolation of typical components in portable electronics. Int. J. Impact Eng. **36**, 1377–1384 (2009)
2. Benning, C.J.: Effect of cell structures in polyethylene foam on shock mitigation. J. Cell. Plast. **5**(1), 40–45 (1969)

Brett Sanborn is a senior member of technical staff at Sandia National Laboratories and previously worked at the US Army Research Laboratory. His research interests include dynamic behavior of materials and experimental impact mechanics.

Chapter 3
Strain Rate Sensitivity of Richtmyer-Meshkov Instability Experiments for Metal Strength

Michael B. Prime

Abstract Recently, Richtmyer-Meshkov instabilities (RMI) have been used for studying strength at strain rates up to at least 10^7/s. RMI experiments involve shocking a metal interface with sinusoidal perturbations that invert and grow subsequent to shock and may arrest because of strength effects. To use RMI strength estimates as calibration data for rate-dependent constitutive models, one must understand the strain rates that apply to the strength estimate, but the strain rate varies spatially and temporally during the instability. In this study, we use a series of numerical simulations to establish the strain rate(s) to which the instability is most sensitive.

Keywords Richtmyer-Meshkov Instability • High-rate strength

3.1 Introduction

Recently, researchers have shown that Richtmyer-Meshkov Instabilities (RMI) are sensitive to strength at strain rates up to at least 10^7/s [1–12]. Figure 3.1 illustrates an RMI experiment in the configuration fielded for recent experiments [3, 4]. As shown in Fig. 3.1, in an RMI experiment the initial perturbations invert after shock, and the subsequent peaks are called spikes and the valleys are called bubbles. Recent work has shown that the peak spike velocity, as shown in Fig. 3.2, is quite sensitive to the deviatoric strength of the sample and offers advantages over using total spike growth as the measure of strength [13]. As with the previous work using total spike growth [3, 4], the peak spike velocity is used to estimate an average strength that best matches the data.

Since strength is generally a function of strain, strain, temperature and pressure, e.g., [14], and those properties vary spatially and temporally during RMI growth and arrest, how useful is an average strength? We begin to explore this question by examining the sensitivity to the strain rate-dependent portion of a constitutive model.

3.2 Modeling Approach

An RMI experiment was modelled using the Abaqus Explicit commercial finite element code [15] based on extensive previous finite element modeling of RMI experiments [13]. A 2-D, plane strain model was used with the model domain including two full wavelengths of the perturbation and 3 mm thickness of copper. Figure 3.3 is zoomed in on the right end of the domain to show the 10 μm zoning and the perturbations with an amplitude $\eta_0 k = 0.5$. The top and bottom surfaces are constrained in the vertical direction to simulate periodic behavior. For simplicity in the sensitivity study, a shock pressure of 30 GPa was applied at $t = 0$ to the left end of the domain. To minimize noise in the predicted velocities, the default linear and quadratic coefficients for bulk viscosity were increased from 0.06 and 1.2 to 0.20 and 1.7. To sufficiently resolve the velocity peak, the SCALE FACTOR parameter was used to reduce Abaqus' default time increment by a factor of five.

M.B. Prime (✉)
MS A142, Los Alamos National Laboratory, Los Alamos, NM, 87545, USA
e-mail: prime@lanl.gov

J. Kimberley et al. (eds.), *Dynamic Behavior of Materials, Volume 1*, Conference Proceedings of the Society for Experimental Mechanics Series, DOI 10.1007/978-3-319-62956-8_3

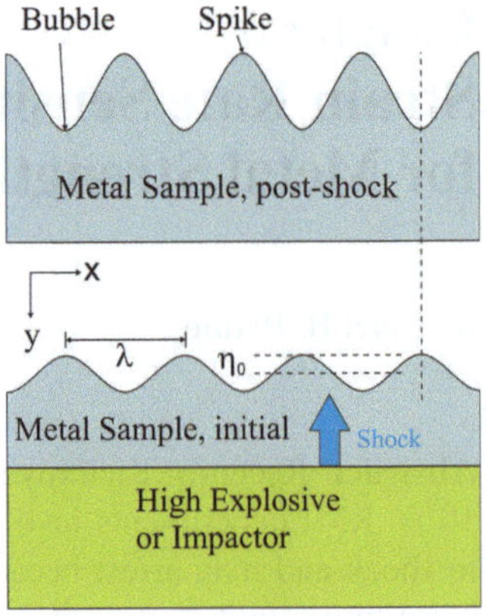

Fig. 3.1 A Richtmyer-Meshkov instability experiment. The perturbed surface of the sample is accelerated by a shock. At later time, the perturbations have inverted

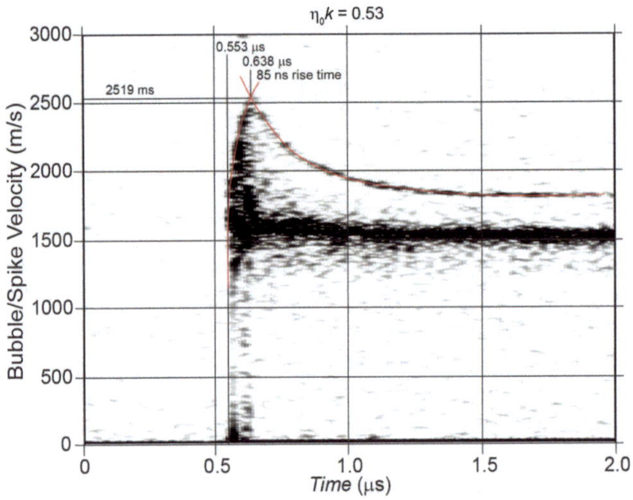

Fig. 3.2 An experimental PDV velocity spectrogram for $\eta_0 k = 0.52$ in copper shows a distinct velocity trace for the spike growth peaking at 2519 m/s

Fig. 3.3 The Abaqus model simulated two full periods of the sine wave perturbation using Cartesian 2D plane strain elements. The figure here shows only perturbation region

The constitutive behavior was modeled using the Zerilli-Armstrong model with parameters for copper [16]. To allow for model variations for the sensitivity study, the model was tabularized for use in Abaqus. For strain rate, Abaqus suggest tabular data be given at regular logarithmic intervals, so we used strain rates of 10^n/s with $n = 4.0, 4.25, 4.5, \ldots 7.75, 8.0$. The "parametric studies" capability in Abaqus was used to sequentially scale the magnitude of the stress-strain curve for a given strain rate to 0.95 of its original value and repeat the simulation. The peak speak velocity for each simulation was extracted and compare to the baseline simulation to estimate the sensitivity of the experiment to a given strain rate.

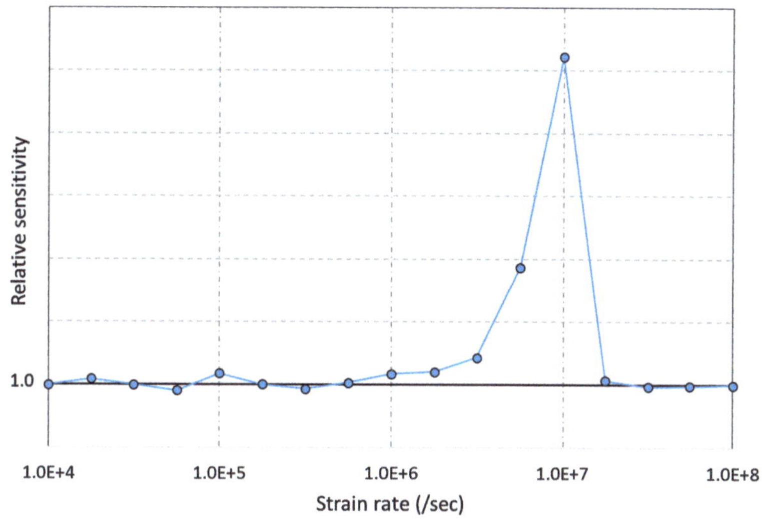

Fig. 3.4 For the simulation with $\eta_0 k = 0.5$ and a 30 GPa shock, the peak spike velocity is sensitive to the strength model over a quite narrow band of strain rates, peaking at 10^7/s

3.3 Results and Discussion

Figure 3.4 shows the encouraging results of the sensitivity study. This particular case is sensitive to a tight range of strain rates, peaking at 10^7/s. Based on the standard logarithmic dependence of strength on strain rate [14], the result indicates that the average strength could be characterized as effectively probing this single strain rate $\pm$ a small band and used as a single point to calibrate a constitutive model.

Further work is needed. Similar sensitivity studies should be performed for strain, temperature, and pressure. As appropriate based on those results, the average strength could be assigned to a point in (strain, strain rate, temperature, pressure) space and used for calibrating a constitutive model that is dependent on all of the variables. Using a number of RMI experiments with different shock loading, perturbation sizes, and initial temperatures, a reasonable span of conditions could be evaluated and extend the usefulness for constitutive model calibration. The ability of the calibrated model to reproduce the full range of experimental data would be the first check on the suitability of such a calibration scheme.

Acknowledgements Los Alamos National Laboratory, an affirmative action/equal opportunity employer, is operated by the Los Alamos National Security, LLC for the National Nuclear Security Administration of the U.S. Department of Energy under contract DE-AC52-06NA25396. By approving this article, the publisher recognizes that the U.S. Government retains nonexclusive, royalty-free license to publish or reproduce the published form of this contribution, or to allow others to do so, for U.S. Government purposes. Los Alamos National Laboratory requests that the publisher identify this article as work performed under the auspices of the U.S. Department of Energy. Los Alamos National Laboratory strongly supports academic freedom and a researcher's right to publish; as an institution, however, the Laboratory does not endorse the viewpoint of a publication or guarantee its technical correctness.

References

1. Piriz, A.R., Cela, J.J.L., Tahir, N.A., Hoffmann, D.H.H.: Richtmyer-Meshkov instability in elastic-plastic media. Phys. Rev. E. **78**(5), 056401 (2008)
2. Piriz, A.R., Cela, J.J.L., Tahir, N.A.: Richtmyer–Meshkov instability as a tool for evaluating material strength under extreme conditions. Nucl. Inst. Methods A. **606**(1), 139–141 (2009)
3. Dimonte, G., Terrones, G., Cherne, F.J., Germann, T.C., Dupont, V., Kadau, K., Buttler, W.T., Oro, D.M., Morris, C., Preston, D.L.: Use of the Richtmyer-Meshkov instability to infer yield stress at high-energy densities. Phys. Rev. Lett. **107**(26), 264502 (2011)
4. Buttler, W.T., Oró, D.M., Preston, D.L., Mikaelian, K.O., Cherne, F.J., Hixson, R.S., Mariam, F.G., Morris, C., Stone, J.B., Terrones, G., Tupa, D.: Unstable Richtmyer-Meshkov growth of solid and liquid metals in vacuum. J. Fluid Mech. **703**, 60–84 (2012)
5. López Ortega, A., Lombardini, M., Pullin, D.I., Meiron, D.I.: Numerical simulations of the Richtmyer-Meshkov instability in solid-vacuum interfaces using calibrated plasticity laws. Phys. Rev. E. **89**(3), 033018 (2014)
6. Mikaelian, K.O.: Shock-induced interface instability in viscous fluids and metals. Phys. Rev. E. **87**(3), 031003 (2013)
7. Plohr, J.N., Plohr, B.J.: Linearized analysis of Richtmyer-Meshkov flow for elastic materials. J. Fluid Mech. **537**, 55–89 (2005)

8. Prime, M.B., Vaughan, D.E., Preston, D.L., Buttler, W.T., Chen, S.R., Oró, D.M., Pack, C.: Using growth and arrest of Richtmyer-Meshkov instabilities and Lagrangian simulations to study high-rate material strength. J. Phys. Conf. Ser. **500**(11), 112051 (2014)
9. Opie, S., Gautam, S., Fortin, E., Lynch, J., Peralta, P., Loomis, E.: Behaviour of rippled shocks from ablatively-driven Richtmyer-Meshkov in metals accounting for strength. J. Phys. Conf. Ser. **717**(1), 012075 (2016)
10. John, K.K.: Strength of tantalum at high pressures through Richtmyer-Meshkov laser compression experiments and simulations. Ph.D. Dissertation, California Institute of Technology, Pasadena (2014)
11. Buttler, W.T., Gray III, G.T., Fensin, S.J., Grover, M., Prime, M.B., Stevens, G.D., Stone, J.B., Turley, W.D.: Yield strength of Cu and a CuPb alloy (1% Pb). AIP Conf. Proc. **1793**(1), 110005 (2017). doi:10.1063/1.4971668
12. Sternberger, Z., Maddox, B.R., Opachich, Y.P., Wehrenberg, C.E., Kraus, R.G., Remington, B.A., Randall, G.C., Farrell, M., Ravichandran, G.: A comparative study of Rayleigh-Taylor and Richtmyer-Meshkov instabilities in 2D and 3D in tantalum. AIP Conf. Proc. **1793**(1), 110006 (2017). doi:10.1063/1.4971669
13. Prime, M.B., Buttler, W.T., Buechler, M.A., Denissen, N.A., Kenamond, M.A., Mariam, F.G., Martinez, J.I., Oró, D.M., Schmidt, D.W., Stone, J.B., Tupa, D., Vogan-McNeil, W.: Estimation of metal strength at very high rates using free-surface Richtmyer-Meshkov instabilities. J. Dyn. Behav. Mater. **3**(2), 189–202 (2017)
14. Preston, D.L., Tonks, D.L., Wallace, D.C.: Model of plastic deformation for extreme loading conditions. J. Appl. Phys. **93**(1), 211–220 (2003)
15. Simulia Abaqus 6.14: Dassault Systèmes, Vélizy-Villacoublay, France (2014)
16. Armstrong, R.W., Zerilli, F.J.: High rate straining of tantalum and copper. J. Phys. D Appl. Phys. **43**(49), 492002 (2010)

Michael B. Prime has been a Technical Staff Member at Los Alamos since 1994. He is a fellow of the ASME and won SEM's AJ Durelli Award in 2012.

Chapter 4
Impact Response of Density Graded Cellular Polymers

Behrad Koohbor, Suraj Ravindran, and Addis Kidane

Abstract Full-field deformation response of density-graded cellular polymers subjected to high velocity impact is investigated experimentally. Recently developed experimental setup consisting of ultra-high speed imaging in conjunction with digital image correlation is used to measure in-situ full-field deformation on density-graded polymeric foam specimens. Loading of the specimens is performed using a direct impact configuration in a modified Hopkinson bar apparatus. Discretely-layered foam specimens made from three distinct layers each with a different bulk density are subjected to direct impact, while their deformation response is studied via ultra-high speed digital image correlation. Formation and propagation of compaction waves from the impact side to the support end of the specimen are observed and analyzed. Spatial distribution of inertia stress is determined from acceleration fields and density of the layers. Full-field stress distribution in the specimen is later used to estimate the stress gradients within compaction waves. Mechanisms associated with the energy dissipation in graded foam specimens are discussed.

Keywords Polymeric foam • Dynamic loading • Digital image correlation • Graded structure • Direct impact

4.1 Introduction

Rigid polymeric foams are used in a variety of applications due to their unique properties such as low density, relatively high specific strength, great heat insulation and excellent energy absorption characteristics. The latter property is the main reason behind the application of these materials in impact loading conditions. It is a well-established fact that the energy absorption and strength of polymeric foams are highly sensitive to their relative density [1]. From a design standpoint, there is always a trade-off between strength, energy absorption and weight of polymeric foams having different bulk densities. To take advantage of all the aforementioned properties, while retaining the structural weight at a low level, the concept of graded foam has been introduced in recent years. The fundamental idea in a graded foam structure is to have an integrated part that is made of both high and low density foams, such that the overall weight of the structure remains low, while the energy absorption characteristics are enhanced due to the presence of high density foam material [2].

Research on the concept of graded foam structures have drawn much attention in recent years. For instance, graded syntactic foams capable of withstanding 60–75% compressive strain before failure were fabricated by Gupta [3]. Graded foam structures with improved energy absorption performance compared with single-phase foams were fabricated by Cui et al. [4]. Structure of stress wave propagation in functionally graded foams with different gradation functions were studied by Kiernan et al. [5]. Zheng et al. [6] investigated one dimensional impact response of density-graded cellular materials with different density distributions. Our objective in the present work is to implement a recently-developed experimental approach [7–9] to characterize full-field deformation response of density graded cellular polymers. Density graded samples with discretized layers each having a different bulk density are prepared in-house and subjected to direct impact loading. High speed imaging in conjunction with full-field DIC measurements are utilized to obtain the distribution of displacement and strain components in the specimen. Acceleration distribution is calculated from full-field displacement and is later used to estimate full-field inertia stress in the specimen.

B. Koohbor (✉) • S. Ravindran • A. Kidane
Department of Mechanical Engineering, University of South Carolina, 300 Main Street, Columbia, SC, 29208, USA
e-mail: koohbor@email.sc.edu

J. Kimberley et al. (eds.), *Dynamic Behavior of Materials, Volume 1*, Conference Proceedings of the Society for Experimental Mechanics Series, DOI 10.1007/978-3-319-62956-8_4

4.2 Materials and Methods

Rigid closed-cell polyurethane foams with three different nominal densities are used to fabricate layered foam specimens in this work. Figure 4.1 shows the quasi-static stress-strain curve for the three foams. Rectangular prism specimens with dimensions $14 \times 14 \times 7$ mm^3 are extracted from these three foam materials. The layers are then bonded to form a three-layered density graded structure of 21 mm total length. Layers are bonded together with a thin layer of highly-flexible strong polyurethane adhesive. Flexibility of the interface is required here to accommodate the slight relative lateral deformation of the foams and to minimize the shear stress developed within the interface [2]. To facilitate DIC measurements, front surface of the graded specimen is polished and then speckled using conventional spray speckling method. As illustrated in Fig. 4.2a, a narrow strip on the front surface of the specimen is intentionally left un-speckled to enable tracking of the layers' interface during deformation time.

A modified Hopkinson bar setup is used to apply impact loading on the specimen. Schematic drawing of our experimental setup is shown in Fig. 4.2b. The specimen is fixed on the left hand-end of the transmitter bar with the use of a lithium grease, which also serves as lubricant. A projectile is directly shot toward the specimen. The impact velocity of the projectile can be controlled by adjusting the pressure inside the gas barrel. In the present work, the projectile velocity upon contact with the specimen was measured as 90 m/s.

In order to measure the full-field deformation response of the specimen during impact time, a single high ultra-high speed HPV-X2 camera (Hadland Imaging) was utilized. This camera facilitates acquisition of 128 images at the full0field resolution of 400×250 pixel^2 upon each trigger. Imaging rate is adjustable, and in this work, was selected to be 500,000 frames per second, to facilitate recording the entire deformation of the sample. The camera was equipped with a 100 mm macro lens (Tokina) providing an optical resolution of 100 μm/pixel. Illumination of the area of interest was provided by a metal arc lamp (Lumen 200).

Images acquired by the camera were post-processed in the commercially available digital image correlation code Vic-2D (Correlated Solutions). In this software, subset and step sizes of 13 pixel and 2 pixel were used for image correlation respectively. Correlation was performed using an incremental algorithm to enable data acquisition at significantly large deformation conditions, i.e. compressive strains >45%. A high order interpolation function (optimized 8-tap) was used to convert discretized point measurements into continuous data. Correlation criterion was chosen to be zero normalized, reported to be insensitive to the scaling of light intensity. Strain, velocity and acceleration fields were numerically calculated

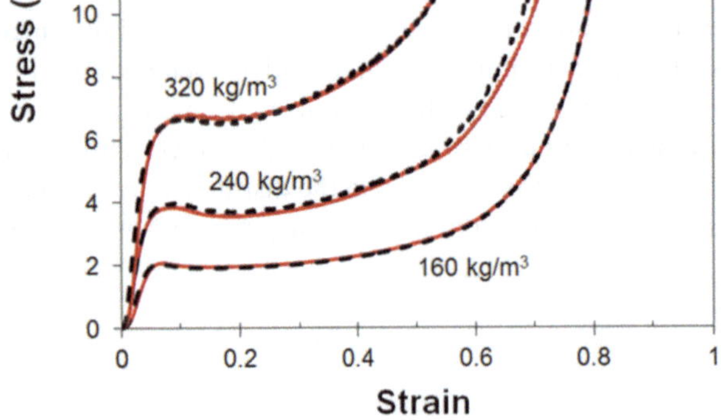

Density (kg/m^3)	Elastic Modulus (MPa)	Yield Stress (MPa)	Nominal Densification Strain
160	54.39±1.61	2.07±0.02	0.57±0.00
240	110.13±1.95	3.93±0.06	0.54±0.02
320	172.99±9.49	6.76±0.05	0.52±0.01

Fig. 4.1 Nominal stress-strain *curves* of the foams with three different densities obtained from two sets of independent quasit-static experiments. Elastic modulus, apparent yield stress and nominal densification strains are tabulated

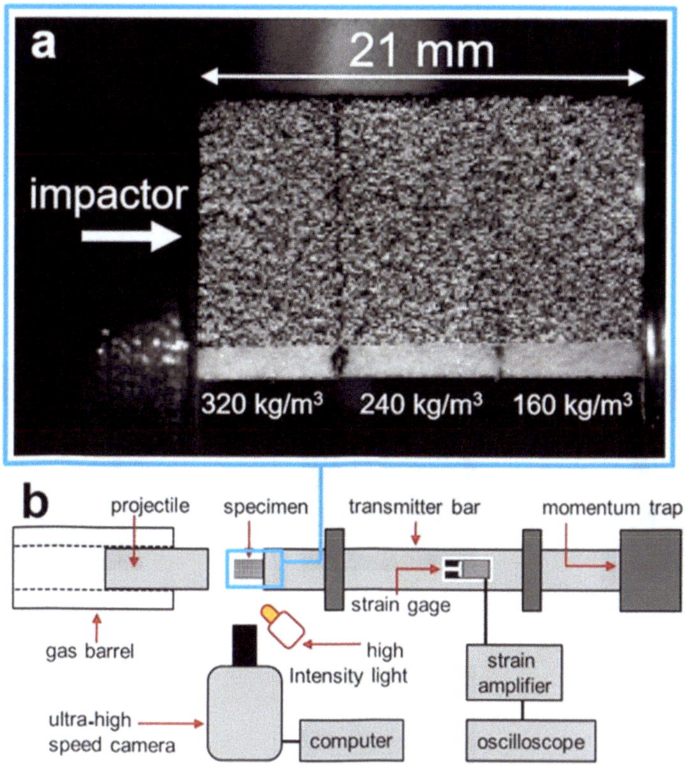

Fig. 4.2 (**a**) Camera view of the layered specimen, (**b**) schematic of the experimental setup consisting of impact setup and ultra-high speed camera

from the full-field displacement distribution. Temporal derivatives, i.e. velocity and acceleration were determined using simple finite differences operations applied to the obtained displacement data. Note that the temporal smoothing of the displacement data was first conducted using 2nd-order polynomials and a 5 μs smoothing window.

4.3 Full-Field Stress Measurement

Our main objective in this work is to adopt a recently introduced [7, 8] approach to measure the full-field axial stress in the impacted specimen. To this goal, let us first consider the general dynamic stress equilibrium equation:

$$\sigma_{ij,j} + \rho b_i = \rho a_i \tag{4.1}$$

where σ_{ij} denotes Cauchy stress tensor; ρ is mass density, b_i is body force in i-direction and a_i is the acceleration in i-direction. Note that the density term in this work is itself a step-wise function of x, since the samples are made of layers with different bulk densities. In uniaxial compression of a cubical specimen, as schematically shown in Fig. 4.3, and in the absence of body forces Eq. (4.1) can be reduced to:

$$\frac{\partial \sigma_x}{\partial x} = \rho a_x \tag{4.2}$$

where subscript x denotes the axial direction. The general solution to this differential equation is given by:

$$\sigma_x(L, t) = \sigma_x(0, t) + \int_{\xi=0}^{\xi=L} \rho(\xi, t)\, a_x(\xi, t)\, d\xi \tag{4.3}$$

which indicates that the total stress: (1) is indeed a location-dependent quantity, and (2) can be reconstructed from the stress measured at the specimen end (i.e. $x = 0$) plus the integral term, which represents inertia stress.

Fig. 4.3 Schematic of a three-layered density graded specimen fixed on one side and subjected to direct impact applied on its opposite end

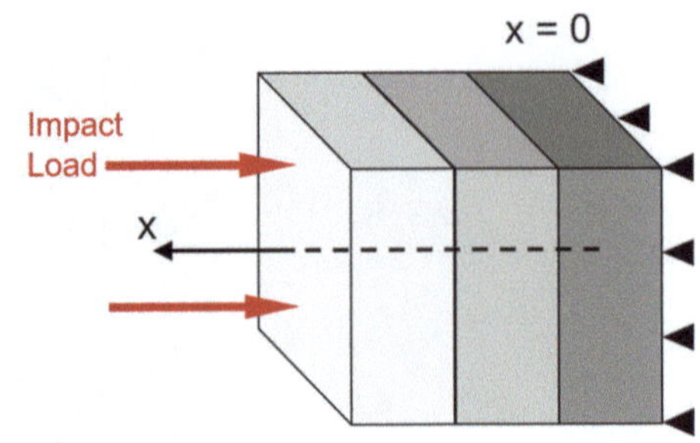

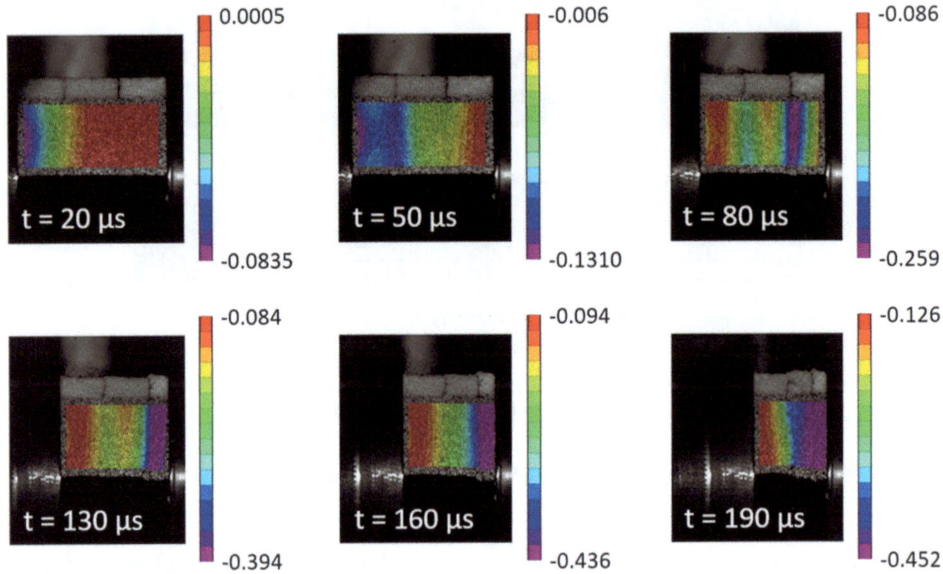

Fig. 4.4 Contour maps showing the distribution of axial strain on surface at different times

Our main goal in this work will be to examine the full-field deformation and stress response of the impacted specimens using the above-mentioned equations. It should also be noted that the material examined in this work is highly compressible, thus, undergoes significant density change during deformation time. The formulation derived here is based on Lagrangian description where acceleration and density are defined in the initial coordinate system. In this regard, the stress must also be based on the First Piola Kirchoff definition. In the case of perfect uniaxial deformation, i.e. lateral and shear deformations are negligibly small, Cauchy and First Piola Kirchoff stress definitions will be equivalent. Therefore, Eq. (4.1) remains valid when using Cauchy stress along with Lagrangian acceleration and density. The assumption of uniaxial deformation is validated in the forthcoming sections.

4.4 Full-Field Deformation Response

Contour maps showing the evolution of axial strain in the graded foam are presented in Fig. 4.4. The state of deformation is shown to be symmetric about the center line and homogeneous in *y*-direction. Also, note that the lateral deformation is minimal during the entire course of deformation. This means that there is an insignificant relative motion of the layers relative to each other, and therefore, development of shear stress at the interface is negligible.

It is clearly shown in Fig. 4.4 that the high-density layer, located on the impacted side of the specimen is first partially compressed during the first 20 μs of the impact. Then, the middle and low density layers located further from the impacted end start and continue to compress. Finally, at $t > 130$ μs, after the complete densification of low density layer (located on the support end), densification switches direction again and propagates back toward the impacted end of the specimen. This behavior is more clearly indicated in Fig. 4.5 where axial strain curves extracted from the line $\bar{y} = 0.5$ (see Fig. 4.5a) are plotted along the specimen axis.

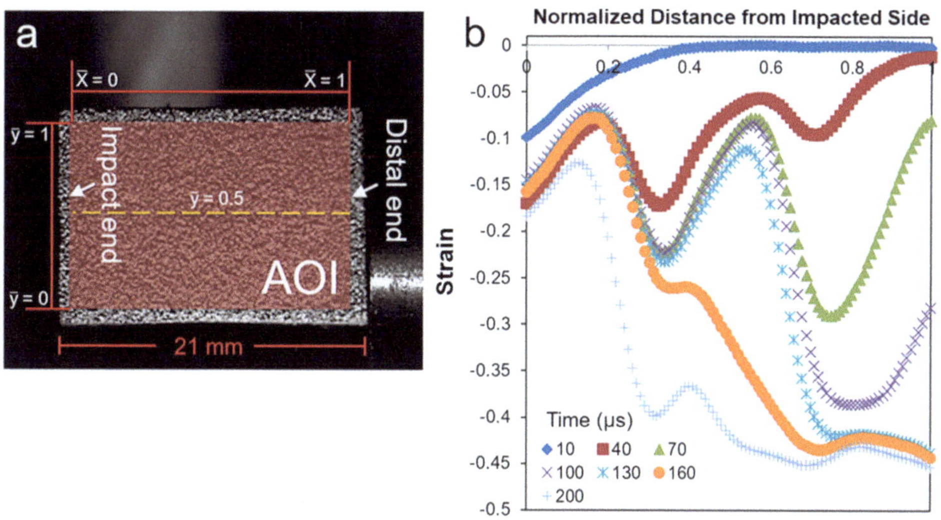

Fig. 4.5 (**a**) Geometry of the area of interest (AOI). (**b**) Variation of axial strain along specimen axis at different times extracted from the line $\bar{y} = 0.5$

Figure 4.5 indicates that the magnitude of axial strain remains nearly constant after $t = 10$ μs. Axial compression then continues to evolve in the lower density layers. Finally, at $t = 130$ μs, when the deformation in the lowest density layer reaches the densification strain, axial deformation followed by densification continue to develop into the middle layer.

4.5 Acceleration and Inertia Stress

Axial acceleration distribution along specimen axis was determined by taking double temporal derivative of the axial displacement. Spatial variation of axial acceleration is plotted in Fig. 4.6 at various times. Acceleration curves in Fig. 4.6 show local peaks that are indicative of the location of the wave front. Tracking the location of acceleration peaks in time, one can estimate elastic and plastic wave speeds in the specimen. In addition, there are several locations on each acceleration curve where a change in the slope of the curve is observed. Such behavior is due to the stress wave passing through an interface, changing its propagation speed as the result of a step-wise density change. Last but not least, it is clearly shown that after $t = 130$ μs, axial acceleration changes sign. This sign change is the consequence of the propagation of compaction wave back into the specimen after it reflects back from the distal end of the specimen.

Distribution of axial inertia stress is calculated using Eq. (4.3) and the acceleration data. Spatial variation of inertia stress at different time step is plotted in Fig. 4.7. The distribution of inertia stress is such that during the first 20 μs after the establishment of impact, high stress regions are formed on the impacted end of the specimen. This high stress region gradually propagates through the entire length of the specimen at longer times. Significant spatial gradients seen on the inertia stress curves can be used to pinpoint the compaction wave location, as well as measurement of compaction wave thickness. Upon reflection of the stress wave from the distal end of the specimen, inertia stress switches sign and becomes tensile. This tensile stress again traverses the entire length of the compressed specimen after $t = 130$ μs.

Our current research on this topic involves the extension of this work such that the stress gradients would be used to estimate the compaction wave thickness during deformation time. Investigation of compaction wave thickness, along with the study of compaction wave speed in different layers of a graded foam facilitated by the experimental methodology described in this work enables a more in-depth insight in the characterization of damping mechanisms in density graded foams. In addition, attempts are being made to conduct similar studies on graded foam structures with various layer thickness and sequence.

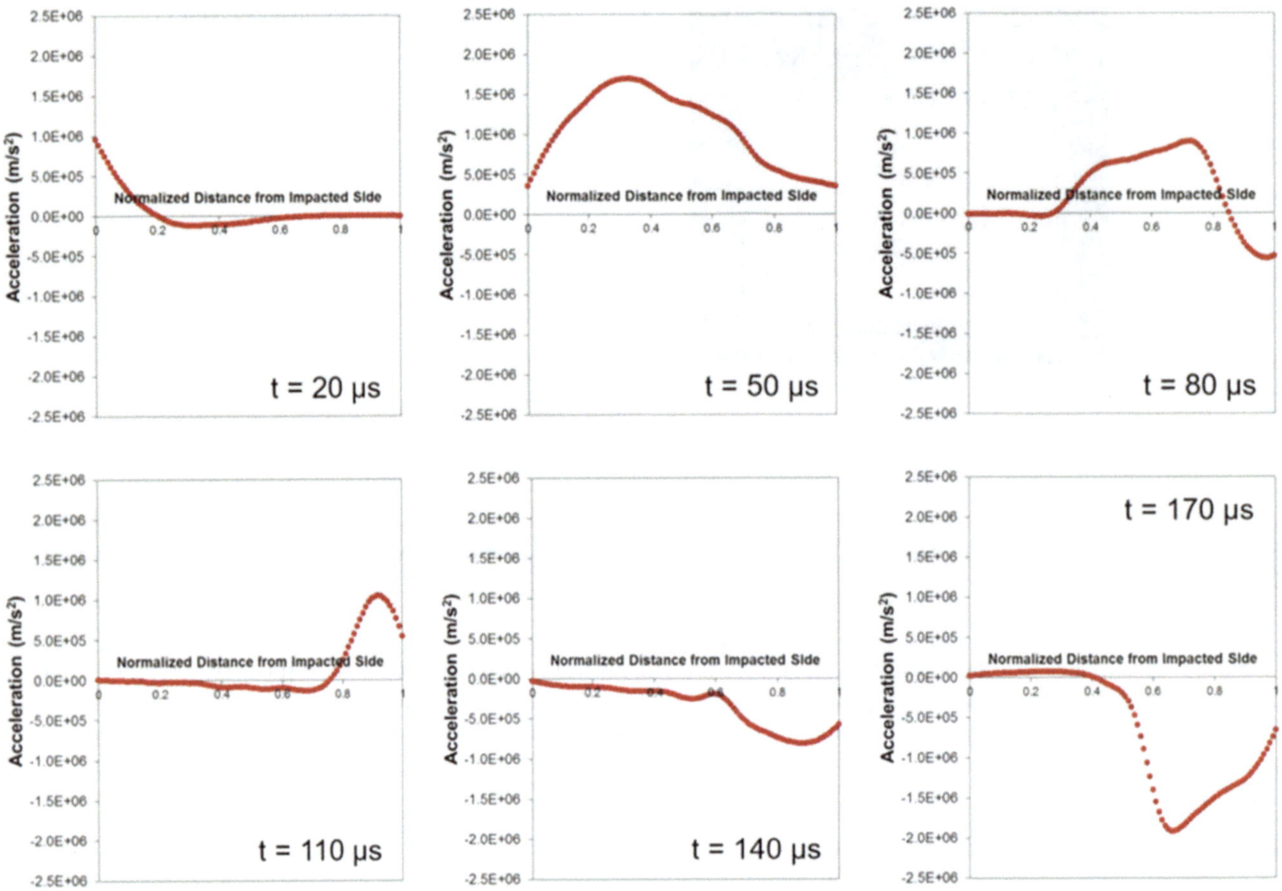

Fig. 4.6 Variation of axial acceleration along specimen axis at different times extracted over the line $\overline{y} = 0.5$ (see Fig. 4.5a)

Fig. 4.7 Variation of axial inertia stress along specimen axis at different times extracted from the line $\overline{y} = 0.5$ (see Fig. 4.5a). Note that a positive stress here means a compressive one

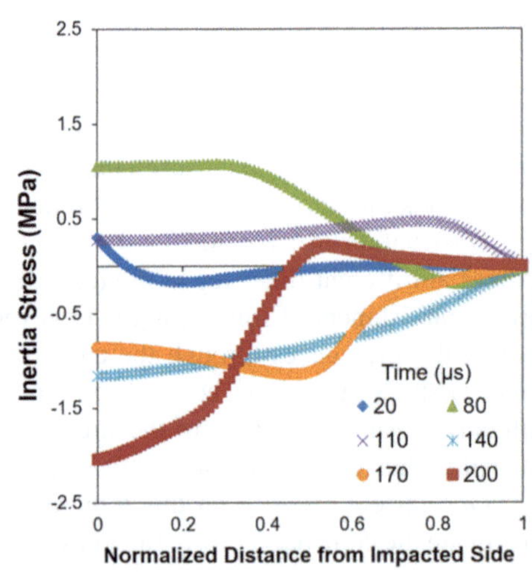

4.6 Summary

Dynamic deformation response of an impacted graded foam specimen fabricated from layers with different bulk densities was studied. An experimental setup consisting of a direct impact apparatus and a single high speed camera was used to enable acquiring images from the specimen during deformation time. Full-field deformation response was studied through 2D digital image correlation. Full-field strain maps extracted from DIC were used to identify the overall compressive response of the graded sample. It was found that deformation is nearly uniaxial with insignificant lateral and shear deformation. Acceleration field was then calculated from full-field displacement and was used to characterize wave propagation in the specimen. Finally, inertia stress distribution in the specimen was determined using acceleration data. Variation of inertia stress was found to have significant gradients, whereas such spatial gradients are attributed to the compaction wave formed in the material.

References

1. Gibson, L.J., Ashby, M.F.: Cellular Solids: Structure and Properties, 2nd edn. Cambridge University Press, Cambridge (1997)
2. Koohbor, B., Kidane, A.: Design optimization of continuously and discretely graded foam materials for efficient energy absorption. Mater. Des. **102**, 151–161 (2016). doi:10.1016/j.matdes.2016.04.031
3. Gupta, N.: A functionally graded syntactic foam material for high energy absorption under compression. Mater. Lett. **61**, 979–982 (2007). doi:10.1016/j.matlet.2006.06.033
4. Cui, L., Kiernan, S., Gilchrist, M.D.: Designing the energy absorption capacity of functionally graded foam materials. Mater. Sci. Eng. A. **507**, 215–225 (2009). doi:10.1016/j.msea.2008.12.011
5. Kiernan, S., Cui, L., Gilchrist, M.D.: Propagation of a stress wave through a virtual functionally graded foam. Int. J. Non Linear Mech. **44**, 456–468 (2009). doi:10.1016/j.ijnonlinmec.2009.02.006
6. Zheng, J., Qin, Q., Wang, T.L.: Impact plastic crushing and design of density graded cellular materials. Mech. Mater. **94**, 66–78 (2016). doi:10.1016/j.mechmat.2015.11.014
7. Koohbor, B., Kidane, A., Lu, W.Y., Sutton, M.A.: Investigation of the dynamic stress-strain response of compressible polymeric foam using a non-parametric analysis. Int. J. Impact Eng. **91**, 170–182 (2016). doi:10.1016/j.ijimpeng.2016.01.007
8. Koohbor, B., Kidane, A., Lu, W.: Effect of specimen size, compressibility and inertia on the response of rigid polymer foams subjected to high velocity direct impact loading. Int. J. Impact Eng. **98**, 62–74 (2016). doi:10.1016/j.ijimpeng.2016.08.006
9. Koohbor, B., Kidane, A., Lu, W.: Characterizing the constitutive response and energy absorption of rigid polymeric foams subjected to intermediate-velocity impact. Polym. Test. **54**, 48–58 (2016). doi:10.1016/j.polymertesting.2016.06.023

Behrad Koohbor received his Ph.D. from the Department of Mechanical Engineering, University of South Carolina.

Chapter 5
Dynamic Mixed-Mode Crack Initiation and Growth in PMMA and Polycarbonate

Balamurugan M. Sundaram and Hareesh V. Tippur

Abstract Mixed-mode dynamic crack initiation and growth in polymethymethacrylate (PMMA) and polycarbonate (PC) are studied experimentally. A simple specimen geometry and loading configuration is used to generate various mode-mixities during dynamic crack initiation and fracture. A Hopkinson pressure bar is used to rapidly load free-standing edge cracked samples in reverse impact configuration. Using eccentric loading relative to the crack line, different mode-mixities at crack initiation are achieved by increasing the initial crack length while keeping all other experimental parameters unchanged. A relatively new full-field optical technique, Digital Gradient Sensing (DGS), along with ultrahigh-speed photography is used to perform full-field measurements. DGS can measure instantaneous angular deflections of light rays due to elasto-optic effects and provides two orthogonal stress gradients under plane stress conditions. The mode-I and -II stress intensity factor histories of PMMA are evaluated via overdeterministic analysis of optically measured data. By quantifying critical stress intensity factors evaluated at crack initiation, dynamic fracture envelopes can be developed.

Keywords Digital Gradient Sensing • Transparent Material • Mixed-mode Fracture • Dynamic Fracture • Polymer Testing

5.1 Introduction

Dynamic mixed-mode fracture is rather common when materials and structures experience a combination of tensile and shear loads simultaneously. Despite decades of research on dynamic fracture mechanics, crack initiation under mixed-mode conditions is yet to be fully understood. This research is an experimental investigation of mixed-mode fracture behaviors of transparent polymethylmethacrylate (PMMA) and polycarbonate (PC). These two polymers are widely used in engineering practice for their characteristics such as optical transparency, stiffness, and resistance to cracking. One of the applications of these materials is in transparent armor when used individually or as part of a layered system which authors have previously studied [1–4]. Interestingly, there are limited studies on these materials in the monolithic state under mixed-mode dynamic loading condition. Most reported works have used global measurements and/or complementary computational simulations to recover fracture parameters such as crack initiation toughness and kink angle. However, this necessitates approximations and creates gaps while interpreting the measured data.

With this background, the current work describes a simple specimen geometry and loading configuration to generate different mode mixities under dynamic loading conditions. The stress intensity factor (SIF) histories are evaluated using an optical method called Digital Gradient Sensing (DGS) [5] in conjunction with ultrahigh-speed photography. As DGS is a stress gradient measurement, a simple 2D integration technique can be used to obtain in-plane stresses [6]. Using the methodology to evaluate SIF histories previously outlined by the authors [1], SIF histories of five configurations for PMMA are evaluated. The further work involves evaluating SIF histories for different the configurations of PMMA and PC and develop dynamic crack initiation toughness envelopes.

B.M. Sundaram • H.V. Tippur (✉)
Department of Mechanical Engineering, Auburn University, Auburn, AL, 36849, USA
e-mail: htippur@eng.auburn.edu

J. Kimberley et al. (eds.), *Dynamic Behavior of Materials, Volume 1*, Conference Proceedings of the Society for Experimental Mechanics Series, DOI 10.1007/978-3-319-62956-8_5

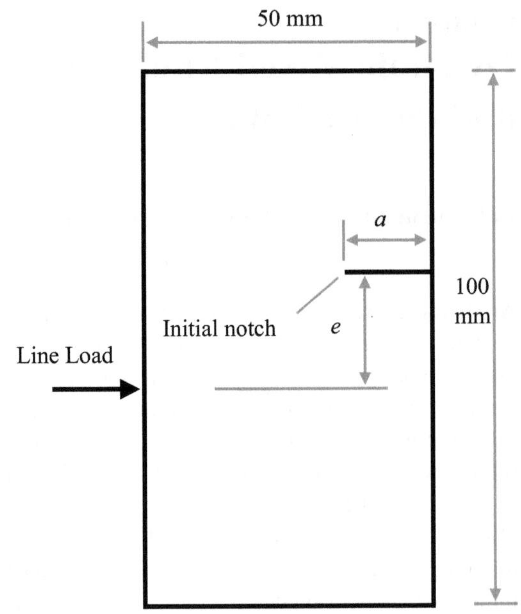

Fig. 5.1 Specimen geometry and loading configuration used in mixed-mode dynamic fracture experiments. Values of a and e used are listed in Table 5.1

Table 5.1 Values of a and e used in Fig. 5.1 to obtain different mixed-mode configurations (1–5)

Mode-I	Config. 1	$a = 6$ mm, $e = 0$ mm
Mixed-Mode	Config. 2	$a = 6$ mm, $e = 15$ mm
	Config. 3	$a = 9$ mm, $e = 15$ mm
	Config. 4	$a = 12$ mm, $e = 15$ mm
	Config. 5	$a = 15$ mm, $e = 15$ mm

5.2 Specimen Geometry and Loading Configurations

SEN plate specimens of both PMMA and PC were machined from sheet stock. The specimens were of dimensions 100 mm × 50 mm × 8.6 mm (see Fig. 5.1). An initial notch was machined into each specimen and was further sharpened using a razor blade. The notch length was varied to obtain different mode-mixities while the eccentricity (e) between the loading point on the opposite face and the crack location was kept constant. Five different configurations, 1–5, were used to produce various mode-mixities at crack initiation. This included the mode-I case in which eccentricity $e = 0$. The different values of a and e used to obtain various configurations are listed in Table 5.1.

5.3 Experimental Details

The schematic of the experimental setup used for dynamic fracture tests is shown in Fig. 5.2. A Hopkinson bar (long-bar) apparatus was used to load the sample [7]. A 6 feet long, 1-inch diameter long-bar with a cylindrical tip was held against an unconstrained specimen placed on an adjustable platform. A 12-inch-long, 1-inch diameter striker held inside the barrel of a gas-gun was used to impact the long-bar. Both the long-bar and the striker were made of AL 7075. The striker was propelled towards the long-bar at a velocity (V) of 13.5 m/using a gas-gun. When the striker contacted the long-bar, a compressive stress wave propagated along the bar before transmission into the specimen in contact. The notch on the opposite edge of the specimen was subjected to tensile and shear stresses simultaneously by the reflected stress waves from the free edges of the specimen.

A target plate decorated with random black and white speckles was placed behind the specimen at a fixed distance (approx. 25 mm) to measure angular deflections of light rays in the region of interest using DGS. The speckle pattern was photographed through the specimen using a Cordin-550 ultrahigh-speed digital camera with 32 sensors. Before loading the specimen, a set of 32 reference images were recorded at a rate of 400,000 frames per second at a resolution of 1000 × 1000 pixels. When the striker contacted the long-bar, a trigger signal initiated recording a second set of 32 images (at the same framing rate) of the fracture event. Thus each reference image had a corresponding image in the deformed

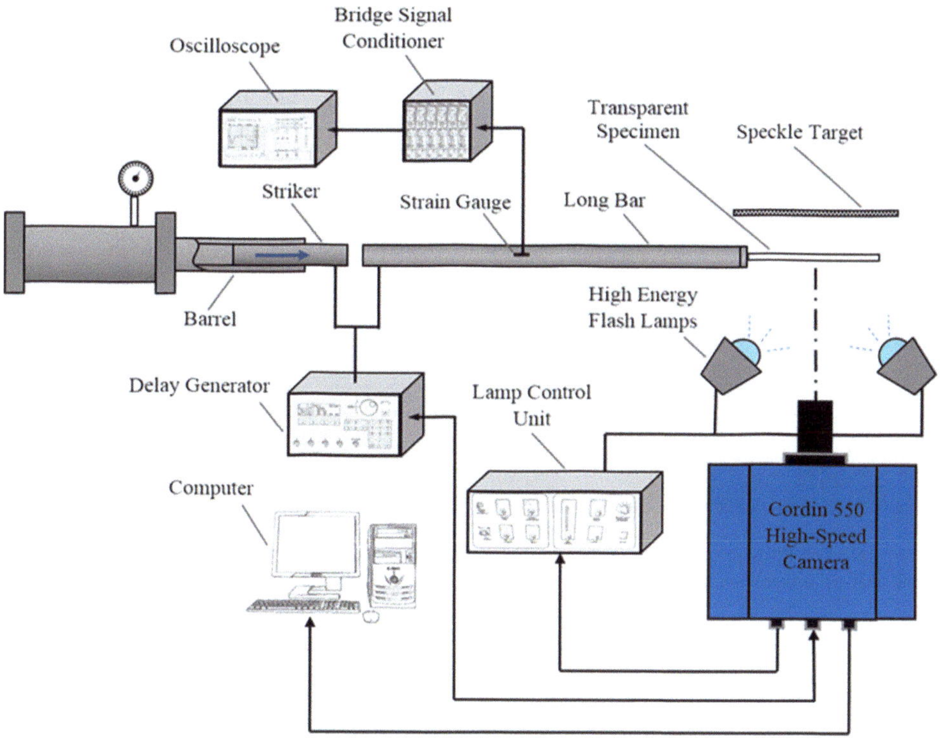

Fig. 5.2 Schematic of the experimental setup used to study mixed-mode fracture of PMMA and PC

state. The corresponding reference and 'deformed' images from same sensor were paired and correlated to obtain in-plane displacements, using which angular deflections of light rays were evaluated as a 2-D array of data in the field of view.

5.4 Mixed-Mode Crack Paths

Figure 5.3a–e shows the photographs of the fractured PMMA specimens of various configurations, 1–5, respectively. The crack growth occurred from right to left in these photographs, as indicated by the arrowhead in Fig. 5.3a. A single crack initiated at the initial crack tip on the right edge in each of the specimens. It can be seen that the crack path changes from mode-I to mixed-mode from Fig. 5.3a to b. From Fig. 5.3b to e the crack initiation angle can be seen to increase from 24.8° to 52.3° even though the overall macroscale path appears nearly the same. The change in crack initiation angles suggests that the crack initiates at different mode-mixities in each configuration.

5.5 Optical Measurements

The recorded reference and deformed images were correlated using ARAMIS image analysis software to obtain contours of constant angular deflections around a propagating crack tip. Other details are avoided here for brevity [8]. The resulting contours in Figs. 5.4 and 5.5 represent the angular deflections of light rays for two different configurations, 1 and 5, at two different time instants ($t = 0$ μs represents the time at which the crack initiated). They represent the stress gradients [5] in two mutually perpendicular directions in the x-z and y-z planes with (x, y) denoting the in-plane coordinates of the specimen.

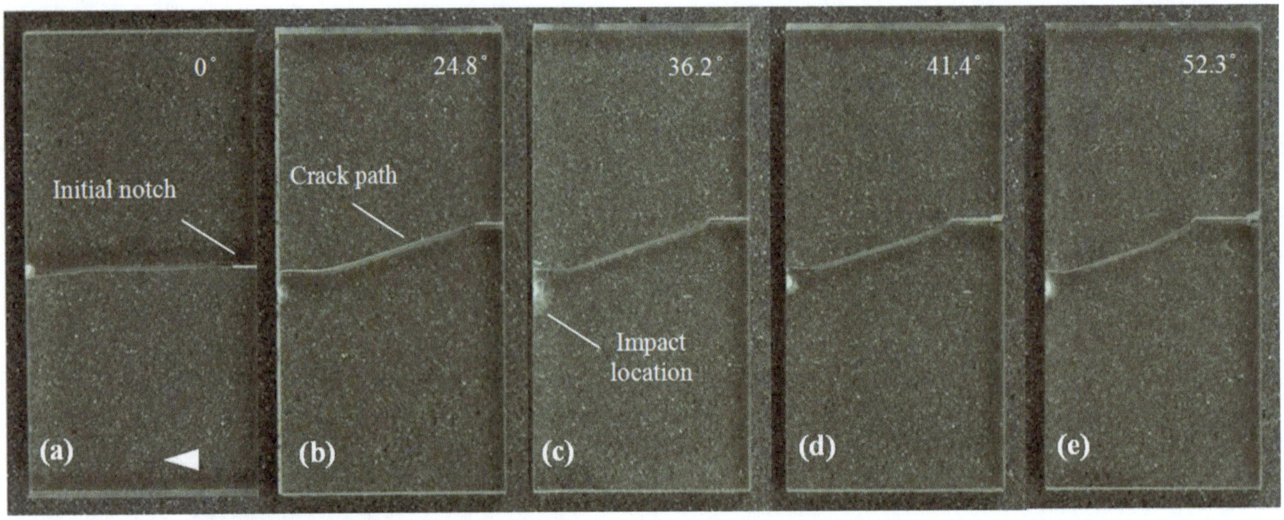

Fig. 5.3 Photographs of fractured PMMA specimens (*upper half*) for all configurations in sequence. The corresponding crack initiation angles measured from enlarged near-tip images is shown in the *right top corner*

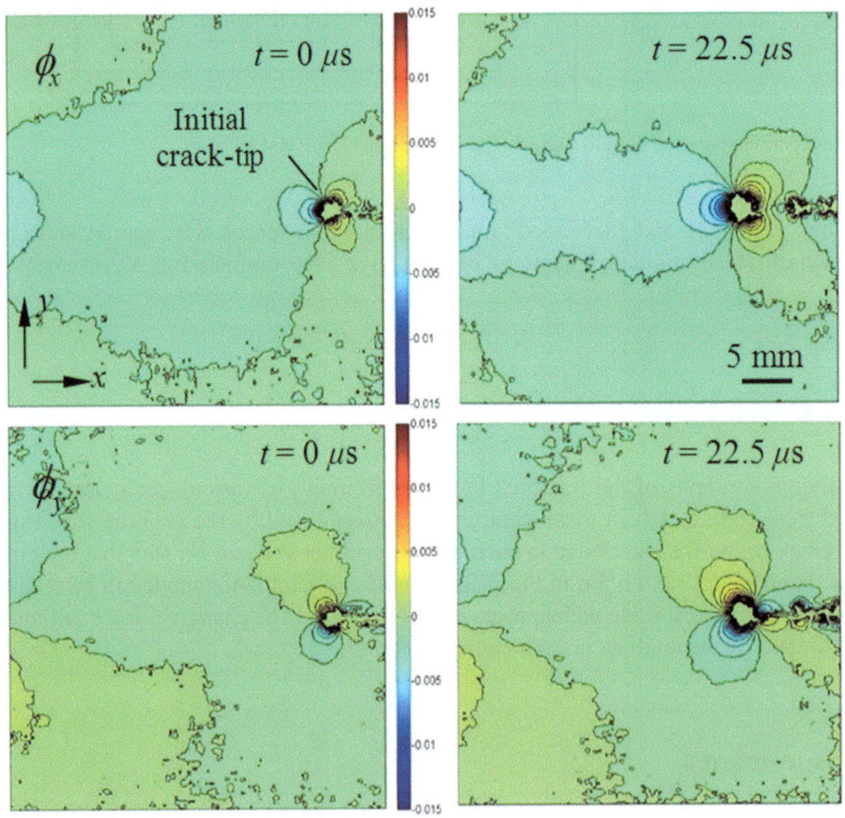

Fig. 5.4 Angular deflection contour plots (contour interval = 5×10^{-4} rad) proportional to stress gradients of $(\sigma_{xx} + \sigma_{yy})$ in the *x*- (*top row*) and *y*-directions (*bottom row*) for a PMMA specimen with $a = 6$ mm and $e = 0$ mm (Note that $t = 0$ in these correspond to the instant at crack initiation)

5.6 Stress Intensity Factor Histories

SIF histories were evaluated using mixed-mode stress gradient crack-tip fields following the methodology described in Ref. [3]. A compilation of mode-I and mode-II stress intensity factor histories evaluated form DGS for all configurations of PMMA are plotted in Fig. 5.6. The time axis of each experiment was shifted such that $t = 0$ corresponds to the instant at

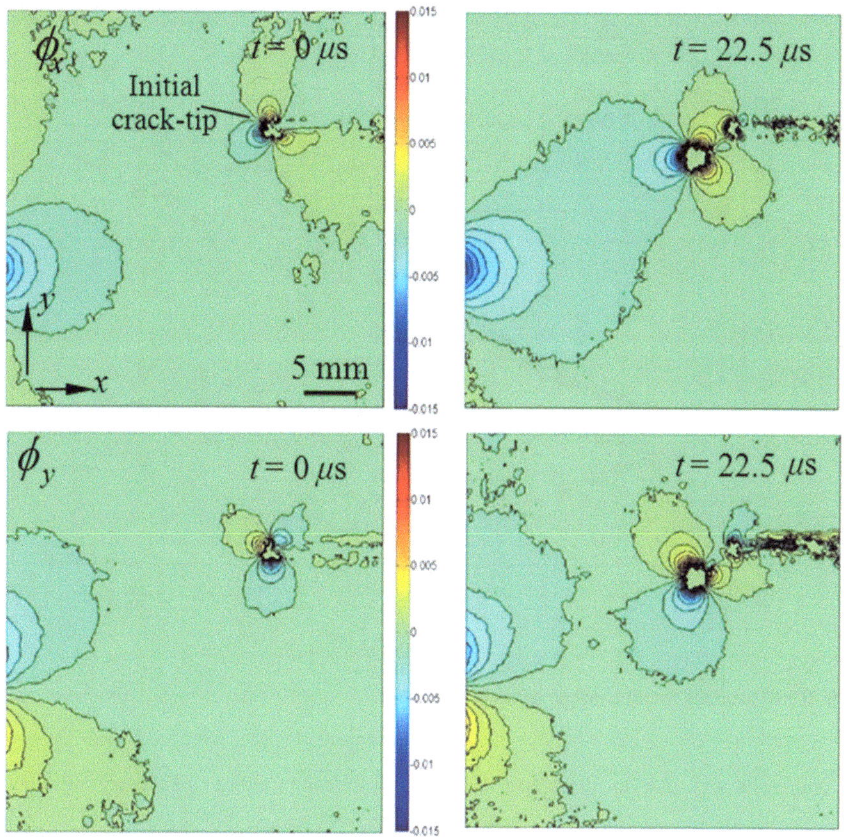

Fig. 5.5 Angular deflection contour plots (contour interval $= 5 \times 10^{-4}$ rad) proportional to stress gradients of $(\sigma_{xx} + \sigma_{yy})$ in the x- (*top row*) and y-directions (*bottom row*) for a PMMA specimen with $a = 15$ mm and $e = 15$ mm (Note that $t = 0$ in these correspond to the instant at crack initiation)

which the crack initiated at the initial notch tip as determined from the recorded images. At crack initiation, there exists both K_I and K_{II} in config. 2 to config. 5 suggesting mixed-mode crack initiation. It can be seen that there is a monotonic reduction in the magnitude of K_{I} at initiation from config. 1 to config. 5. The K_{II} histories, on the other hand, show an opposite trend with a monotonic increase in magnitude of K_{II} at initiation from config. 1 to config. 5. Further, K_{II} values approximately reach a plateau prior to crack initiation. In the post-initiation period, K_{I} histories continue to increase while K_{II} precipitously drop in all the cases. An oscillatory decaying trend in K_{II} about zero is also evident suggesting crack growth preference to mode-I following crack initiation transient. Similar experiments were carried out using PC specimens to study its mixed-mode dynamic fracture behavior and the SIF histories for all the configurations of PC can be evaluated. Using the measured SIF histories, the dynamic fracture envelope of both PMMA and PC can be developed [9].

5.7 Summary

An eccentric reverse impact specimen geometry involving different crack lengths has been investigated for producing dynamic crack initiation under different mixed-mode conditions. The results show that the crack initiation angle increases with crack length when all other geometric and loading parameters remain fixed. The crack tip fields have been quantified using the Digital Gradient Sensing method with ultrahigh-speed photography before, at, and after crack initiation. Using the measured fields, the mixed-mode SIF histories are being evaluated in order to produce dynamic crack initiation toughness envelopes and study post-initiation behaviors of both these polymers.

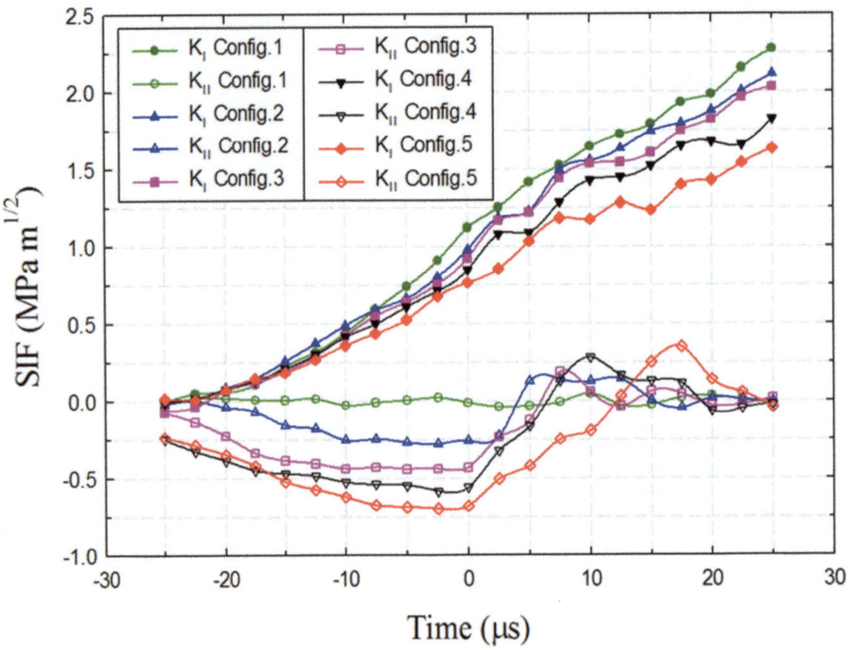

Fig. 5.6 Mode-I and mode-II SIF histories for all configurations of PMMA

Acknowledgement The authors would like to thank the U.S. Army Research Office for supporting this research through grant W911NF-16-1-0093.

References

1. Sundaram, B.M., Tippur, H.V.: Dynamic crack growth normal to an interface in bi-layered materials: an experimental study using digital gradient sensing technique. Exp. Mech. **56**, 37–57 (2016)
2. Sundaram, B.M., Tippur, H.V.: Dynamics of crack penetration vs. branching at a weak interface: an experimental study. J. Mech. Phys. Solids. **96**, 312–332 (2016)
3. Sundaram, B.M., Tippur, H.V.: Dynamic crack propagation in layered transparent materials studied using digital gradient sensing method: part II. In: Dynamic Behavior of Materials, Vol. 1, Proceedings of the Society for Experimental Mechanics, pp. 25–31 (2016)
4. Sundaram, B.M., Tippur, H.V.: Dynamic crack propagation in layered transparent materials studied using digital gradient sensing method. In: Dynamic Behavior of Materials, Vol. 1, Proceedings of the Society for Experimental Mechanics, pp. 197–205 (2015)
5. Periasamy, C., Tippur, H.V.: A full-field digital gradient sensing method for evaluating stress gradients in transparent solids. Appl. Opt. **51**, 2088–2097 (2012)
6. Miao, C., Sundaram, B.M., Huang, L., Tippur, H.V.: Surface profile and stress field evaluation using digital gradient sensing method. Meas. Sci. Technol. **27**(9), 095203 (2016)
7. Sundaram, B. M., Tippur, H. V.: Full-field measurement of contact-point and crack-tip deformations in soda-lime glass. Part-II: Stress wave loading. Int. J. Appl. Glass Sci. (2017). https://doi.org/10.1111/ijag.12289
8. Sundaram, B. M., Tippur, H. V.: Full-field measurement of contact-point and crack-tip deformations in soda-lime glass. Part-I: Quasi-static loading. Int. J. Appl. Glass Sci. (2017). https://doi.org/10.1111/ijag.12278
9. Sundaram, B.M., Tippur, H.V.: Dynamic mixed-mode fracture behaviors of PMMA and polycarbonate. Eng. Fract. Mech. **176**, 186–212 (2017)

Chapter 6
Dynamic Response of Alumina Ceramics Under Brazilian Disc Test Conditions

Ping Hong Lin, Liren Tsai, and N.S. Liou

Abstract In the joint replacement surgery, the alumina ceramic are widely used as femoral heads for total hip replacements as an alternative to metal parts. Alumina ceramic materials has good mechanical properties, corrosion resistance, wear resistance, and biocompatible properties. However, the alumina ceramics typically exhibit brittle behavior thus limiting their performance under certain conditions. Due to some machining difficulties, there is almost no direct measurement of dynamic tensile properties of ceramics in the strain rate ranges between 500 s^{-1} and 5000 s^{-1}. To understand dynamic tensile response of alumina ceramics, Flattened Brazilian Disc method with Split Hokinson Pressure Bar was utilized. In this research, the dynamic fracture process of the alumina ceramic at various tensile loading rates was studied. The effects of sample size were also monitored and analyzed. This paper explores the feasibility of applying the tensile behavior of alumina ceramics to total hip arthroplasty using a Brazilian Disc method.

Keywords Spilt Hopkinson Pressure Bar • Brazilian Disc • Alumina ceramic

6.1 Introduction

The development of alumina ceramics [1] is an important civilization of mankind. Therefore, from ancient times to the present it has a long history, with the development and history of the evolution of new technologies. In recent years, alumina ceramic materials continue to develop new products. It is a very wide range of Alumina ceramic materials have good mechanical properties, corrosion resistance, abrasion resistance and biocompatibility applications, such as: industrial applications, biomedical, precision materials, and even it can also be made of fibrous material. Alumina ceramics is a common raw material. It has good conductivity, high hardness, low density, wear resistance, but also a high temperature resistant materials. It is brittle nature, the advantage is able to withstand great compressive stress, but the disadvantage is not able to withstand the tensile, bending load and cannot withstand plastic deformation.

From the past to the present, total hip arthroplasty [2] for the treatment of arthritis in elderly patients. In biomedical, joint replacement surgery often use alumina ceramic as a material. Alumina ceramic materials have good mechanical properties, corrosion resistance, abrasion resistance 3.5 and biocompatibility and prolonged effects on human implants [5]. Therefore, it is used as a substitute for metal parts. Hip replacement surgery uses ceramic materials, but rarely explores its tensile properties. In this study, the dynamic tension properties of ceramic materials were observed at strain rates between 500^{-1} and 5000^{-1}.

P.H. Lin • L. Tsai (✉)
National Kaohsiung University of Applied Sciences, Department of Mechanical Engineering, Kaohsiung, Taiwan
e-mail: liren@gm.kuas.edu.tw

N.S. Liou
Southern Taiwan University of Science and Technology, Department of Mechanical Engineering, Tainan, Taiwan

J. Kimberley et al. (eds.), *Dynamic Behavior of Materials, Volume 1*, Conference Proceedings of the Society for Experimental Mechanics Series, DOI 10.1007/978-3-319-62956-8_6

6.2 Equation

6.2.1 *One-Dimensional Wave Equation*

1. The SHPB experiment is mainly based on the one-dimensional wave theory of stress wave motion Eq. (6.1):

$$\frac{\partial^2 u}{\partial x^2} = \frac{1}{c^2} \cdot \frac{\partial^2 u}{\partial t^2} \tag{6.1}$$

2. The specimen is subject to strain rate and strain of the equation. The equation is derived by H. Kolsky [6]. Is show by the following Eqs. (6.2) and (6.3):

$$\dot{\varepsilon} = -\frac{c}{ls} \cdot (\varepsilon_i - \varepsilon_r - \varepsilon_t) \tag{6.2}$$

$$\varepsilon = -\frac{c}{ls} \cdot \int_0^t \cdot (\varepsilon_i - \varepsilon_r - \varepsilon_t)\, dt \tag{6.3}$$

Where $\dot{\varepsilon}$ is obtained strain rate of the specimens, and ε is obtained strain of the specimens. The strain signals are the incident wave (ε_i), the reflected wave (ε_r) and the transmitted wave (ε_t) respectively. Then c is the elastic wave velocity of the bar material, l_s is the original gage length of the specimen.

When the test specimen reaches the dynamic equilibrium, it can be assumed that the stress at the front of the specimen is σ_{front} and the stress at the front end of the specimen is σ_{back}.

$$\sigma_{front} = \sigma_{back} \tag{6.4}$$

The stress balance is achieved by Eq. (6.4), and the strain and strain rate formula of the specimen are obtained. The following Eqs. (6.5) and (6.6) are shown:

$$\varepsilon = -2\frac{c}{ls} \cdot \int_0^t \varepsilon_r(t)\, dt \tag{6.5}$$

$$\dot{\varepsilon} = -2\frac{c}{ls} \cdot \varepsilon_r \tag{6.6}$$

6.2.2 *Brazilian Disc Method Equation*

Usually the tensile strength of the Brazilian disc is quasi-static test. Subsequently M. Iqbal and other scholars [7] have verified that the following equation can be used for dynamic tensile tests:

Brazilian disc experiment has been used in rock for many years. The static rocking test was carried out using the rock at the International Society of Rock Mechanics (ISRM) [8] and American scholar tests concrete tests [9]. According to the Brazilian disc, the tensile strength is calculated in Eq. (6.7).

$$\sigma_t = \frac{2P_{Max}}{\pi t DB} \tag{6.7}$$

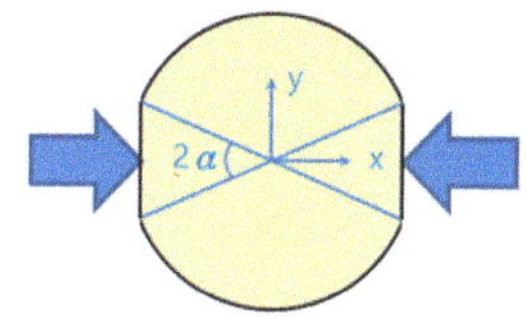

Fig. 6.1 The end of the specimen is subjected to a balanced load

6.2.3 *Flattened Brazilian Disc Method Equation*

The two ends of the plane are mainly to reduce and reduce the stress concentration. Both ends of the specimen are balanced and capable of accurate contact. The main specimen fracture process ensures that the central position begins to crack and travels along the load diameter. The Brazilian disc test encounters stress concentration problems, so the formula is not valid. The main specimen is selected as the appropriate value of 2α, the specimen is subjected to the average radial load as shown in Fig. 6.1. According to Wang and scholars [10] using Flattened Brazilian Disc (FBD) dynamic tension tests. The following Eq. (6.8) is shown:

$$\sigma_t = K\frac{2P_{Max}}{\pi DB} \tag{6.8}$$

Where σ_t is tensile strength of the specimen, P_{max} is subjected to the maximum load of the specimen. B is the specimen thickness, D is the specimen diameter, K is a 2α for load angle, and it is an important relationship coefficient. In calculating K, it can be substituted into Eq. (6.9) using an approximation:

Where

$$k = \frac{\left(\cos^3\alpha + \cos\alpha + \frac{\sin\alpha}{\alpha}\right)}{8\left(\cos\alpha + \frac{\sin\alpha}{\alpha}\right)}\frac{\alpha}{\sin\alpha} \tag{6.9}$$

6.3 Experiment Instrument Description and Procedure

6.3.1 *Instrument Description*

This study uses a spilt Hopkinson pressure bar, through several amendments to improve, therefore, it can be used for dynamic experimental aspects of the test equipment. The main observation of materials under high strain rate response measurement and change their mechanical properties. In the experiment, the mechanical impedance of the rod and the specimen is not matched and the one-dimensional wave theory is applied. The mechanical properties of the test specimen under high strain rate response are analyzed and studied. In the SHPB, they consist of a Pressure Chamber, a Striker, Incident Bar, and a Transmitter Bar, as shown in Fig. 6.2.

6.3.2 *Experimental Procedure*

In this study, dynamic tests were performed using split Hopkinson pressure bars (SHPB) and Brazilian disc (BD) method tests as shown in Fig. 6.2. First, place the Incident bar and the Transmitter bar, which is clamped on the two-bar end. Second, set and adjust the signal amplifier (Amplifier). Let semiconductor strain gauge reaches the voltage balance. Third, adjust the pressure chamber to start inflating. The striker is blown by the compressed gas. Fourth, the tensile load voltage is measured by a semiconductor strain gauge. Finally, the Wheatstone Bridge modifies the initial signal and amplifies the amplifier signal to the oscilloscope for recording. Figure 6.2 is represented. Then, the stress-strain curve of the dynamic tensile test is calculated from the formula.

In the dynamic tensile test method is roughly divided into two test methods. The first way is to use direct tension testing. The second way is to use indirect tension test [11]. For brittle materials such as ceramics, rocks are difficult to carry out direct tension experiments and are subject to some limitations and are therefore carried out using the Brazilian disc method.

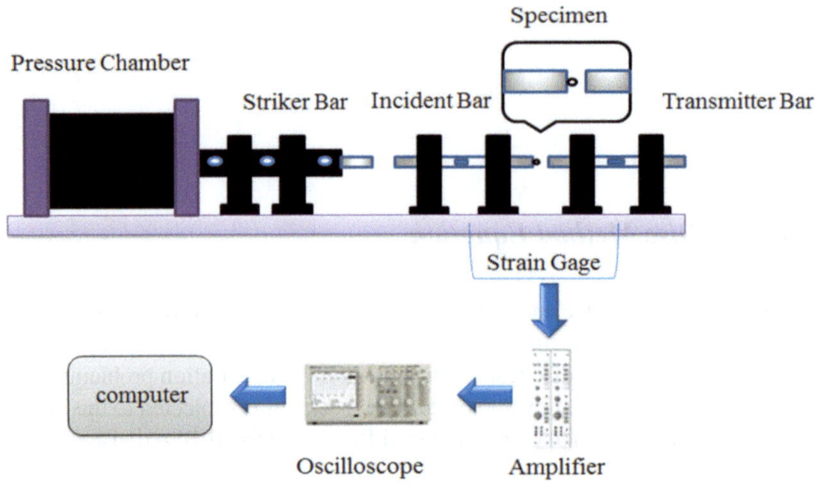

Fig. 6.2 Spilt Hopkinson pressure bar equipment and Brazilian disc method

Table 6.1 Physical properties of SKD11

Young's modulus E (GPa)	Density ρ (g/cm^3)	Poisson's ratio υ
209.09	7.85	0.3

Table 6.2 Experimental bar (unit: mm)

Bar diameter	Incident bar	Transmitter bar	Striker
10 (0.0393″)	1500 (59.055″)	1500 (59.055″)	400 (15.748″)

Brazilian discs are an experimental method of indirect tension. Therefore, it overcomes the direct tension test problem (for example, the rod and the specimen are difficult to combine, the fixture shape is difficult to process). This experiment was carried out using the Brazilian disc method.

6.4 Experimental Specimen and Bar Size

6.4.1 Experimental Bar

In this research experiment, the experimental bar material is mainly used in alloy tool steel (SKD11). SKD11 has a very high wear resistance and high hardness, therefore, used it as an experimental material. The experimental bars are Young's modulus E = 209.9 GPa and Density ρ = 7.85, Poisson's ratio υ = 0.3, Bar diameter are 10-mm (0393″), respectively, as shown in Tables 6.1 and 6.2. And bar length is striker 400-mm (15.748″) and incident bar 1500-mm (59.055″), transmitter bar 1500-mm (59.055″), respectively.

6.4.2 Specimen

In this study, experimental specimen material mainly used in alumina ceramic material, using it to explore total hip arthroplasty material dynamic tensile test. This experiment was carried out using flat face Brazilian disc method. The main problem of stress concentration can be reduced. Thus, the specimen was cut into two end faces. The specimen diameters are 6 and 8 mm (0.236″ and 0.314″) in diameter and 2-mm (0.078″) in thickness, as shown in Figs. 6.3 and 6.4.

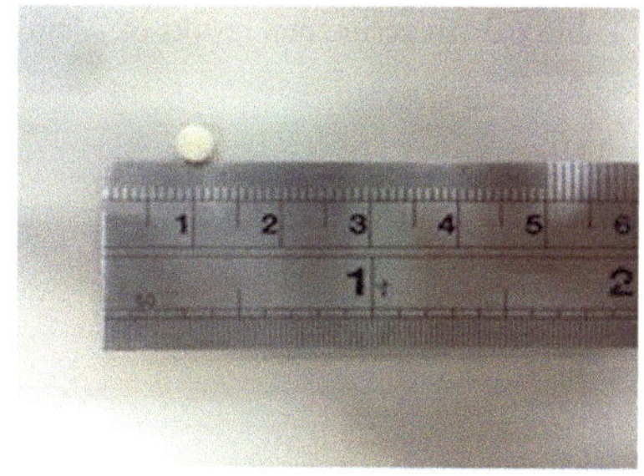

Fig. 6.3 Specimen diagram

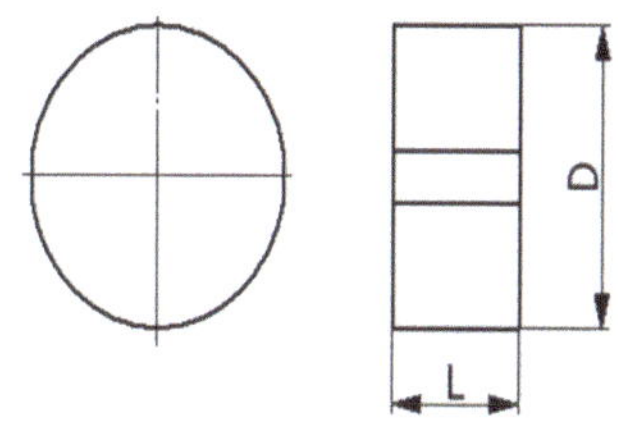

Fig. 6.4 Specimen size diagram (unit: mm)

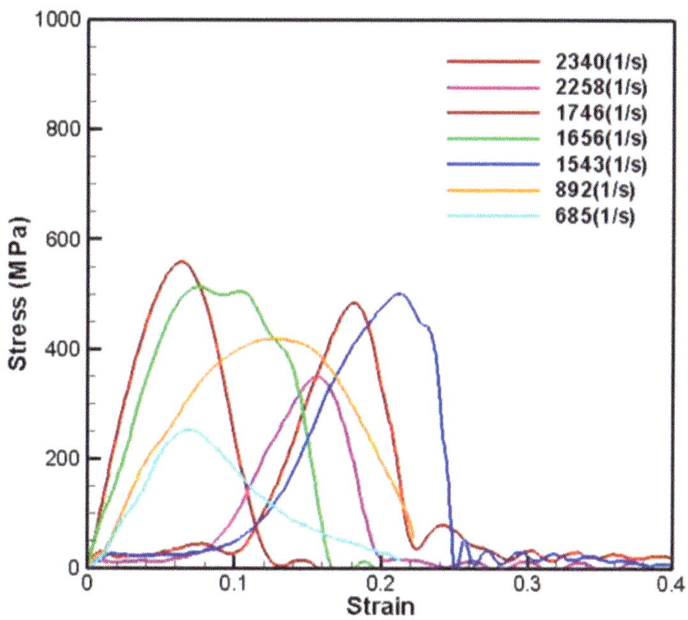

Fig. 6.5 Stress – strain curve

6.5 Results and Discussion

The operation must be corrected before the experiment, and the correction positions are the end faces of the incident bar and the transmitter bar, respectively. Thus ensure the integrity of their ends and the wave forwarding integrity. In this study, the SHPB experiment and the Brazilian disk method approach were investigated. Stress-strain curves of alumina ceramics were shown in Fig. 6.5. The trends of the stress-strain curves were found to be quite similar within the tested strain rate rage. Figure 6.5 shows the stress-strain curves of alumina ceramic with seven different strain rates: 2340 s^{-1}, 2258 s^{-1}, 1746 s^{-1}, 1656 s^{-1}, 1543 s^{-1}, 1706 s^{-1}, 892 s^{-1} , 685 s^{-1}. Found Table 6.3, elastic modulus between 260 and 792. Found Table 6.3, Ultimate strength between 252 and 558.

Figure 6.8 shows the stress-strain curves of alumina ceramic with six different strain rates: 2153 s^{-1}, 2085 s^{-1}, 2011 s^{-1}, 1810 s^{-1}, 1439 s^{-1}, 945 s^{-1}. Found Table 6.3, elastic modulus between 224 and 621. Found Table 6.4, Ultimate strength between 281 and 578.

Table 6.3 Experimental results of Alumina ceramic, ∅6 (mm)

Strain rate (1/s)	Young's modulus (GPa)	Ultimate strength (MPa)
2340	321.9	558
2258	481.1	349
1746	739.2	484
1656	500	514
1543	518.4	501
892	382.8	418
685	260.4	252

Table 6.4 Experimental results of Alumina ceramic, ∅8 (mm)

Strain rate (1/s)	Young's modulus (GPa)	Ultimate strength (MPa)
2153	621.1	578
2085	561.1	500
2011	224.6	286
1808	529.9	452
1439	441.1	446
945	302.3	281

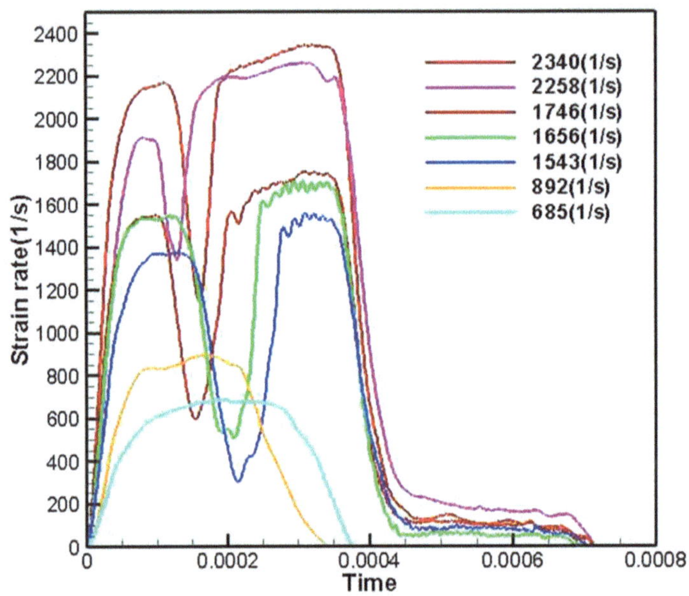

Fig. 6.6 Strain rate – time curve

The trend of stress-strain curves in the range of strain rates tested in Figs. 6.5 and 6.6 is very similar. And the strength of the limit strength looks quite similar.

As shown in Figs. 6.5 and 6.7, they mainly begin to exhibit linearity to reach the maximum ultimate strength to produce damage. Here, they indicate that they are pressed by pressure, and cracks are mainly generated from the point of contact between the test piece and the rod. They are damaged from the loading point, and then expand to the center of the disc. Thus, they observed the same trend from the figure.

When comparing the high strain rate performance between alumina ceramics diameter of 6 and 8-mm, as shown in Figs. 6.5 and 6.7, it is quite obvious that alumina ceramic diameter of 8-mm is more sensitive to the strain rates and possess higher and maximum ultimate strength. Figures 6.5 and 6.7 show the tensile strength values of alumina ceramics increasing with the increase of the loading rate. Therefore, Diameter of 8-mm alumina ceramic possess better mechanical properties to withstand dynamic tension loading.

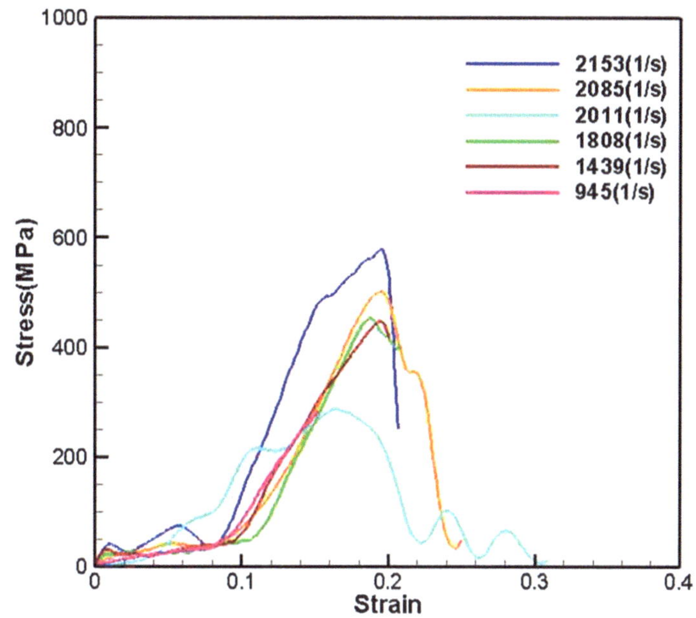

Fig. 6.7 Stress – strain curve

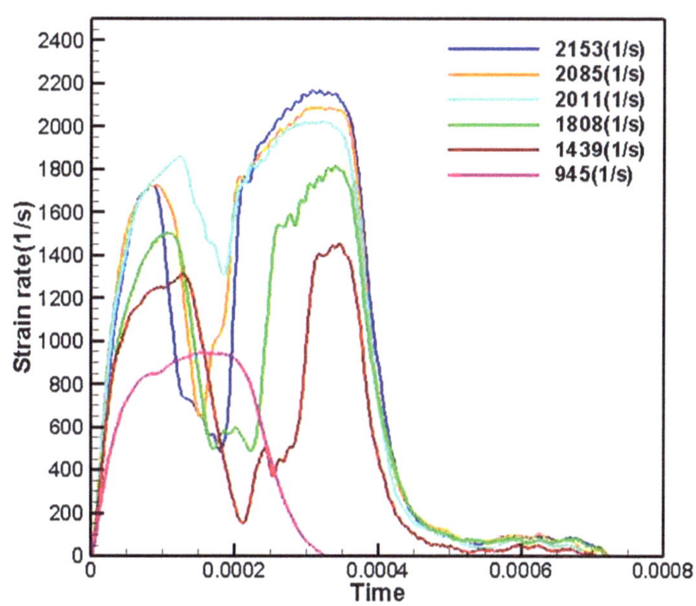

Fig. 6.8 Strain rate – time curve

At present, we use the parameters in addition to the size of the specimen to do the same, from the above two experimental results that the diameter of 8-mm specimen tensile strength slightly higher than 6-mm, but the experiment is in the initial stage, the lack of more data to compare, The future will be into the use of more different diameters of test pieces to analyze.

Acknowledgments This research was supported by the Taiwan National Science Council under grant no.105-2221-E-151-015, and National Kaohsiung University of Applied Sciences, Taiwan.

References

1. Gitzen, W.H.: Alumina as a Ceramic Material. Wiley (1970)
2. Pivec, R., Johnson, A.J., Mears, S.C., Mont, M.A.: Hip arthroplasty. The Lancet. **380**(9855), 1768–1777 (2012)
3. Hsu, S.M., Shen, M.C.: Ceramic wear maps. Wear. **200**, 154–175 (1996)
4. Willmann, G.: Ceramics for total hip replacement – what a surgeon should know. Orthopedics. **21**(2), 173–177 (1998)
5. Vallet-Regi, M.: Bio-ceramics with Clinical Applications. Wiley, Chichester (2014)
6. Kolsky, H.: An investigation of the mechanical properties of materials at very high rates of loading. Proc. Phys. Soc. **B62**, 676 (1949)
7. Iqbal, M., Mohanty, B., Xia, K.: Dynamic tensile strength and mode-I fracture toughness in granitic rocks, pp. 1–8, Orlando, Florida, USA (2008)
8. ISRM: Suggested methods for determining tensile strength of rock materials. Int. J. Rock Mech. Min. Sci. Geomech. Abstr. **15**, 99–103 (1978)
9. ASTM C496: Standard test method for splitting tensile strength of cylindrical concrete specimens. In: ASTM, Standards, 0.042, pp. 336–341. ASTM, Philadelphia (1984)
10. Wang, Q.Z., Li, W., Xie, H.P.: Dynamic split tensile test of flattened Brazilian disc of rock with SHPB setup. Mech. Mater. **41**(3), 252–260 (2009)
11. Zhou, Y.X., Xia, K., Li, X.B., Li, H.B., Ma, G.W., Zhao, J., Zhou, Z.L., Dai, F.: Suggested methods for determining the dynamic strength parameters and mode-I fracture toughness of rock materials. Int. J. Rock Mech. Min. Sci. **49**, 105–112 (2012)

Chapter 7
Digital Gradient Sensing Method to Visualize and Quantify Crack-Tip Deformations in Soda-Lime Glass Under Static and Dynamic Loading

Balamurugan M. Sundaram and Hareesh V. Tippur

Abstract Transparent ceramic materials such as soda-lime glass pose unique challenges for visualizing and quantifying deformations to characterize fracture and failure. The relatively low fracture toughness coupled with high stiffness and elastic wave speeds create spatio-temporal challenges as deformations tend to be confined to a very small region around an almost mathematically sharp crack that tends to propagate at ~1500 m/s. Soda-lime glass show weak birefringence to be able to use conventional photoelasticity. Interferometric methods often need elaborate coherent optics and are less attractive in terms of experimental simplicity. Digital image correlation (DIC) techniques, on the other hand, require creating speckles on the specimen surface which makes it difficult to locate a propagating crack-tip for evaluating fracture parameters precisely. Also, achieving a good measurement resolution at conventional magnifications using currently fielded high-speed cameras is difficult. Motivated by these overlapping factors, the feasibility of Digital Gradient Sensing (DGS) method to measure crack-tip deformations near stationary and growing cracks in soda-lime glass is investigated in this work. Both quasi-static and dynamic problems associated with crack initiation and growth are successfully demonstrated.

Keywords Digital Gradient Sensing • Transparent Ceramics • Soda-lime Glass • High-Speed Photography • Dynamic Fracture

7.1 Introduction

Soda-lime glass (plate glass) is a common structural material. Its superior optical transparency, scratch resistance, high stiffness, good chemical resistance and low-cost make it very attractive for structural applications. Further, understanding the dynamic fracture behavior of transparent ceramics in general is important for life critical applications such as automotive windshields, blast resistant structural cladding, etc. Performing full-field, non-contact measurement of deformations in glasses is rather challenging due to high spatial and temporal resolution requirements. The lack of a simple, full-field, non-contact optical technique to study glasses (and ceramics in general) needs attention.

In this context, the present research examines the feasibility of a full-field optical technique called Digital Gradient Sensing (DGS) [1] to tackle this problem. In DGS the elasto-optic effects exhibited by transparent soda-lime glass subjected to non-uniform state of stress are quantified by evaluating the angular deflections of light rays propagating through the material. The angular deflection measurements represent two orthogonal in-plane stress gradients under plane stress conditions. These stress gradients can be integrated to obtain the stresses themselves [2]. DGS has been extended here to measure deformations in soda-lime glass and evaluate crack-tip stress intensity factors (SIFs) [3, 4]. The deformation fields have been mapped and analyzed under both static and dynamic loading conditions.

7.2 Digital Gradient Sensing Method

In transmission-mode DGS method [1] random black and white speckles coated on a planar surface ('target') is photographed through an optically transparent specimen. A white light is used to illuminate the target. The speckle pattern is first photographed through the specimen in a no-load/undeformed state to obtain a reference image. When loaded, the specimen

B.M. Sundaram • H.V. Tippur (✉)
Department of Mechanical Engineering, Auburn University, Auburn, AL, 36849, USA
e-mail: htippur@eng.auburn.edu

J. Kimberley et al. (eds.), *Dynamic Behavior of Materials, Volume 1*, Conference Proceedings of the Society for Experimental Mechanics Series, DOI 10.1007/978-3-319-62956-8_7

experiences non-uniform stresses that affect the local refractive index. Additionally, the specimen thickness changes non-uniformly due to the Poisson effect. A combination of these two effects, known as the *elasto-optic effect*, cause the light rays to deflect from their initial path when they propagate through the specimen. The speckle pattern is photographed again through the specimen in this loaded/deformed state. The local deviations of light rays can be quantified by correlating speckle images in the deformed and undeformed states to obtain in-plane speckle displacement components on the target plane. The angular deflections of light can be computed if the distance between the specimen plane and the target plane is known (see Ref. [1]). The local angular deflections are related to the gradients of in-plane normal stresses as,

$$\phi_{x;y} = C_\sigma B\left[\partial\left(\sigma_x + \sigma_y\right)/\partial x; y\right], \tag{7.1}$$

where C_σ is the elasto-optical constant of the materials and B is its thickness.

7.3 Quasi-static Crack-Tip Fields

A 130 mm × 50 mm × 4.65 mm soda-lime glass SENB specimen of was used to perform three-point-bend test (see Fig. 7.1). An initial notch of 10 mm length and 2 mm width was machined on one edge of the specimen. Initial notch was intentionally kept wide to avoid premature failure. Instron 4465 testing machine was used to load the specimen in displacement control mode. The schematic of the experimental setup used is shown in Fig. 7.1. A planar target plate decorated with random black and white speckles was placed behind the specimen at a distance of 712 mm. A Point Grey Grasshopper3 digital camera placed in front of the specimen (at a distance of 445 mm) focused on the target plate through the specimen was used to record the speckle images during experiments. Two CFL lamps were used to illuminate the target plate uniformly. A reference image was recorded through the crack-tip vicinity at no-load condition. As the load increased gradually, the camera recorded images at 15 frames per minute. The images in the deformed state of the specimen were correlated with the one in the reference state using ARAMIS® image analysis software. The measurements were converted to angular deflections on the specimen plane using paraxial approximations and mapping function based on the principle of pin-hole camera. Thus obtained angular deflection fields in two orthogonal in-plane directions for a select load level of 880 N is shown in Fig. 7.2. A heavy white line is overlaid on the contour plot to represent the initial notch. Using the methodology described in Ref. [1], the SIFs were evaluated.

The mode-I SIFs evaluated from the load-cell measurements and the sample geometry [3] can be expressed as,

$$K_I = \frac{F\,S}{Bw^{3/2}}\,\frac{3(\xi)^{1/2}\left[1.99 - \xi\,(1-\xi)\left\{2.15 - 3.93\,(\xi) + 2.7(\xi)^2\right\}\right]}{2\,(1+2\xi)\,(1-\xi)^{3/2}}, \quad \xi = \frac{a}{w} \tag{7.2}$$

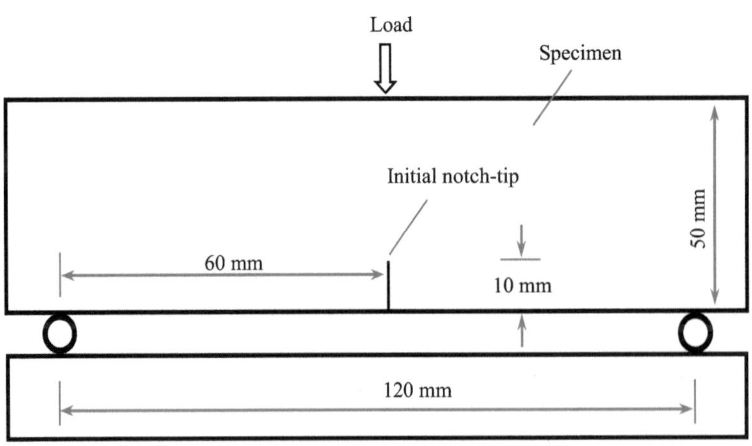

Fig. 7.1 Specimen geometry and the SEN three-point-bend setup used

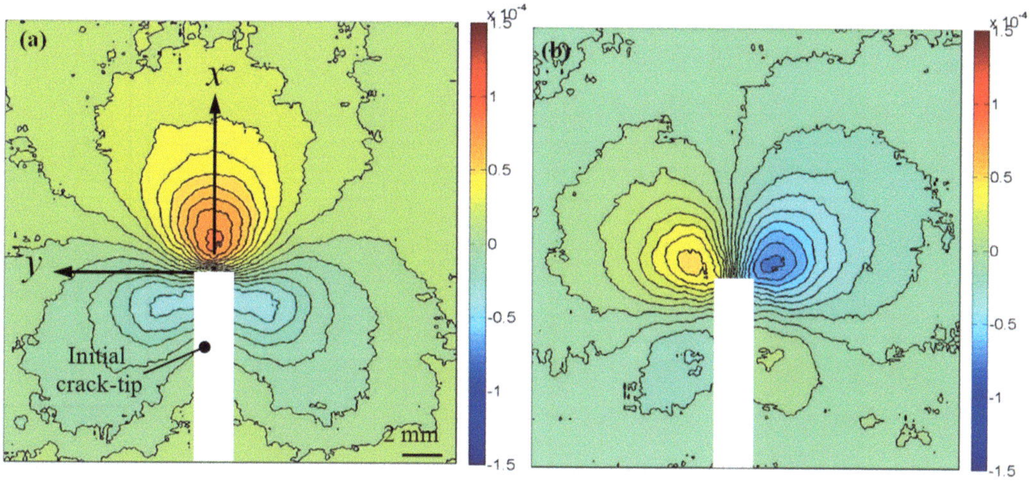

Fig. 7.2 Angular deflection contour plots (contour interval = 8×10^{-6} rad) proportional to stress gradients of $(\sigma_x + \sigma_y)$ in the (**a**) x- and, (**b**) y-directions near the notch-tip for a soda-lime glass specimen under quasi-static three-point bending load

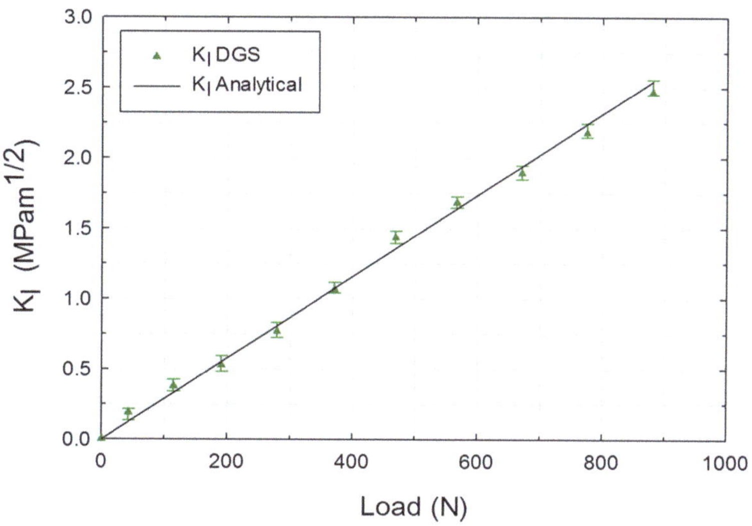

Fig. 7.3 Comparison of mode-I stress intensity factors from DGS data and Eq. (7.1)

where F is the applied load, S is the span, B is the thickness of the specimen, w is the width of the specimen and a is the initial crack length. The mode-I SIFs obtained from the regression analysis of measured DGS data at various load levels are plotted in Fig. 7.3 along alongside the ones obtained analytically from Eq. (7.1). A good agreement between both can be seen. Further, DGS can also be used to visualize and quantify contact-point deformations in soda-lime glass (see Ref. [5] for details).

7.4 Dynamic Crack-Tip Fields

A rectangular SEN specimen of dimension 150 mm × 100 mm of thickness 5.65 mm was used for conducting dynamic experiment. An initial notch of 12 mm length and 300 μm width was machined on one edge of the specimens. The specimen dimension and loading configuration is shown in Fig. 7.4. An area loading on the edge opposite to the edge with the initial notch was used to dynamically load a free standing specimen.

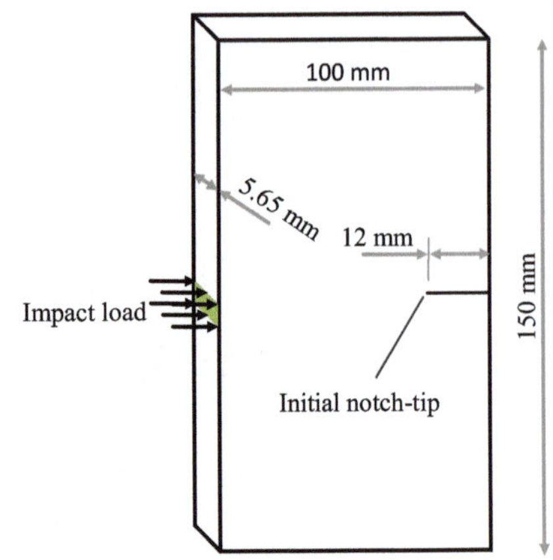

Fig. 7.4 Specimen geometry and loading configuration used to perform dynamic fracture experiments

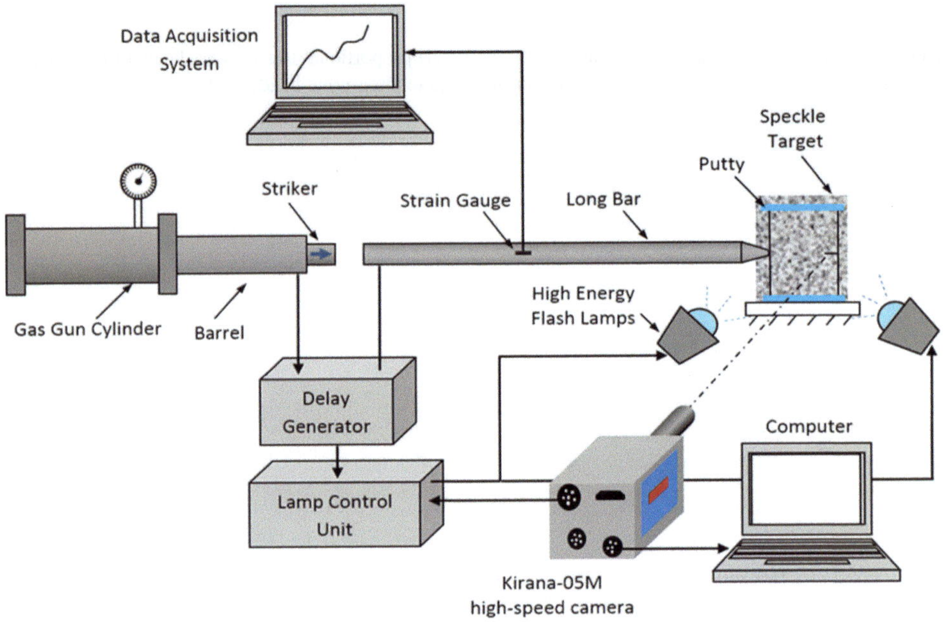

Fig. 7.5 Schematic of the experimental setup used for performing dynamic fracture experiments on soda-lime glass

The schematic of the experimental setup used for dynamic fracture tests is shown in Fig. 7.5 [6]. A Hopkinson pressure bar was used to load the sample. A 6 feet long, 1-inch diameter long-bar with a blunt wedge shaped tip was held against the unconstrained SEN specimen. A 12-inch long, 1-inch diameter striker held inside the barrel of a gas-gun was used to impact the long-bar. Both the long-bar and the striker were of the same diameter and made of maraging steel to eliminate the impedance mismatch between them. The striker was launched towards the long-bar using a gas-gun. When the striker contacted the long-bar, a compressive stress wave propagated along the bar before getting transmitted into the specimen in contact generating stress wave loading.

Figure 7.6 shows the angular deflections in the x- (top row) and y- (bottom row) directions at various time instants. The time $t = 0$ corresponds to the instant at which the crack initiates at the original crack-tip. Here the crack propagates from right to left of the specimen as indicated by the arrow head in the first contour plot. A white line is overlaid to the plots to depict the crack path along the x-axis. The crack follows a self-similar path during the window of observation consistent with mode-I loading configuration. Further, these crack tip fields can be analyzed to extract fracture parameters including crack speed and the dynamic SIF histories [7].

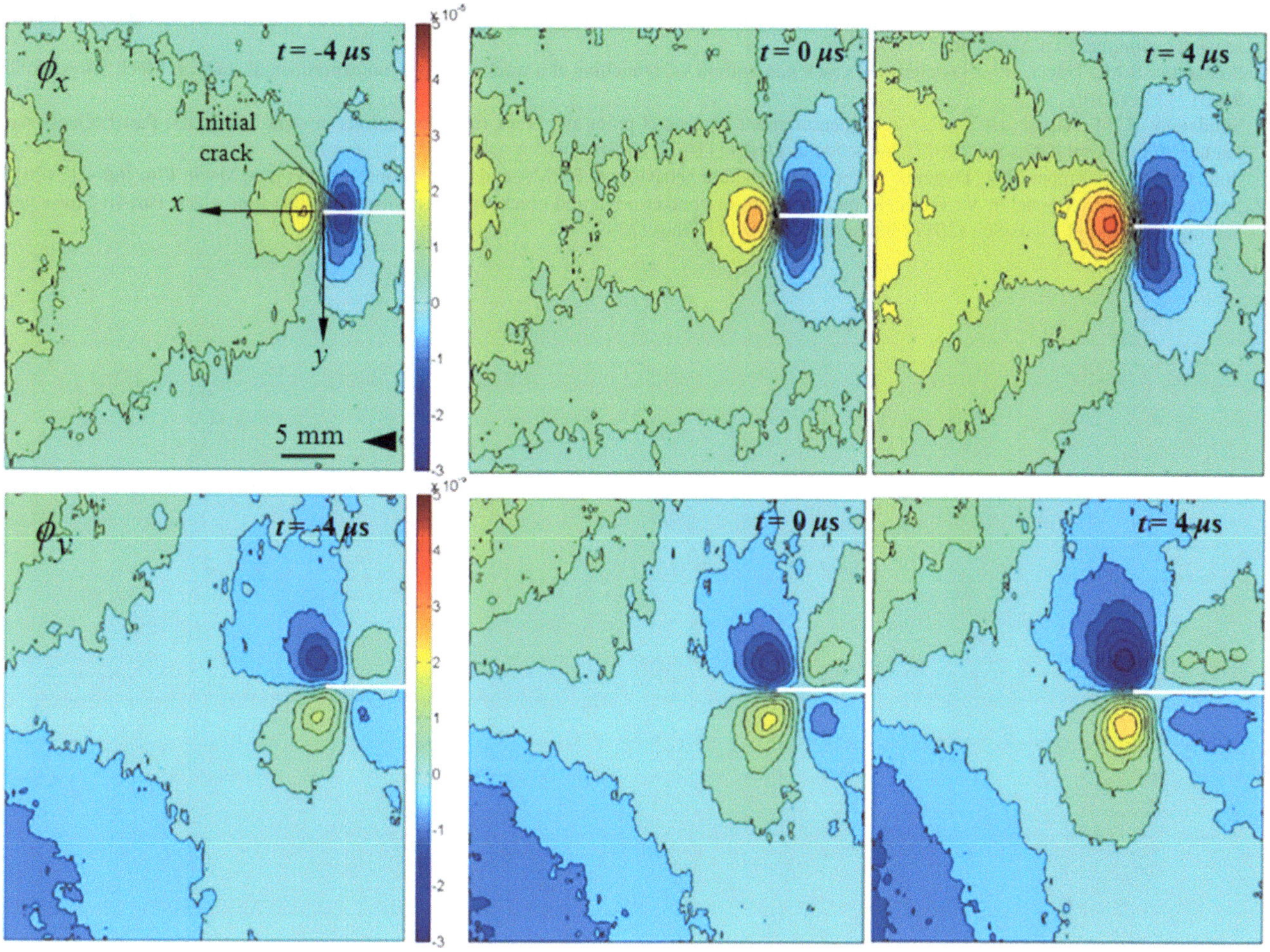

Fig. 7.6 Angular deflection contour plots (contour interval = 4×10^{-6} rad) for a soda-lime glass SEN specimen subjected to dynamic edge loading

7.5 Summary

In this research the feasibility of transmission-mode DGS to study fracture mechanics of soda-lime glass has been demonstrated under static and dynamic loading conditions. By optically mapping the crack-tip deformations in soda-lime glass – a relatively stiff, very brittle, low fracture toughness material – the possibility of investigating transparent ceramics optically is revealed. The static mode-I SIFs evaluated by analyzing the notch-tip deformations in three-point bend SEN specimens agree well with the analytical counterparts. The SIF histories under dynamic loading conditions can be evaluated by analyzing the optical fields shown in Fig. 7.6.

Acknowledgement The authors would like to thank the U.S. Army Research Office for supporting this research through grant W911NF-16-1-0093 and W911NF-15-1-0357 (DURIP).

References

1. Periasamy, C., Tippur, H.V.: A full-field digital gradient sensing method for evaluating stress gradients in transparent solids. Appl. Opt. **51**, 2088–2097 (2012)
2. Miao, C., Sundaram, B.M., Huang, L., Tippur, H.V.: Surface profile and stress field evaluation using digital gradient sensing method. Meas. Sci. Technol. **27**(9), 095203 (2016)

3. Sundaram, B.M., Tippur, H.V.: Dynamic crack growth normal to an interface in bi-layered materials: an experimental study using digital gradient sensing technique. Exp. Mech. **56**, 37–57 (2016)
4. Sundaram, B.M., Tippur, H.V.: Dynamics of crack penetration vs. branching at a weak interface: an experimental study. J. Mech. Phys. Solids. **96**, 312–332 (2016)
5. Sundaram, B. M., Tippur, H. V.: Full-field measurement of contact-point and crack-tip deformations in soda-lime glass. Part-I: Quasi-static loading. Int. J. Appl. Glass Sci. (2017). https://doi.org/10.1111/ijag.12278
6. Sundaram, B.M., Tippur, H.V.: Dynamic mixed-mode fracture behaviors of PMMA and polycarbonate. Eng. Fract. Mech. **176**, 186–212 (2017)
7. Sundaram, B. M., Tippur, H. V.: Full-field measurement of contact-point and crack-tip deformations in soda-lime glass. Part-II: Stress wave loading. Int. J. Appl. Glass Sci. (2017). https://doi.org/10.1111/ijag.12289

Chapter 8
Construction of Phase Diagrams of Mg-Zn with Selected Rare Earth (R.E) Elements

C.G. Fountzoulas

Abstract Rare earth alloys are a relatively new group of materials. Magnesium is the lightest structural metal, which is used in a variety of markets and applications, each one using the unique physical and mechanical properties of the element and its alloys. Addition of RE can refine the grains and weaken the anisotropy to improve its ductility and produce thermally stable precipitates with high melting point (above 620 °C), an improvement over precipitates of the Mg-Zn system. The high temperature strength and creep resistance of Mg alloys, such as ZK60, which is Mg alloyed with other elements often aluminum, zinc, manganese, silicon, copper, rare earths and zirconium, are improved by the addition of various amounts of rare earth elements (i.e. 0.5, 1.5, and 2.5 wt%). The Phase Diagrams and thermodynamic properties of these alloys are necessary for the modeling and development of materials and processes for innovative applications involving rare-earth based materials. However, there are very few known phase diagrams of Mg and magnesium alloys with rare earth elements. In this work we study the construction of the phase diagram of Mg-Zn alloyed with Pr, a light R.E. element, Y, and Er, a heavy R.E. element respectively using the Thermo-Calc [1] commercial software.

Keywords Phase diagram • Mg-rare earth elements alloys • Thermo-Calc • Modeling

8.1 Introduction

Aluminum and magnesium are the lightest metallic materials used for structural applications with densities of 2.70 g/cm^3 and 1.738 g/cm^3 respectively. Aluminum and magnesium are the third and the eighth most abundant element on the earth's crust, 8.1% and 2.1% respectively [2]. It is an alkaline earth element (Group II) that crystallizes in a hexagonal structure (hcp-A3). Magnesium alloys have excellent strength-to-weight ratio, good fatigue and impact strengths, and relatively large thermal and electrical conductivities [3]. Magnesium compounds are used as refractory material in furnace linings for metal production such as iron and steel, and nonferrous metals, glass, and cement [3]. Due to its small density it has numerous applications in the aeroplane and missile construction. Magnesium components are widely used in industry and agriculture. Other uses include: removal of sulphur form iron and steel, photoengraved plates in the printing industry; reducing agent for the production of pure uranium and other metals from their salts; flashlight photography, flares, and pyrotechnics [3]. Magnesium alloys are one of the most promising light-weight materials for automotive, aerospace, consumer electronic (computer-camera, and cell phone), and biomedical applications due to its biodegradability [4]. It is being used in the automotive industries in steering column parts, shift actuators, valve covers and housings, brackets, and intake manifold blades [3]. However, poor ductility and secondary processing induced crystallographic asymmetry (texturing) caused by its hexagonal closed pack (HCP) crystal structure constitute the major limitations of Mg. According to Tekumalla et al. [5], its limited ductility is due to few available slip systems and the activation difficulty of non-basal slip at room temperature. However, development of new magnesium based alloy have overcome this limitation. Rare earth metals (RE) addition is promising, by weakening the texture and improving the deformability of Mg [5]. In addition, the hard eutectic phases formed also improve the strength of the alloy [5]. It is believed that the strengthening of Mg by the addition of RE is caused by a solid solution strengthening mechanism and a precipitation hardening mechanism [5].

According to Tekumall et al. [5], the Mg-RE based alloys are useful in automotive industry as superior light metal-alloys in cast or wrought condition. The Mg-RE alloys are also used in biomedical applications as biodegradable implant materials

C.G. Fountzoulas (✉)
U.S. Army Research Laboratory, WMRD, RDRL-WMM-B, Aberdeen Proving Ground, Aberdeen, MD 21005-5069, USA
e-mail: htippur@eng.auburn.edu

J. Kimberley et al. (eds.), *Dynamic Behavior of Materials, Volume 1*, Conference Proceedings of the Society for Experimental Mechanics Series, DOI 10.1007/978-3-319-62956-8_8

that help in healing of the tissues and leaving no implant residues. To be used in such applications, the Mg-RE alloys must be fabricated economically with simple processing. The Mg-RE phase diagrams indicate that each RE behaves uniquely when it is added as a dominant alloying element [4]. In this work we study the construction of the phase diagram of Mg alloyed with 1, 2, 3, 4, and 5% Pr, Y, and Er, respectively alloyed with 0–100% Zn, using the material models of Thermo-Calc commercial software library.

8.2 Modeling and Simulations

The commercial software Thermo-Calc was used to construct the phase diagrams of the above mentioned alloys, using only the existing thermodynamic models of the software library. Figure 8.1a shows the phase diagram of Mn-Zn system constructed using the Thermo-Calc library. Figure 8.1b shows a detailed phase diagram of Mg-Zn [4] constructed with the aid of CALPHAD software optimization [6], which is part of the Thermo-Calc software, using the heat of formation of the intermetallic compounds obtained from experimental, and first principle calculations. Figure 8.1a shows a peritectic invariant metallurgical reaction at 16.70 Zn w/o and 620 °C that does not appear in Fig. 8.1b. Figure 8.1a shows phase α3, which corresponds to solid solution of Zn in Mg. Figure 8.1a shows above 15% Zn a Zn (HCP) single phase, However, Figure 8.1b shows various Mg-Zn phases.

Figure 8.2a–d shows the phase diagrams of Mg-Zn alloys containing 1%, 2%, 3% and 5% Pr respectively. The presence of praseodymium (Pr) has affected the peritectic invariant reaction, stoichiometry, temperature and appearance. The stoichiometry and temperature of the peritectic has been changed from 16.70% Zn and 620 °C (Fig. 8.1a) to 20% Zn and temperature 618 °C to 612 °C with increasing Pr stoichiometry {1%–5%]. The simulations showed that two new phases appeared, a BCC and DHCP microstructures, which qualitative resembles, although remotely, the various phases appear in Fig. 8.1a, of unknown composition.

Figure 8.3a–d shows the phase diagrams of Mg-Zn alloys containing 1%, 2%, 3% and 5% Y.

Figure 8.3a–d shows the phase diagrams of Mg-Zn alloys containing 1%, 2%, 3% and 5% yttrium (Y) respectively. The presence of yttrium has also affected the stoichiometry and temperature of the peritectic invariant reaction. The peritectic stoichiometry varied from 16% to 18% 20% 21% and 25% with increasing amount of Y. The peritectic temperature varied as 614 °C, 610 °C, 602 °C, 600 °C and 593 °C with increasing Y composition. No new phases appeared, although a "knee" at higher compositions than the peritectic composition and lower temperatures lower than the peritectic invariant reaction (Fig. 8.3d, 63% Zn and 593 °C), a possible indication of different metallurgical reaction.

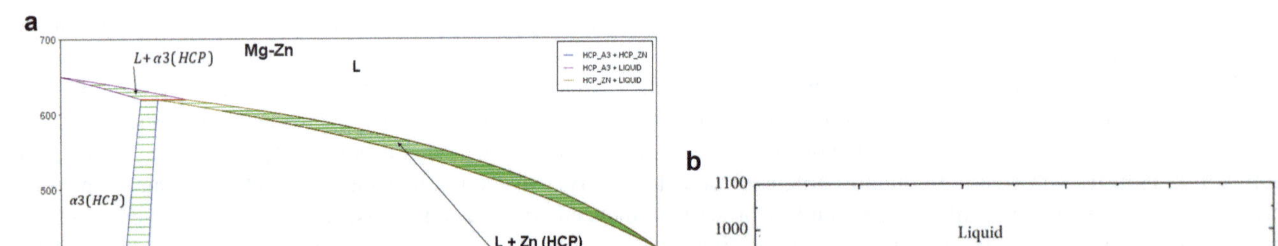

Fig. 8.1 Mg-Zn phase diagram: (**a**) using thermodynamic models of Thermo-Calc library; (**b**) derived from Islam et al. [4]

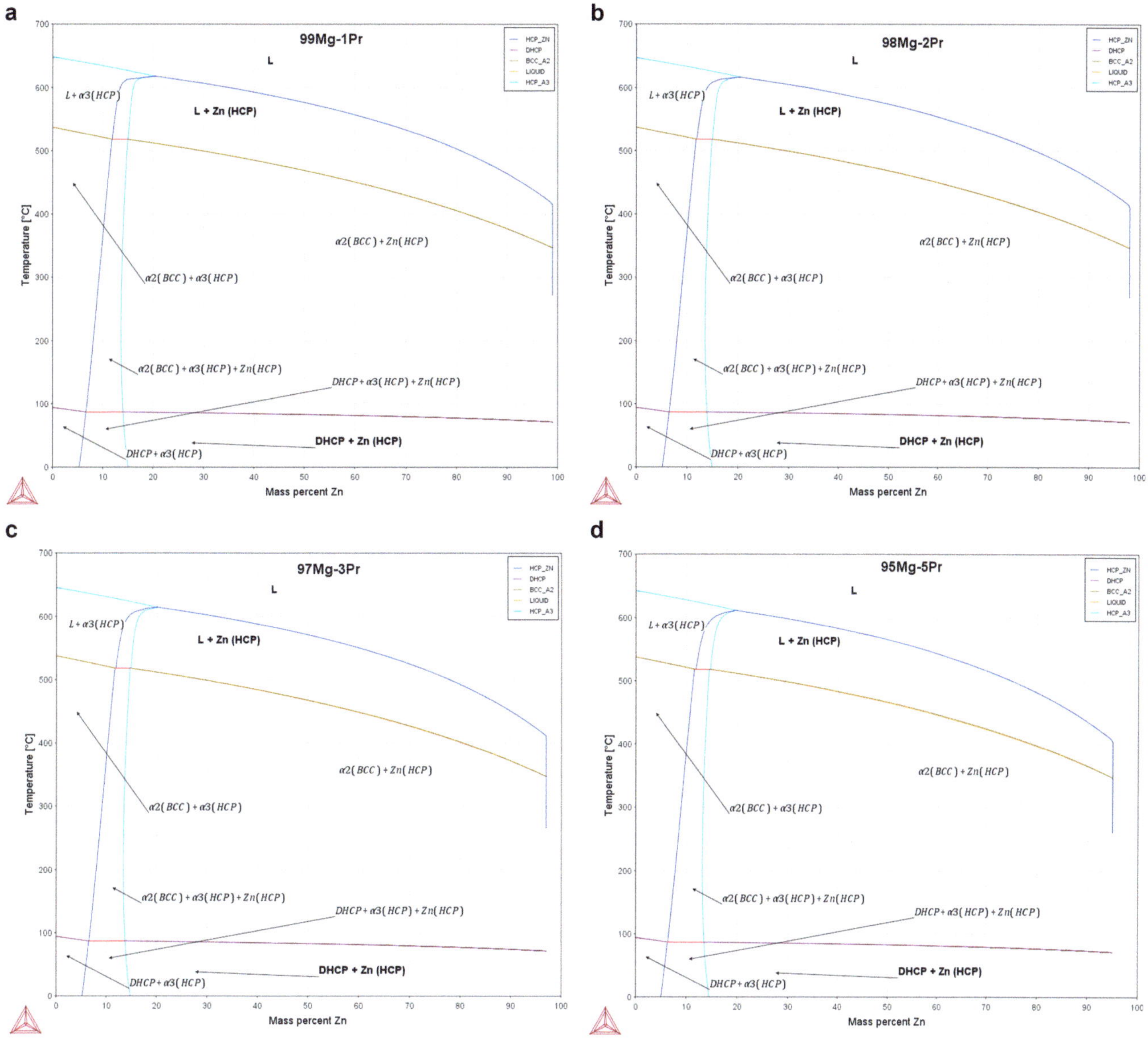

Fig. 8.2 Phase diagram of Mg-Zn-Pr: (**a**) 1% Pr; (**b**) 2% Pr; (**c**) 3% Pr; (**d**) 5% Pr

Figure 8.4a–d shows the phase diagrams of Mg-Zn alloys containing 1%, 2%, 3% and 5% erbium (Er) respectively. No new phases appear. The peritectic stoichiometry varied from 14.75% to 16%, 17%, 19% and 20% with increasing amount of Er. The peritectic temperature varied as 612 °C, 610 °C, 607 °C, 607 °C and 602 °C with increasing Y composition. No new phases appeared, although a "knee" at higher compositions than the peritectic composition and lower temperatures lower than the peritectic invariant reaction (Fig. 8.4d, 62% Zn and 539 °C), a possible indication of different metallurgical reactions.

8.3 Conclusions

The phase diagrams of Mg-Zn alloys with the addition of 1%, 2%, 3% 4% and 5% (w/o) praseodymium (Pr), yttrium (Y) and erbium (Er) were constructed with the aid of the thermodynamic models in the library of the commercial software Thermo-Calc. The Mg-Zn phase diagram constructed with the aid of the Thermo-Calc and it was compared with the published by

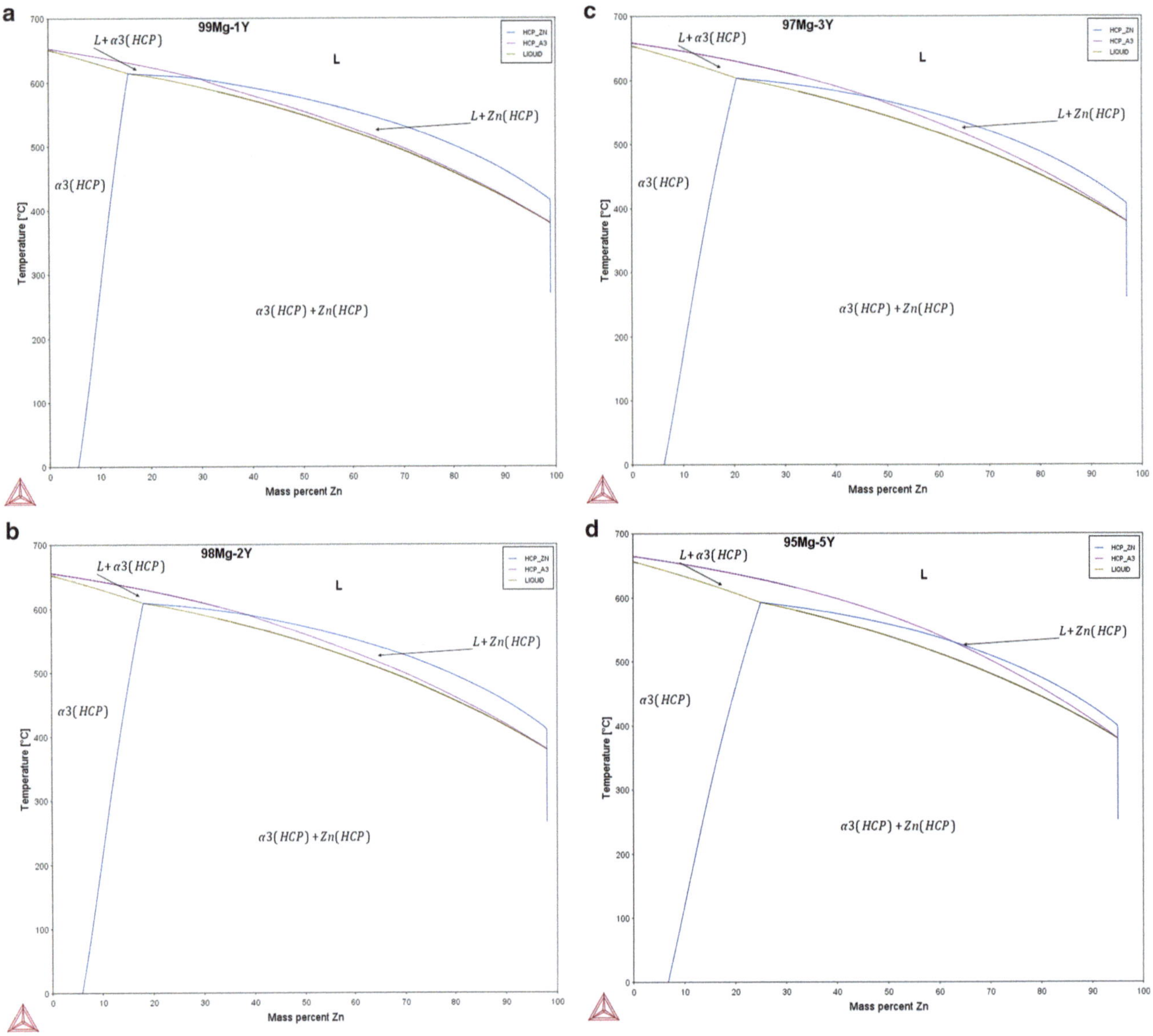

Fig. 8.3 Phase diagram of Mg-Zn-Y: (**a**) 1% Y; (**b**) 2% Y; (**c**) 3% Y; (**d**) 5% Y

Islam et al. [4]. The latter was constructed with the aid of CALPHAD software optimization [6], which is part of the Thermo-Calc software, using the heat of formation of the intermetallic compounds obtained from experimental, and first principle calculations. The Mg-Zn phase diagram we constructed was different than the constructed by Islam et al. [4], whose interim phases did not appear in ours.

However, it was shown that the addition of the rare earth metals, Pr, Y and Er affect the composition and temperature of the invariant metallurgical reaction (peritectic) and the stoichiometry of the solid solution of the R.E. element in the Mg-Zn alloy system. Detailed phase diagrams of the Mg-Zn-R.E systems can be constructed with the aid of Thermo-Calc, by modifying the thermodynamic models of the Thermo-Calc by using available experimental data and improved thermodynamic model, such as Gibbs free energy of the R.E. metals.

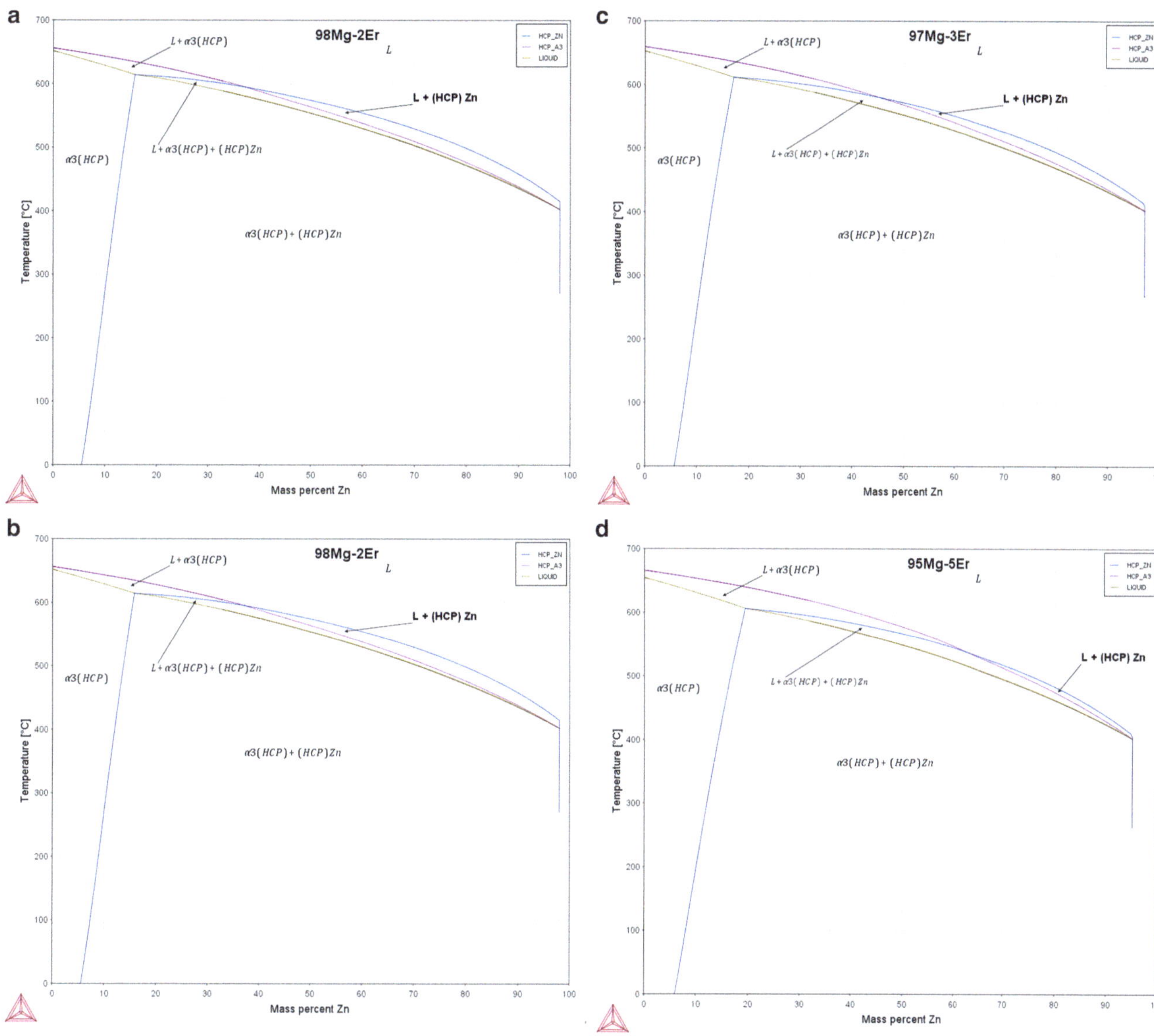

Fig. 8.4 Phase diagram of Mg-Zn-Er: (**a**) 1% Er; (**b**) 2% Er; (**c**) 3% Er; (**d**) 5% Er

References

1. http://www.thermocalc.com/
2. http://hyperphysics.phy-astr.gsu.edu/hbase/Tables/elabund.html
3. http://www.lenntech.com/periodic/elements/mg.htm
4. Mezbahul-Islam, M., Mostafa, A.O., Medraj, M.: Essential magnesium alloys binary phase diagrams and their thermochemical data. J. Mater. **2014**, 33, Article ID 704283 (2014)
5. Tekumalla, S., Seetharaman, S., Almajid, A., Gupta, M.: Mechanical properties of magnesium-rare earth alloy systems: a review. Metals. **5**, 1–39 (2015). doi:10.3390/met5010001
6. http://www.calphad.org/

C.G. Fountzoulas received his Ph.D. in Materials Science and Engineering and M.S. in Mechanical Engineering from Northeastern University, Boston, MA, and Diploma in Mining and Metallurgical Engineering from National Technical University, Athens, Greece. He taught at Northeastern and Tufts University, Boston, MA and he is currently an adjunct associate professor at School of Engineering, Widener University, Chester, PA. He is the lead modeling expert for transparent armor technologies for the Army Research Laboratory. He has over 8 years of experience in predicting ballistic performance of laminates using glass, plastic, polymer, and ceramic materials by using the AUTODYN, WORKBENCH, THERMOCALC.

Chapter 9
High Strain Rate Transverse Compression Response of Ballistic Single Fibers

Subramani Sockalingam, Daniel T. Casem, Tusit Weerasooriya, and John W. Gillespie Jr.

Abstract Ballistic impact onto fiber-based personnel protection armor systems induce multiaxial loading in the affected impact zone that includes axial tension, axial compression, transverse compression and transverse shear. The influence of the transverse compression response of ballistic fibers at high rates of loading is not well understood. In this study, high strain rate transverse compression response of aramid Kevlar KM2 and ultra-high molecular weight polyethylene (UHMWPE) Dyneema SK76 single fibers are investigated. Micron scale single fibers are transversely compressed at high loading rates in a small diameter Kolsky bar with optical interferometry to measure strains in the bars. The fibers exhibit a nonlinear inelastic behavior in transverse compression. Comparison of high rate and quasi-static response indicate a smaller contact area growth and a stiffer material response at higher rates of loading for both types of fibers.

Keywords Kevlar • UHMWPE • Ballistic Impact • Transverse Compression • Fibers

9.1 Introduction

Kevlar KM2 and Ultra-high molecular weight polyethylene (UHMWPE) Dyneema SK76 fibers are used in personnel protection ballistic impact applications [1, 2] in the form of multi-layered flexible textile fabrics and laminated composites. These polymeric fibers exhibit a fibrillar structure [3] and lend themselves to such applications due to their superior specific axial tensile strength and specific tensile modulus. The fibers experience multi-axial loading [4, 5] including axial tension, axial compression, transverse compression and transverse shear during impact. While axial specific toughness and longitudinal wave speed are important fiber properties contributing to the ballistic performance [6], the role of transverse properties, the effect of transverse loading on the longitudinal strength degradation and the associated macro, micro and sub-structural damage mechanisms during impact are not well understood. Under quasi-static (QS) transverse compression, both Kevlar KM2 and Dyneema SK76 fibers exhibit a nonlinear inelastic nominal stress-strain behavior [7, 8] with a very small elastic limit of about 1%. This inelastic behavior influences the intensity of fiber bounce and the evolution of axial strains and subsequent fiber failure during impact [9]. The Kevlar KM2 fibers also show a reduction in tensile strength when subjected to prior transverse compression [5, 10, 11]. In this work we build upon our previous QS work to investigate the high strain rate (HSR) transverse deformation behavior of Kevlar KM2 and Dyneema SK76 single fibers.

S. Sockalingam (✉)
SmartState Center for Multifunctional Materials and Structures, University of South Carolina, Columbia, SC, USA

Department of Mechanical Engineering, University of South Carolina, Columbia, SC, USA

Center for Composite Materials, University of Delaware, Newark, DE, USA
e-mail: sockalin@cec.sc.edu

D.T. Casem • T. Weerasooriya
US Army Research Laboratory, Aberdeen Proving Ground, Aberdeen, MD, USA

J.W. Gillespie Jr.
Center for Composite Materials, University of Delaware, Newark, DE, USA

Department of Mechanical Engineering, University of Delaware, Newark, DE, USA

Department of Materials Science and Engineering, University of Delaware, Newark, DE, USA

Department of Civil and Environmental Engineering, University of Delaware, Newark, DE, USA

J. Kimberley et al. (eds.), *Dynamic Behavior of Materials, Volume 1*, Conference Proceedings of the Society for Experimental Mechanics Series, DOI 10.1007/978-3-319-62956-8_9

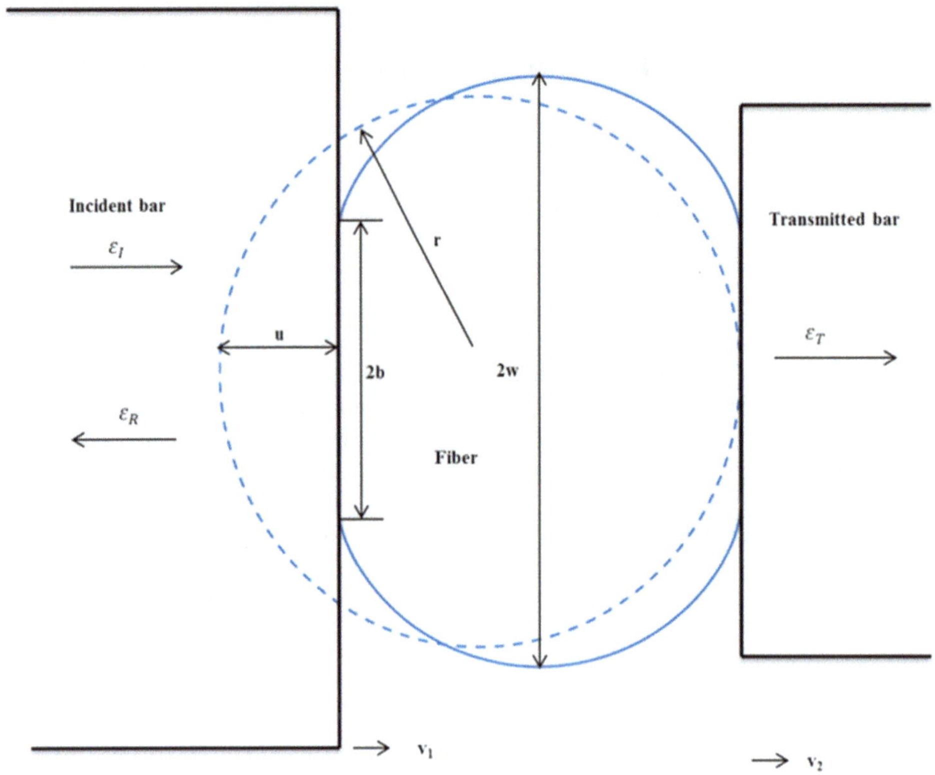

Fig. 9.1 Schematic of single fiber HSR transverse compression (not to scale)

9.2 Experimental Set Up

The HSR experimental set up involves compressing a single fiber in a smaller diameter Kolsky bar set up as shown in the schematic in Fig. 9.1. The steel incident and transmitted bar diameters are 3.175 mm and 0.283 mm respectively. A semiconductor strain gage on the incident bar is used to record the input pulse. Optical instrumentation using a normal displacement interferometer (NDI) is used at the free end of the transmitted bar to record the particle velocity [12]. The compressive load per unit length (F) and displacement (u) of the fiber are measured in real time. Compressed width ($2w$) and contact width ($2b$) is measured post-test at different applied load levels. The fibers are compressed at nominal strain rates (velocity divided by original fiber diameter) in the range of 10,000–90,000 1/s and the corresponding particles velocities are in the range of 0.20–1.10 m/s. The high strain rate achieved in Kolsky bar is seven orders of magnitude higher than the QS strain rate of 0.0059 1/s. The applied strain rate is varied by controlling the pressure applied to accelerate the striker bar. However, in this set up it is difficult to control the applied maximum nominal strain at a given strain rate. A more detailed explanation of the experimental set up is given in [12].

9.3 Results and Discussion

The experimental nominal stress $\overline{\sigma} = \frac{F}{d}$ nominal strain $\overline{\varepsilon} = \frac{u}{d}$ (d is the undeformed fiber diameter) response for Kevlar KM2 and Dyneema SK76 fibers is shown in Fig. 9.2. Both Kevlar KM2 and Dyneema SK76 fibers exhibit a nonlinear inelastic response with negligible strain recovery during unloading. The nonlinearity is attributed to both geometric stiffening due to growing contact area and material softening.

Figure 9.3 shows confocal microscopy images of both types of fibers subjected to the highest strain rates. It is seen that the fibers are compressed uniformly along the length. The length (L) of the compressed fiber is measured to be the same as the diameter of the transmitted bar.

The normalized effective contact width defined as $2w_{eff} = \frac{2w+2b}{2}$ as a function of maximum applied nominal strain is shown in Fig. 9.4. The general trend for both fibers is that the contact area growth at higher rates is found to be smaller than the quasi-static rate of loading. This may be attributed to the smaller time scales associated with higher rates such that there is not sufficient time for the inter-fibrillar damage to occur resulting in smaller width growth. This observation also indicates that inter-fibrillar interactions and associated adhesive stress may be strain rate dependent.

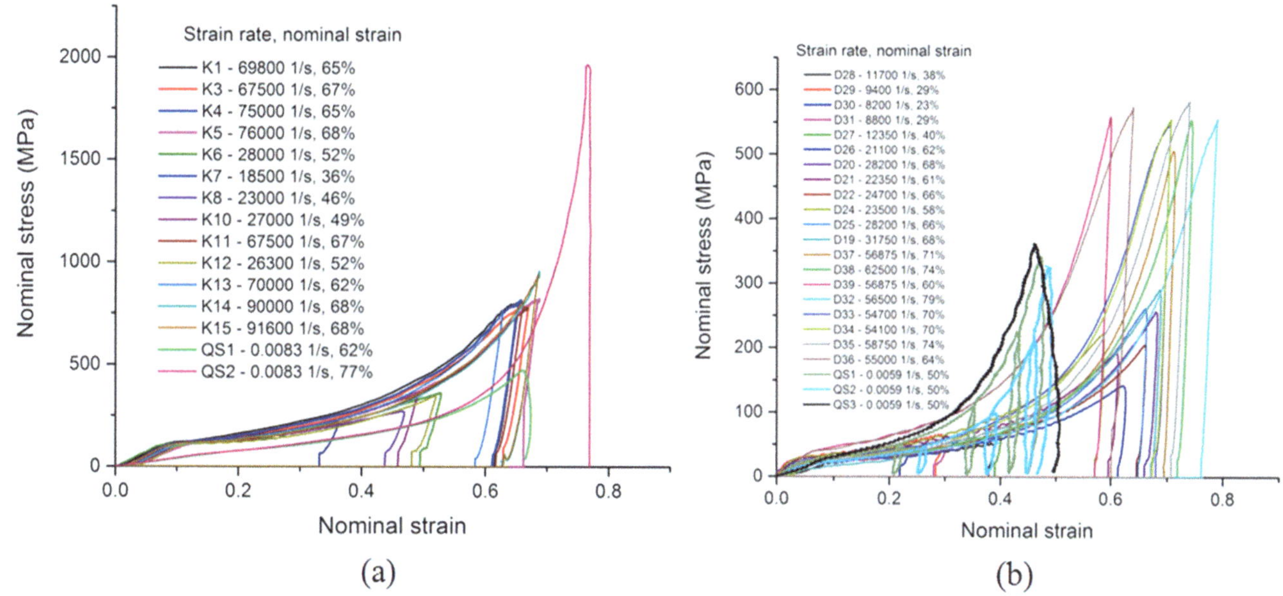

Fig. 9.2 HSR experimental nominal stress strain (**a**) Kevlar KM2 (**b**) Dyneema SK76

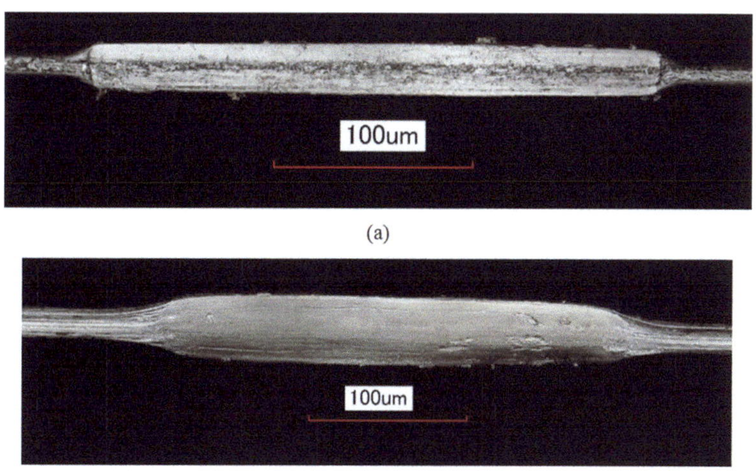

(b)

Fig. 9.3 Confocal microscopy images of deformed fibers at high strain rates (**a**) Kevlar KM2 at 75,000 1/s to 65% nominal strain (**b**) Dyneema SK76 at 54,117 1/s to 70% nominal strain

The intrinsic material behavior is isolated from the total response using the measured contact area by defining true stress as $\sigma = \frac{F}{2w_{eff}}$. The original fiber cross section is assumed to be a circle and the compressed fiber cross section is assumed to be a rectangle. The compressed cross sectional area is estimated as $2w_{eff} \times (2r - u)$. The cross sectional area remains approximately constant at different nominal strain levels and appears to be independent of the strain rate for Kevlar KM2. Whereas for Dyneema SK76, an increase in cross sectional is observed at quasi-static rate of loading which is attributed to the higher degree of fibrillation through atomic force microscopy (AFM) analysis [8]. At higher rates, the cross sectional area is found to slightly decrease with increase in the applied nominal strain. Further AFM investigation is needed to better understand this trend in Dyneema SK76 fibers.

The true strain is obtained by equating the internal energy to the external work done given by Eq. (9.1). The true strains are computed incrementally using trapezoidal rule to evaluate the integrals in Eq. (9.1). The normalized compressed cross sectional area is found to be approximately constant as shown in Fig. 9.5b.

$$\int_{A_0}^{A} \int_{0}^{\varepsilon_{eff}} \sigma_{eff} d\varepsilon_{eff} \, dA = \int_{0}^{u} F \, du \tag{9.1}$$

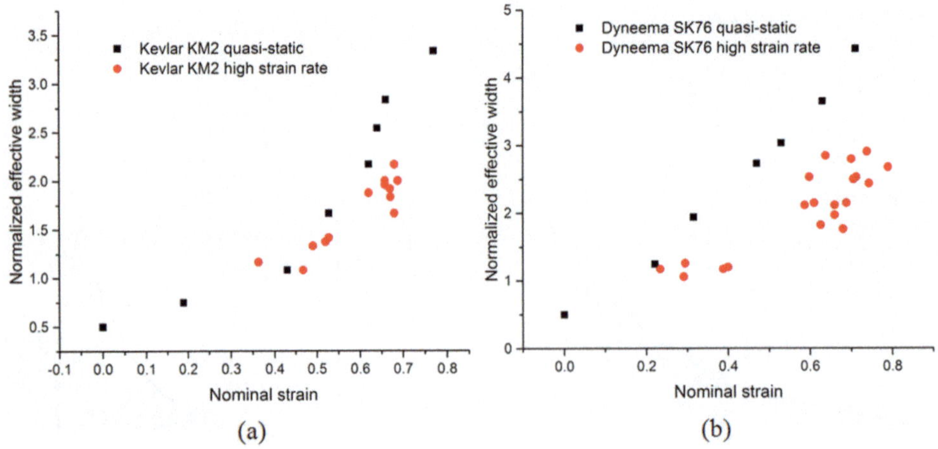

Fig. 9.4 Normalized effective width (**a**) Kevlar KM2 (**b**) Dyneema SK76

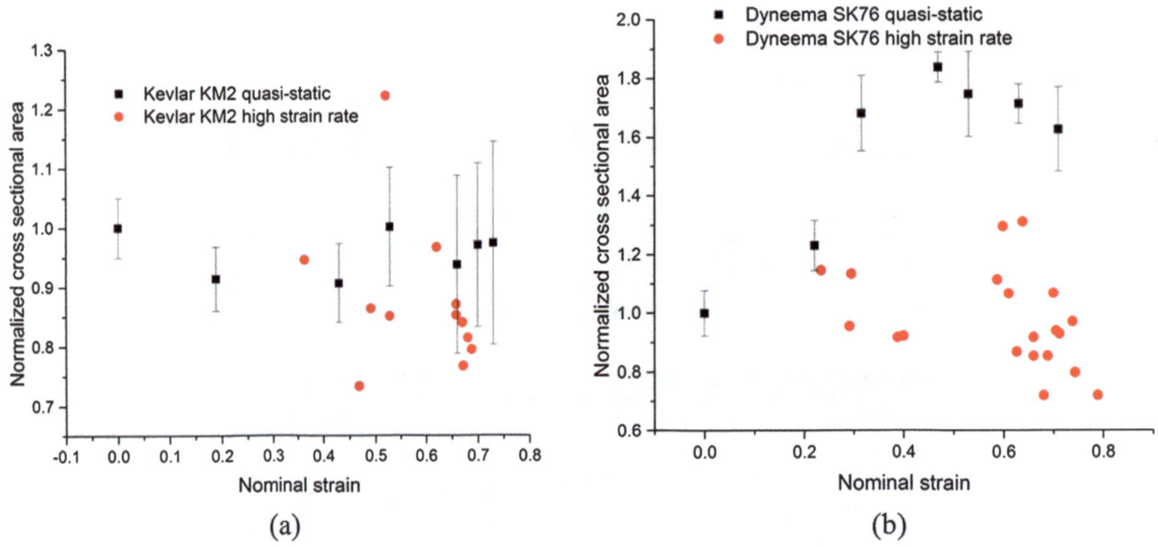

Fig. 9.5 Normalized cross sectional area (**a**) Kevlar KM2 (**b**) Dyneema SK76

The true stress strain response obtained by imposing a constant volume is shown in Fig. 9.6. Comparison to representative QS response indicates a stiffer material response in both Kevlar KM2 and Dyneema SK76 at higher rates. The Kevlar KM2 data may be grouped into three different strain rates – 24,700, 71,200 and 84,100 1/s. While there is only marginal difference in the true stress-strain response among these HSR groups, the response is significantly stiffer compared to the QS. Similarly, Dyneema SK76 data may be grouped into – 9550, 25,400, 55,800 and 58,750 1/s. Dyneema SK76 true stress-strain response is more scattered compared to Kevlar KM2. This may be due to the higher diameter variation (15.0–20.0 μm) in Dyneema SK76 fibers compared to the Kevlar KM2 which has a consistent diameter of 12.0 μm.

9.4 Conclusions

This paper presented the HSR transverse compression response of Kevlar KM2 and UHMWPE Dyneema SK76 ballistic fibers in a smaller diameter Kolsky bar. The fibers exhibit nonlinear inelastic behavior under large compressive strains under HSR transverse compression in the range of 10,000–90,000 1/s. The true stress strain behavior of the fiber is determined by removing the geometric nonlinearity using the measured contact area. Both types of fibers display a smaller contact area growth and a stiffer response at higher rates compared to QS rate of loading. Further investigation is required to better understand the fibrillation mechanisms in these fibers during HSR transverse compression.

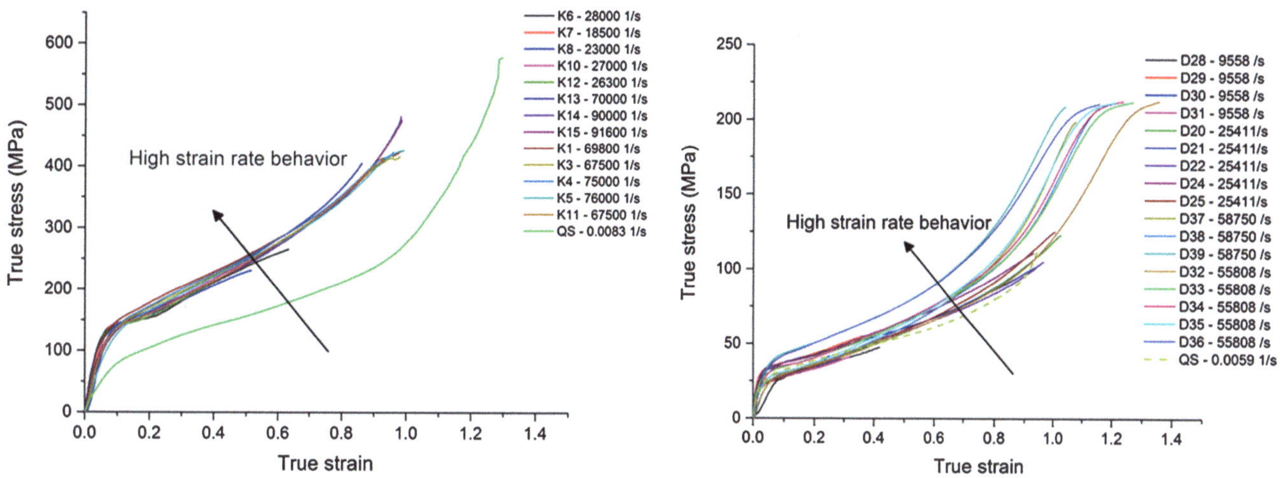

(a) (b)

Fig. 9.6 True stress strain behavior (**a**) Kevlar KM2 (**b**) Dyneema SK76

Acknowledgements Research was sponsored by the Army Research Laboratory and was accomplished under Cooperative Agreement Number W911NF-12-2-0022. The views and conclusions contained in this document are those of the authors and should not be interpreted as representing the official policies, either expressed or implied, of the Army Research Laboratory or the U.S. Government. The U.S. Government is authorized to reproduce and distribute reprints for Government purposes notwithstanding any copyright notation herein. The author SS wish to acknowledge the funding from the MEDE program through a subcontract from the University of Delaware.

References

1. Krishnan, K., Sockalingam, S., Bansal, S., Rajan, S.: Numerical simulation of ceramic composite armor subjected to ballistic impact. Compos. Part B. **41**(8), 583–593 (2010)
2. Sockalingam, S., Chowdhury, S.C., Gillespie Jr., J.W., Keefe, M.: Recent advances in modeling and experiments of Kevlar ballistic fibrils, fibers, yarns and flexible woven textile fabrics – a review. Text. Res. J. 1–27 (2016). doi:10.1177/0040517516646039
3. McDaniel, P.B., Deitzel, J.M., Gillespie, J.W.: Structural hierarchy and surface morphology of highly drawn ultra high molecular weight polyethylene fibers studied by atomic force microscopy and wide angle X-ray diffraction. Polymer. **69**, 148–158 (2015)
4. Sockalingam, S., Gillespie Jr., J.W., Keefe, M.: Dynamic modeling of Kevlar KM2 single fiber subjected to transverse impact. Int. J. Solids Struct. **67–68**(0), 297–310 (2015)
5. Sockalingam, S., Gillespie Jr., J.W., Keefe, M.: Influence of multiaxial loading on the failure of Kevlar KM2 single fiber. Text. Res. J. 1–16 (2016). doi:10.1177/0040517516681961
6. Cunniff, P.M.: Dimensionless parameters for optimization of textile-based body armor systems. In: Proceedings of the 18th International Symposium on Ballistics, pp. 1303–1310 (1999)
7. Sockalingam, S., Bremble, R., Gillespie Jr., J.W., Keefe, M.: Transverse compression behavior of Kevlar KM2 single fiber. Compos. A: Appl. Sci. Manuf. **81**, 271–281 (2016)
8. McDaniel, P.B., Sockalingam, S., Deitzel, J., Gillespie Jr., J.W., Keefe, M., Bogetti, T.A., Casem, D.T., Weerasooriya, T.: The effect of fiber meso/nanostructure on the transverse compression response of ballistic fibers. Compos. A: Appl. Sci. Manuf. **94**, 133–145 (2017)
9. Sockalingam, S., Gillespie Jr., J.W., Keefe, M.: Role of inelastic transverse compressive behavior and multiaxial loading on the transverse impact of Kevlar KM2 single fiber. Fibers. **5**(1), 9 (2017). doi:10.3390/fib5010009
10. Chowdhury, S.C., Sockalingam, S., Gillespie, J.W.: Molecular dynamics modeling of the effect of axial and transverse compression on the residual tensile properties of ballistic fiber. Fibers. **5**(1), 7 (2017)
11. Cheng, M., Chen, W., Weerasooriya, T.: Experimental investigation of the transverse mechanical properties of a single Kevlar® KM2 fiber. Int. J. Solids Struct. **41**(22–23), 6215–6232 (2004)
12. Casem, D., Grunschel, S., Schuster, B.: Normal and transverse displacement interferometers applied to small diameter Kolsky bars. Exp. Mech. **52**(2), 173–184 (2012)

Subramani Sockalingam is Assistant Professor in Mechanical Engineering at the University of South Carolina. His research interests are in the development of advanced materials that are lightweight and stronger through a multiscale experimental and computational approach for aerospace, automotive and defense applications.

Chapter 10
WIAMan ATD Polymeric Material Characterization for Under-Body Blast Environment Simulation

Mostafiz R. Chowdhury and Dawn M. Crawford

Abstract Material characterization is one of the critical challenges facing the anthropomorphic test device (ATD) modeling community because of the lack of adequate material models and inconsistency between material models available in different computational Finite Element codes. The WIAMan modeling and simulation team systematically characterized the material properties using experimental data for 8 polymeric materials used in key components in the WIAMan ATD Tech Demonstrator. Two modeling teams independently developed finite-element model of the WIAMan ATD in vertical accelerative loading using 2 different modeling platforms (LS-DYNA and Velodyne). The WIAMan FEM has successfully supported the design and development of the WIAMan ATD, the world's first biofidelic ATD for vertical accelerative loading. The approaches to derive the material parameters, constitutive material models, and the fitting algorithms are briefly discussed in this paper. A brief comparison of the whole-body simulation results of an ATD response to those of the corresponding test data due to vertical accelerative loading shows strong confidence in the predictive capability of the WIAMan ATD FE model.

Keywords Under-body blast • Finite element model • WIAMan ATD • Polymeric material characterization

Historically, the Live-Fire Test and Evaluation (LFT&E) Community has been using the Hybrid III physical manikin for vehicle survivability and occupant safety assessment [1]. The Hybrid III in its best form (originally intended for automotive crash safety), is inadequate, has usability issues, and lacks the biofidelity to determine the risk of vehicle mounted occupant injury in an under-body blast (UBB) environment [2–4]. The Warrior Injury Assessment Manikin (WIAMan) Engineering Office (EO), sponsored by the Director, Operational Test and Evaluation (DOT&E), and executed by the Army, has been developing a biofidelic anthropomorphic test device (ATD) to enable the LFT&E community to address the deficiencies associated with the use of the Hybrid III ATD to assess the risk of mounted Soldiers' injuries subjected to UBB [5].

A finite-element (FE) virtual modeling tool has been developed to support the design and development of the WIAMan ATD. Material characterization is one of the critical challenges facing the ATD modeling community because of the lack of adequate material models and inconsistency between material models available in different FE codes. The WIAMan modeling and simulation (M&S) team systematically characterized the material properties using experimental data for eight polymeric materials used in key components in the WIAMan ATD Technology Demonstrator (TD) [6]. Figure 10.1 shows a schematic of polymeric material definitions in the WIAMan ATD. Two modeling teams independently simulated the WIAMan ATD in vertical accelerative loading simulations using two different modeling platforms (LS-DYNA and Velodyne). The approaches to derive the material parameters, constitutive material models, and the fitting algorithms are briefly discussed in this enhanced abstract. A brief summary of the whole-body simulation results of an ATD response due to vertical accelerative loading are also presented to illustrate the predictive capability of the derived material models.

10.1 WIAMan Finite Element Model

The success of an FE model depends primarily upon the model's ability to predict the necessary physics and interactions of the simulated environment. Material characterization is one of the critical challenges facing the ATD modeling communities because of the lack of adequate material models and inconsistency between material models available in different FE codes

M.R. Chowdhury (✉) • D.M. Crawford
WIAMan Engineering Office, Army Research Laboratory, APG, Aberdeen, MD, USA
e-mail: mostafiz.r.chowdhury.civ@mail.mil

J. Kimberley et al. (eds.), *Dynamic Behavior of Materials, Volume 1*, Conference Proceedings of the Society for Experimental Mechanics Series, DOI 10.1007/978-3-319-62956-8_10

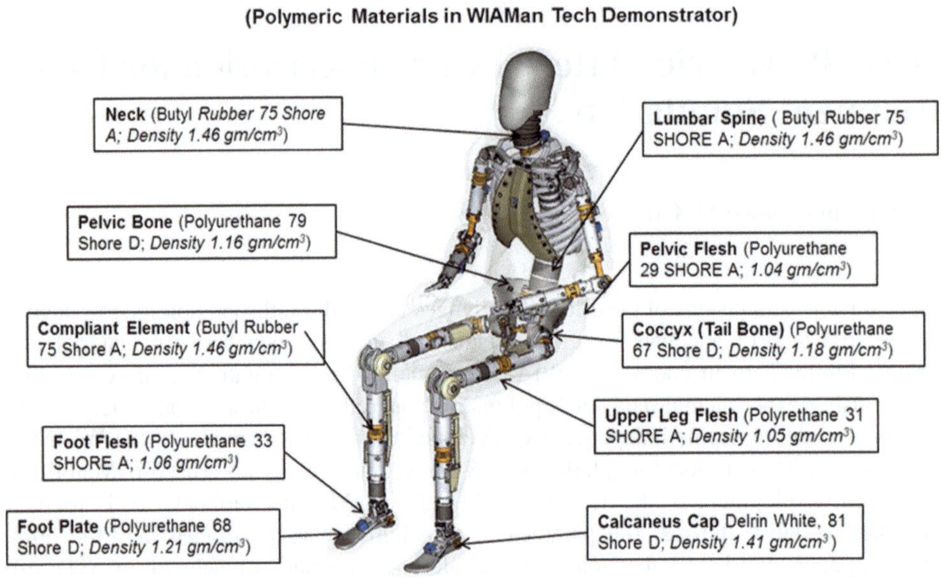

Fig. 10.1 WIAMan ATD TD polymeric material definition

[4, 7]. The WIAMan M&S effort arguably addressed these critical ATD modeling challenges in their undertakings. The WIAMan M&S team used experimentally derived material data to construct material models for the FE models developed using two independent modeling platforms [8, 9].

Material characterization was used to determine material model parameters specific to the two finite element modeling codes (LS-DYNA and VELODYNE) for eight polymeric materials used in key components in the WIAMan ATD TD. There are multiple constitutive models in each of the FEM codes used in this effort. LS-DYNA has a variety of material models that can model polymeric materials behavior using an empirical or rational approach [9–11]. VELODYNE has their own material models mostly inclusive of those available in the LS-DYNA code. The constitutive material models that these two codes used in their numerical simulations were derived from the same experimental data. Thus, the inconsistency between material models used by two different codes is of little concern since the simulation results were compared and verified against the same set of test data with a common reference threshold. The use of two FE codes in solving an identical simulation environment also provided an opportunity to test the hypothesis that vertical accelerative response simulation of the ATD is independent of solver types given that the material models are derived from the same source. The approaches to derive the material parameters, constitutive material models and the fitting algorithms are available in the literature [12]. Finally a brief summary of the whole-body simulation results (both models) of an ATD response due to vertical accelerative loading are compared for the same experimental data to document the predictive capability of the material models.

10.2 LS DYNA Material Models

The LS-DYNA material characterization was performed with multi-rate compression and tension testing on most of the plastic and rubber materials used in the dummy. Material models were created based on all of the available characterizations. Typically, an Ogden or Bergstrom-Boyce (BB) material model was used for soft materials such as the flesh [13]. The material models were subsequently validated in the subassembly level validation.

Table 10.1 features the polymer materials used in the LS-DYNA WIAMan model and lists where these materials are used; what material model is used to represent them; the strain rate range used to fit the material models; and the material library identification number.

Table 10.1 WIAMan LS-DYNA polymer materials

Material name	Anatomic location(s)	LS-DYNA material model	Strain rate range fitted	Material library ID
Butyl Rubber 75 Shore A	Compliant elements; spine rubber	*MAT_OGDEN_RUBBER	T: – C: 1–850	58
Proflex 30	Pelvis flesh	*MAT_BERGSTROM_BOYCE_RUBBER	T: 0.003–300 C: 0.01–950	59
E1030AL	Flesh-other	*MAT_BERGSTROM_BOYCE_RUBBER	T: 1–90 C: 1–850	60
XE1031	Foot flesh	*MAT_OGDEN_RUBBER	T: – C: 1–600	61
FD70	Foot plate	*MAT_SIMPLIFIED_JOHNSON_COOK	T: 1–75 C: 1–150	62
Delrin	Calcaneus cap	*MAT_SIMPLIFIED_JOHNSON_COOK	T: 1–50 C:40–600	63
TC892	Pelvic bone	*MAT_SIMPLIFIED_JOHNSON_COOK	T: 1–80 C: 1–550	64
Rencast 6425	Tailbone	*MAT_SIMPLIFIED_JOHNSON_COOK	T: 1–300 C: 1–160	65
ISODAMP	Heel plug	*MAT_OGDEN_RUBBER	T: 1–300 C: 1–160	68

Note: *T* is tension and *C* is compression

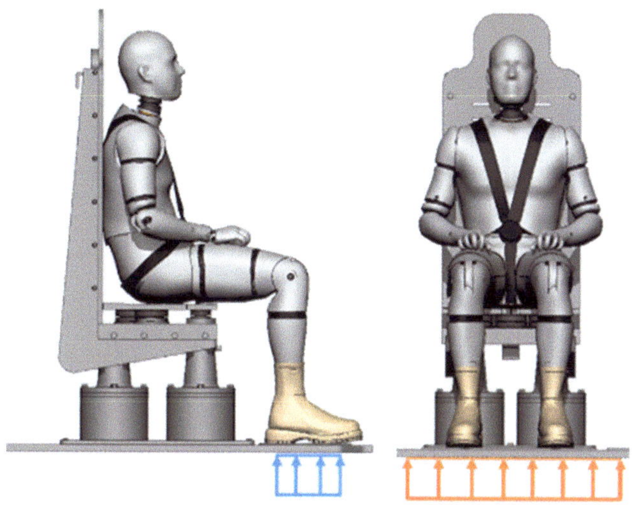

Fig. 10.2 VALTS test simulation schematic

10.3 Whole-Body ATD Model Simulation Results

An illustrative example of the system-level, whole-body response prediction is shown in Figs. 10.2 and 10.3. The whole-body test results came from the Johns Hopkins University Applied Physics Laboratory (APL) Vertically Accelerated Load Transfer System (VALTS) tests that impacted the seat and the floor at prescribed velocity profiles associated with a peak velocity and a time to peak (TTP) based on the target input conditions for sub-injurious UBB loadings used for the WIAMan biomechanical research investigations. Total mass of the WIAMan TD FE model was 79.8 kg for the LS-Dyna case compared to the mass of the physical prototype of 82.7 kg, which included external sensor wiring not present in the model. Figure 10.1 shows a whole-body system-level test configuration in which the ATD with no personal protective equipment (PPE) was tested at the APL VALTS for a loading profile having an amplitude of 4 m/s at a 5-ms TTP. Both floor and seat were subjected to the amplitude and the TTP with different loading profiles discussed in the simulation results section. Figure 10.3 compares the experimental and simulation responses of the WIAMan ATD for the lower extremity at the heel and mid-tibia sensor regions.

10.4 Conclusion

The WIAMan ATD FE model used the experimentally derived multi-rate compression and tension material test data from the same source to extract material model parameters for their respective analytical ATD models for eight polymeric materials. Material models were created based on a number of available characteristic models. Typically, an Ogden or BB material model was used for the soft materials. The material models were subsequently validated in the subassembly level validation.

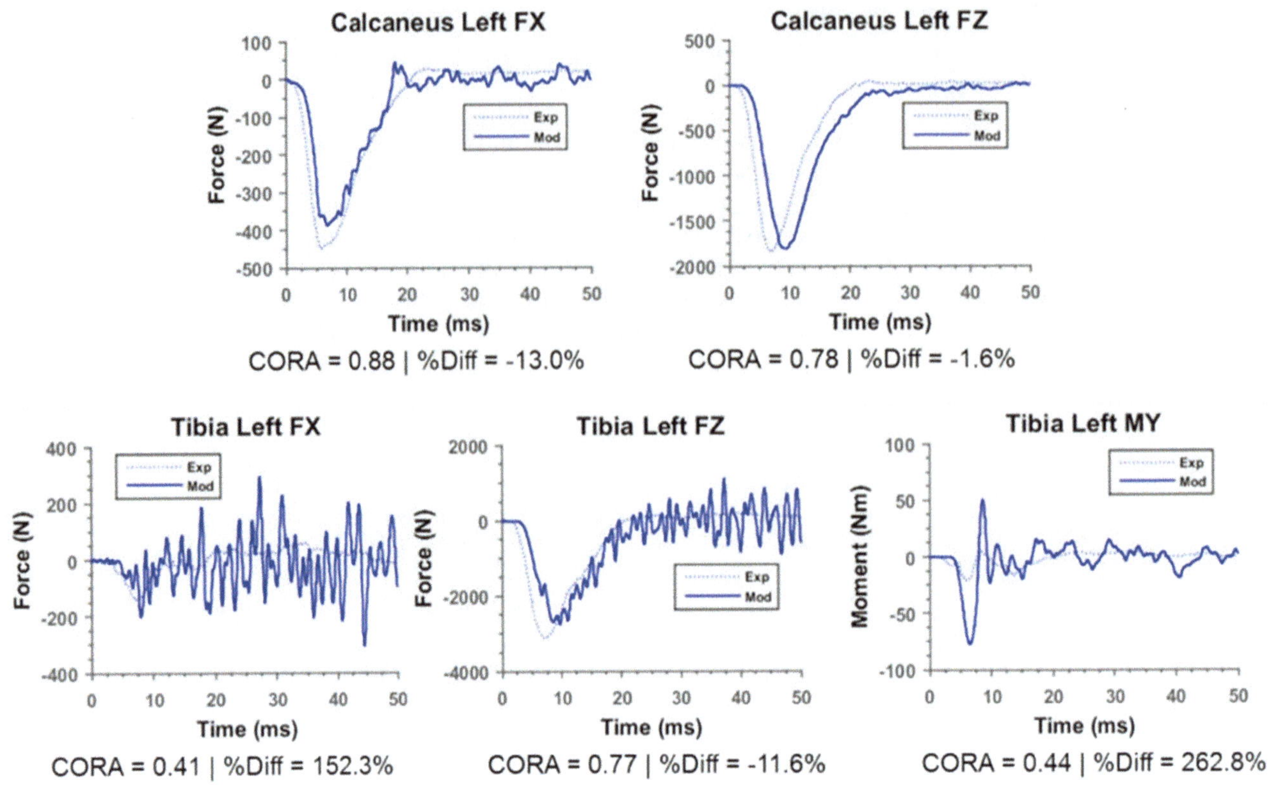

Fig. 10.3 Matched-pair validation results for the lower extremity at the heel and mid-tibia sensor regions for the simulation and the experiment

There is room for improvement in the material model fitting process; however, because of the good comparison (acceptable to a degree that serves the purpose of evaluating the evolving ATD design iterations) of experimental and simulation results for whole-body test cases, the model parameters extracted in the effort were deemed appropriate. Intra- and inter-comparisons of model simulation results for both FE codes with the identical experimental test cases indicated good correlations. The use of two FE codes in solving an identical simulation environment also provided an opportunity to extract two data points for the same simulation environment thus providing an increased degree of confidence in the model predictability. Finally, a quick comparison of the whole-body simulation results of an ATD response to those of the corresponding test data shows strong confidence in the predictive capability of the WIAMan ATD FE model.

References

1. Warrior Injury Assessment Manikin (WIAMan): https://www.fbo.gov/notices/144703326476d3ec9c18778ab0c312df (2016)
2. Hoppel, C., et al.: Workshop on Numerical Analysis of Human and Surrogate Response to Accelerative Loading: Final Report. Aberdeen Proving Ground (MD): Army Research Laboratory (US), May Report No.: ARL-SR-287 (2014)
3. DoD Blast Injury Research Program Coordinating Office: Science & Technology Efforts & Programs, Ft. Detrick, MD. US Army Medical Research & Material Command, Prevention, Mitigation, and Treatment of Blast Injuries, FY14 Report to the Executive Agent (2014) https://blastinjuryresearch.amedd.army.mil/assets/docs/ea_report/FY14_Report_to_the_Executive_Agent.pdf
4. Kleinberger, M., et al.: Second Workshop on Numerical Analysis of Human and Surrogate Response to Accelerative Loading. Aberdeen Proving Ground (MD): Army Research Laboratory (US); Workshop Proceedings (2016)
5. Chancey, C., McEntire, J., Scherer, R., Frounfelker, P., Tegtmeyer, M.: Program for Warrior Injury Assessment Manikin (WIAMan). Aberdeen Proving Ground (MD): Army Research Laboratory (US) (2011) https://blastinjuryresearch.amedd.army.mil/docs/WIAMan_Project_Plan_Industry_Day_V3.pdf
6. Crawford, D.M., Chowdhury, M.R., Pietsch, H.A.: Mechanical Properties of Polymers Used for Anatomical Components in the Warrior Injury Assessment Manikin (WIAMan) Technology Demonstrator. White Sands Missile Range (NM): Army Research Laboratory (US); 2016 Jul. Report No.: ARL-TR-7728 (2016)
7. Sanborn, B., Moy, P., Mrozek, R., Weerasooriya, T.: Effect of Strain Rates on the Compressive Response of Neck Rubber From Humanetics HIII 50th Percentile Male Dummy Under Different Loading Sequences. Aberdeen Proving Ground (MD): Army Research Laboratory (US); 2013 Feb. Report No.: ARL-TR-6336 (2013)

8. Bell, C., Lister, K., Nochetto, H., Robeck, C., Yow, C.: User's Manual and Validation, WIAMan Velodyne Model – TDv1.2, Manual Release v1.0 6/30/2016. CORVID Technologies, Mooresville (2016)
9. Vavalle, N., Shanaman, M., Lomicka, C., Boyle, M., Armiger, R.: User's Manual and Validation Summary; WIAMan ATD Tech Demonstrator LS-Dyna FEM v1.0.0. Johns Hopkins Applied Physics Lab, Laurel (2016)
10. LS-DYNA Support, Material Models for Polymers. Livermore Software Technology Corporation, Livermore (2016) http://www.dynasupport.com/howtos/material/material-models-for-polymers
11. Jeong, D., Kwok, P., Canha, J.: Characterization of deformable materials in the THOR dummy, Injury Science Research Proceeding of the 28th International Workshop, pp. 147, Atlanta (2000)
12. Chowdhury, M., Crawford, D., Shanaman, M., Boyle, M., Armiger, R., Bell, C., Lister, K., Shirley, A.: Polymeric Materials Models in the Warrior Injury Assessment Manikin (WIAMan) Anthropomorphic Test Device (ATD) Tech Demonstrator, Aberdeen Proving Ground (MD): Army Research Laboratory (US); ARL-TR-7927 (2017)
13. Bergstrom, J.: Mechanics of Solid Polymers Theory and Computational Modeling. William Andrew, San Diego (2015) http://www.worldcat.org/title/mechanics-of-solid-polymers-theory-and-computational-modeling/oclc/913869177

Mostafiz R. Chowdhury is currently leading the Modeling and Simulation Product Team in the WIAMan Engineering Office at ARL to develop analytical Finite-Element Model of the World's first anthropomorphic test device (ATD) for under body blast. He worked as a research structural engineer at the Engineer Research and Development Center (ERDC) in Vicksburg before joining ARL in 2001. He has served in a number of different assignments within the Army, including positions in the office of the Deputy Assistant Secretary of the Army for Research and Technology (DASA(R&T)) and the Office of the Assistant Secretary of Defense – Research & Engineering (ASD(R&E)).

He has a PhD in Structural Engineering and has taught engineering courses at various universities. At ARL he has implemented validation and verification of seated manikin finite element (FE) models with the vertical drop test data that guided the end-to-end simulation of blast induced vehicle dynamic simulation for crew survivability evaluation.

Chapter 11
Investigation of Transmission of a Shock Wave Through Thin Films

Hongjoo Jeon and Veronica Eliasson

Abstract The aim of the present experimental study is to find physical laws of shock wave transmission through thin films. A horizontal shock tube was used to produce a planar shocks with incident shock Mach numbers of $M_s = 1.20$, 1.34, 1.39, and 1.46 onto thin polyester films. The test section of the shock tube was constructed such that it could hold either a single or multiple layers of 12.7, 25.4, or 50.8 μm thick films. High-speed schlieren photography and high-frequency pressure transducers were used to evaluate the shock wave interactions with the films and the subsequent film breakup. Following incident shock wave impingement, the transmitted and reflected shock waves displayed different strengths, depending on the incident shock Mach numbers and the film thicknesses.

Keywords Shock wave propagation • Transmission of a shock wave • Thin films • Shock wave attenuation • Schlieren photography

11.1 Introduction

A polyester film has excellent thermal, mechanical, and chemical properties so it is widely used for many purposes including packaging for barrier protection. In particular, the polyester film is used in food packaging for fresh products, such as raw meat, poultry, fish, and shellfish. Therefore, it will be crucial to find a method to effectively kill infectious bacteria after food packaging during storage conditions. Loske et al. [1] investigated possible applications of shock waves in food preservation and demonstrated significant reduced number of microorganism using repeated shock waves in certain foods. In addition, Gnanadhas et al. [2] showed that shock waves could be used to treat biofilm infections on medical devices. In the current study, the transmission of a shock wave through polyester (ductile) or aluminum (brittle) films is quantitatively investigated with different incident Mach numbers and polyester film thicknesses.

11.2 Experimental Setup

Figure 11.1 shows a schematic side view of a horizontal shock tube. The horizontal shock tube has a driver, a driven section, and a test section. The driver and driven sections have a length of 648 mm and 1860 mm so that a 2-D planar shock wave can be observed in the test section [3, 4]. Different thicknesses of plastic membrane were used to create incident shock Mach numbers of $M_s = 1.20$, 1.34, 1.39, and 1.46. Four pressure transducers (S1 and S4: PCB 113B21, S2 and S3: PCB 113B31, flush mounted with 100 mm distance) are equipped on the top of the test section to measure overpressure and incident shock Mach numbers, M_s, and transmitted shock Mach numbers, M_t. A z-folded schlieren setup is used to visualize shock waves together with a high-speed camera (Phantom V711) and a continuous light source (Cree XLamp, XP G-2 LEDs, Cool White) [3, 4].

H. Jeon
Department of Aerospace and Mechanical Engineering, University of Southern California, Los Angeles, CA 90089-1191, USA
e-mail: hongjooj@usc.edu

V. Eliasson (✉)
Department of Structural Engineering, University of California, San Diego, La Jolla, CA 92093-0085, USA
e-mail: eliasson@ucsd.edu

Department of Aerospace and Mechanical Engineering, University of Southern California, Los Angeles, CA 90089-1191, USA

J. Kimberley et al. (eds.), *Dynamic Behavior of Materials, Volume 1*, Conference Proceedings of the Society for Experimental Mechanics Series, DOI 10.1007/978-3-319-62956-8_11

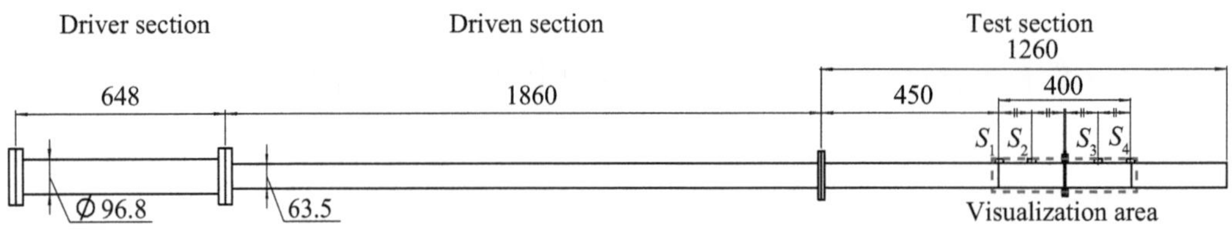

Fig. 11.1 Schematic side view of a horizontal shock tube (Dimensions in mm)

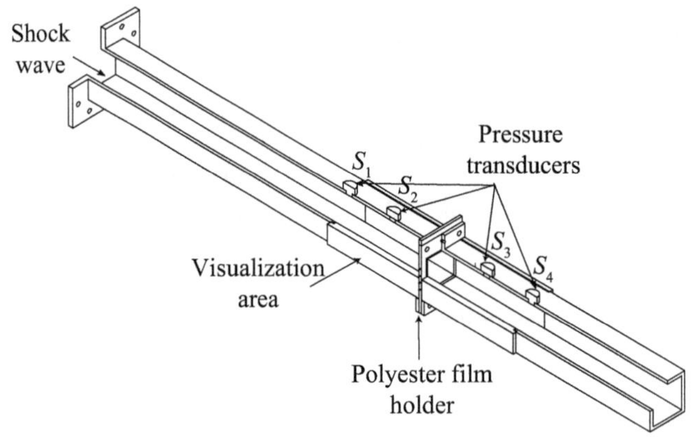

Fig. 11.2 Schematic detailed view of the test section

Figure 11.2 shows a schematic detailed view of the test section. In the current study, a unique polyester film holder was designed to study the shock wave transmission through a polyester film. A single layer of polyester films ($h = 12.7$, 25.4, or 50.8 μm thick) was mounted using O-rings and bolted joints to prevent gas leakage between the test section and the polyester film holder.

11.3 Results

Figure 11.3 represents a series of high-speed schlieren photographs of shock wave transmission through a 12.7 μm thick polyester film at the incident Mach number of 1.46. The time $t = 0$ corresponds to the arrival of the incident shock wave at the sensor S_1. In Fig. 11.3a, the incident shock wave approaches the polyester film holder, which is located in the center. After the incident shock wave impacts the polyester film, the transmitted and reflected shock wave are generated by the impedance mismatch, see Fig. 11.3d, e. The pressure behind the incident and reflected shock waves expanded the polyester film and finally the film broke up after a threshold pressure, see Fig. 11.3f.

Based on the arrival time of shock waves at pressure transducers, the incident and transmitted shock Mach numbers are calculated and the results are presented in Fig.11.4. Higher incident shock Mach numbers, M_s, resulted in higher transmitted shock Mach numbers, M_t, and the transmitted shock Mach number was decreased with increasing thickness of the films. It was expected that the transmitted shock wave through a brittle material (aluminum film) could be stronger compared to a ductile material (polyester film) since brittle materials absorb relatively little energy when subjected to stress. However, it is interesting that the transmitted shock Mach number through aluminum films observed lower than that through polyester films except the 25.4 μm thick aluminum film with $M_s = 1.39$ and 1.46. In these cases, the transmitted shock wave was merged with another shock wave caused by early breakup of aluminum films. Therefore, it can be concluded that the shock wave transmission is more depending on the material density rather than brittle and ductile behavior.

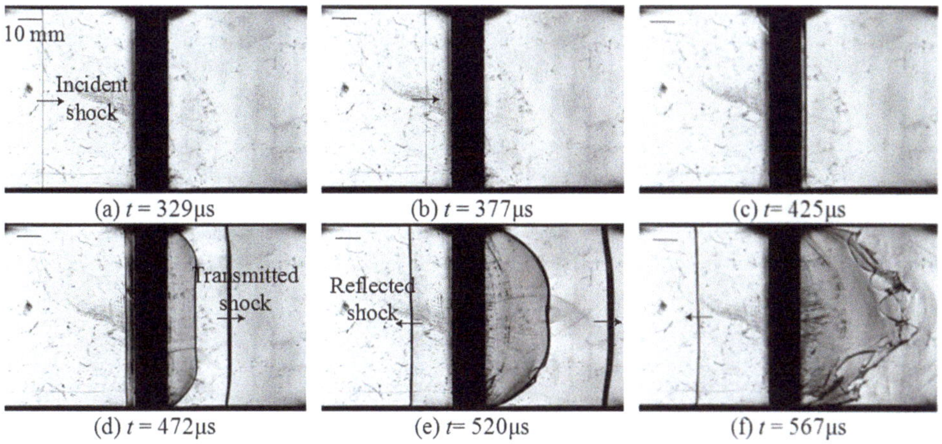

Fig. 11.3 A series of high-speed schlieren images of shock wave interaction with a 12.7 μm thick polyester film at the incident Mach number of 1.46 (304 × 184 pixels) (**a**) $t = 329$ μs (**b**) $t = 377$ μs (**c**) $t = 425$ μs (**d**) $t = 472$ μs (**e**) $t = 520$ μs (**f**) $t = 567$ μs

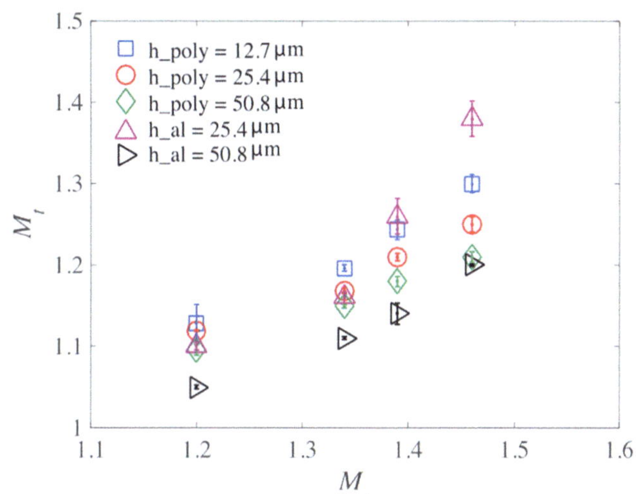

Fig. 11.4 Incident shock Mach numbers versus transmitted shock Mach numbers with different thicknesses of polyester (*poly*) and aluminum (*al*) films

Acknowledgements This study is supported by the National Science Foundation under grant No. CBET-1437412.

References

1. Loske, A.M., Prieto, F.E., Zavala, M.L., Santana, A.D., Armenta, E.: Repeated application of shock waves as a possible method for food preservation. Shock Waves. **9**(1), 49–55 (1999)
2. Gnanadhas, D.P., Elango, M., Janardhanraj, S., Srinandan, C.S., Datey, A., Strugnell, R.A., Gopalan, J., Chakravortty, D.: Successful treatment of biofilm infections using shock waves combined with antibiotic therapy. Sci. Rep. **5**, 17440 (2015)
3. Jeon, H., Eliasson, V.: Shock wave interactions with liquid sheets. Exp. Fluids. **58**(4), 1–11 (2017) (accepted for publication)
4. Jeon, H., Gross, J.R., Estabrook, S., Koumlis, S., Wan, Q., Khanolkar, G.R., Tao, X., Mensching, D.M., Lesnick, E.J., Eliasson, V.: Shock wave attenuation using foam obstacles: does geometry matter? Aerospace. **2**(2), 353–375 (2015)

Hongjoo Jeon is a Ph.D. student of Department of Aerospace and Mechanical Engineering at University of Southern California since 2014. His research activities have focused on shock wave dynamics, such as attenuation, propagation, interactions, with diverse materials and geometries. His research themes include understanding the dynamics of incident and reflected shock waves.

Chapter 12
Experimental Testing and Computational Analysis of Viscoelastic Wave Propagation in Polymeric Split Hopkinson Pressure Bar

M. Bustamante, D.S. Cronin, and D. Singh

Abstract The use of polymeric bars in the traditional Kolsky or Split Hopkinson Pressure Bar (SHPB) has been suggested by several authors as a means of improving coupling to low impedance materials and to increase incident wave rise time to assist in achieving dynamic equilibrium when testing soft materials. However, one aspect that must be addressed in this application is viscoelastic wave propagation leading to wave attenuation and dispersion. The amount of dispersion and attenuation depends on the bar material selection and incident wave signal. Viscoelastic wave propagation has been successfully addressed in Polymeric SHPB through experimental determination of the wave propagation coefficients, and has been investigated through analytical techniques; however, there is no widely accepted method for computationally modeling these events, which would benefit test apparatus design and optimization.

To address this challenge, experimental impact tests were undertaken on a polymeric SHPB utilizing 2.5 m long acrylic (PMMA) bars with various strikers (0.712 m bar, 0.459 m bar, steel sphere) and impact velocities. The response was measured using the incident bar strain gauge signal and a Photon Doppler Velocimetry apparatus to measure the incident bar end velocity. The apparatus was computationally modeled using a commercial explicit finite element code and a viscoelastic Ogden material model. Parameter identification was undertaken using optimization software.

The experimental test results were repeatable for all strikers and impact velocities, with the amount of attenuation and dispersion being highest for the sphere impact, compared to the striker bars. The computational model and optimization was applied to the incident bar strain gauge data to identify the material coefficients. The dominant material properties were dependent on the nature of the striker, but were able to accurately represent the incident and reflected wave signals. The resulting models were then validated using the PDV data using cross-correlation, demonstrating numerical robustness.

Keywords Polymeric Split Hopkinson Pressure Bar apparatus • Impact test • Viscoelastic bar • Computational model • Cross-correlation

12.1 Introduction and Background

The testing and characterization of materials at high deformation rates (e.g. $\sim$1000 s^{-1}) is commonly undertaken using the Kolsky [1] or Split Hopkinson Pressure Bar apparatus [2], enabling uniform deformation rates and relatively large deformations. However, the testing of low impedance materials such as polymers, elastomers, and biological tissues presents some challenges due to the low stiffness and low wave speed in these materials [3]. These challenges manifest as low signal to noise ratios on the transmitter bar and difficulty achieving dynamic equilibrium of the test sample.

Two common approaches for testing low impedance materials include the use of traditional metallic bars with a pulse shaper to lower the rise time of the incident wave and embedded load cells at the bar ends [4]; and the use of polymeric bars in a Polymeric Split Hopkinson Pressure Bar (PSHPB) apparatus (Fig. 12.1). The benefit of using polymeric bars is that they provide a longer signal rise time enabling dynamic equilibrium in the test sample without pulse shaping, and they improve the transmitter bar signal to noise ratio due to lower bar impedance, which is on the order of soft tissues [4] and elastomers. However, one challenge with the PSHPB apparatus is the need to account for viscoelastic attenuation and dispersion of the incident, reflected, and transmitted stress waves, as the waves travel from the strain gauge station at the mid-length of the bar to the bar end. Polymethyl methacrylate (PMMA) is often used in PSHPB [5] and is known to be viscoelastic based on measured material properties [6], and dynamic materials testing using this bar material [7]. Experimentally, this challenge has been successfully resolved through experimental calibration of the bars using the method proposed by Bacon [8] and

M. Bustamante • D.S. Cronin (✉) • D. Singh
Department of Mechanical & Mechatronics Engineering, University of Waterloo, Waterloo, ON, Canada
e-mail: duane.cronin@uwaterloo.ca

J. Kimberley et al. (eds.), *Dynamic Behavior of Materials, Volume 1*, Conference Proceedings of the Society for Experimental Mechanics Series, DOI 10.1007/978-3-319-62956-8_12

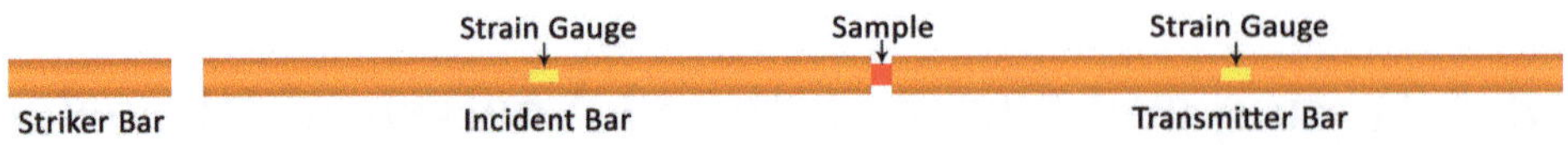

Fig. 12.1 Polymeric Split Hopkinson Pressure Bar (PSHPB) apparatus schematic

Table 12.1 Free-end incident bar tests

Striker type	Striker velocity (m/s)	Averaged velocity (m/s)
Short	4.770	4.82[b]
	4.791	
	4.898	
	6.235	6.24[b]
	6.220	
	6.269	
	7.091	7.07[b]
	7.043	
	7.063	
Long	4.609	4.58[b]
	4.585	
	4.551	
	5.563	5.55[b]
	5.553	
	5.535	
	6.686	6.64[b]
	6.598	
	6.645	
Sphere	7.554	7.57[b]
	7.578	
	9.168[a]	9.13
	9.119	
	9.110	
	10.330	10.36[b]
	10.380	
	10.380	

[a]**Experimental test used for optimization model**
[b]Experimental tests used for model validation

validated for the apparatus described in this study by van Sligtenhorst [3] and Salisbury [7]. Analytical methods (e.g. [9]) have demonstrated that, with some reasonable assumptions, the response of viscoelastic impact and bar response could be approximated analytically; however, the presented data could not be implemented directly in a computational model. A computational model of the apparatus is beneficial to investigate different apparatus configurations, evaluate challenges with alternate configurations, and provide guidance for new experimental tests. The aim of this study was to generate an initial set of experimental data, identify the material parameters of the bars using inverse methods, and validate the model with additional impact scenarios.

12.2 Methods

The experimental testing was undertaken using an existing PSHPB apparatus (Fig. 12.1), using only the incident bar (25.4 mm diameter, 2438 mm length) and three striker bars: long striker (PMMA, 25.4 mm diameter, 712 mm length), short striker (PMMA, 25.4 mm diameter, 459 mm length) and a sphere (steel, 25.4 mm diameter). The striker bars were launched at three different impact velocities and the strain gauge (CEA-13-250UW-120, Micro-Measurements) signal on the incident bar was amplified (2210B Signal Conditioning Amplifier, Vishay) and recorded (BNC-2110, National Instruments) at a frequency of 2 Mhz. Three repeats were undertaken for each test. High speed imaging was used to measure the velocity of the striker and the incident bar end velocity was measured using a Photon Doppler Velocimetry (PDV) (1550 nm wavelength, Ohio Manufacturing Institute) apparatus. A total of 26 tests were undertaken (Table 12.1).

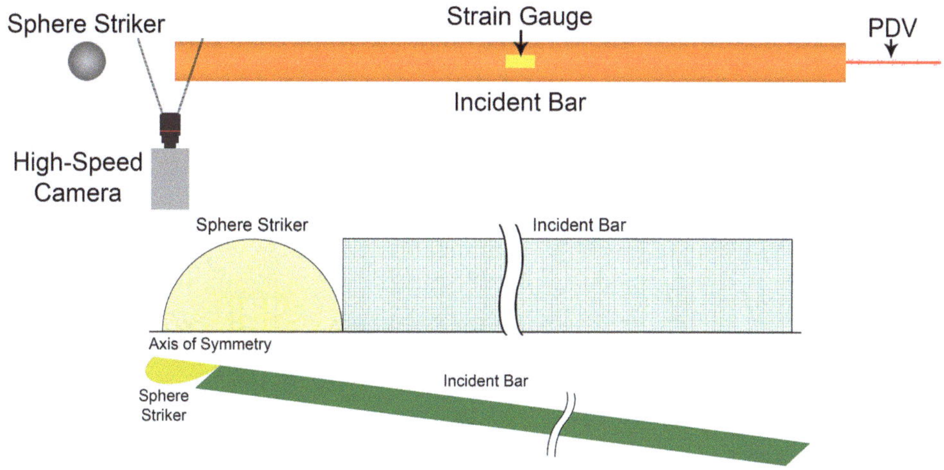

Fig. 12.2 Schematic of incident bar, striker and instrumentation (*top*) and corresponding axisymmetric computational model (*bottom*)

Table 12.2 PMMA incident bar material properties, Ogden constitutive model

μ_1 (GPa)	3.776		
α_1	1		
G_1 (GPa)	0.0065	β_1	0.001
G_2 (GPa)	0.0492	β_2	4.79
G_3 (GPa)	0.0418	β_3	29.0
G_4 (GPa)	0.1353	β_4	12,840

An axisymmetric finite element model of the apparatus was created for analysis in a commercial finite element code (LS-Dyna, R7.1.2) (Fig. 12.2). The PMMA bar was modeled using an Ogden material model, with elements ~0.40 mm in size, based on a convergence analysis. An initial sensitivity study was undertaken to evaluate the Ogden material parameters and range of parameters, followed by an optimization study using a response surface methodology (LS-OPT Version 5.2, LSTC, Livermore, CA). The optimization used a polynomial sampling and D-optimal methodology with a space filling algorithm. Convergence was evaluated using a tolerance criterion using multi-objective function including: Mean Squared Error (MSE), Curve Matching, and Peak Values. The model predictions were compared to an independent set of data based on peak values and shape to verify the model predictions for a wide range of impact conditions.

12.3 Results

The experimental test results (Fig. 12.3) show the incident and reflected waves from the strain gauge as well as the bar end velocity. The experiments were very repeatable with impact velocities varying by +1.6% to −1.3% from the average for a given impact condition. Furthermore, the wave processing technique from Salisbury [10], based on free-end calibration, provided an excellent prediction of the wave shape, arrival time, and bar end velocity. The results from the sphere impact at 18 psi firing pressure (9.168 m/s impact, Table 12.1) were used for the optimization model.

The optimization analysis was preceded by several initial analyses to identify parameter sensitivity, evaluate variations in impact velocity on the response outcome, and to identify appropriate limits for the parameters. The final optimization analysis converged after approximately eight iterations, but was run as long as 20 iterations (Fig. 12.4). In all cases, the analysis was robust and converged to the same material parameter values (Table 12.2). The remaining experiments (excluding the sphere at 18 psi) were simulated using the computational model (Fig. 12.3, long striker $V_{average}$ = 4.58 m/s), and were compared to the experiments using cross-correlation (Table 12.3) [11].

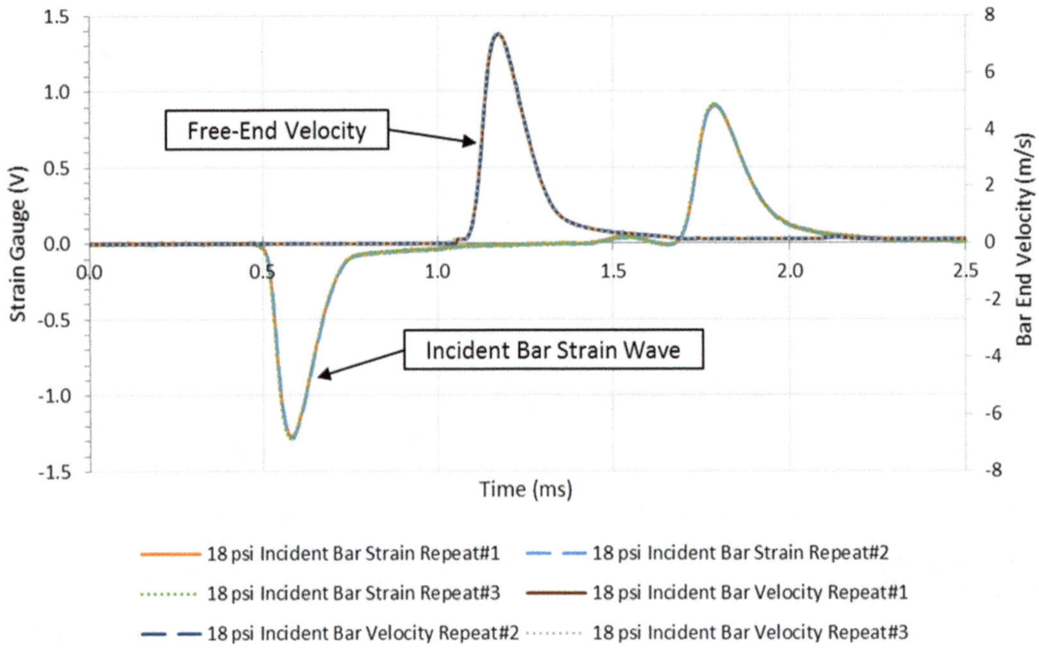

Fig. 12.3 Typical experimental result for bar free-end test, demonstrating test repeatability (9.13 m/s average impact velocity)

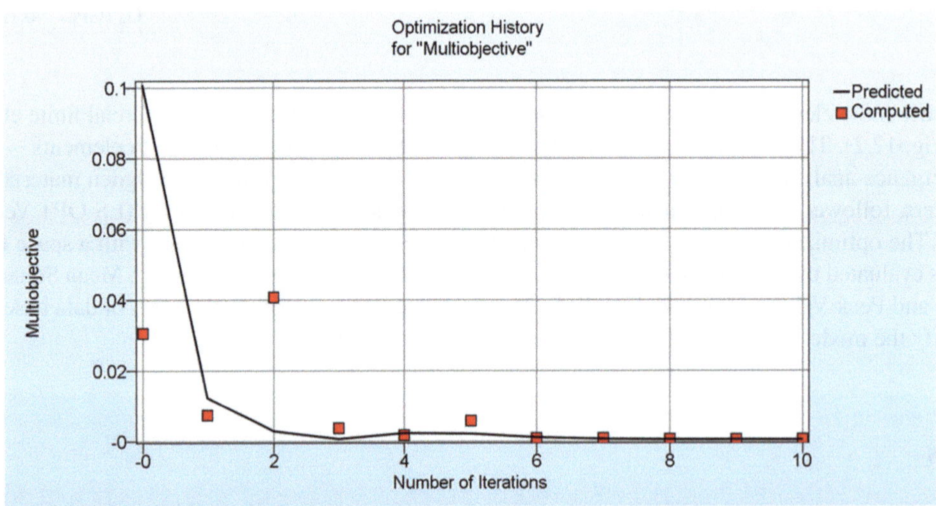

Fig. 12.4 Optimization analysis convergence, predicted and computed responses

Table 12.3 Computational model and experiment comparison

Measurement data	Striker type	Cross-correlation rating
Strain data	Short	0.988
	Long	0.970
	Sphere	0.997
Free-end velocity	Short	0.984
	Long	0.990
	Sphere	0.993

12.4 Discussion and Conclusions

Twenty six experimental impact tests were undertaken comprising three different strikers, three different impact velocities for each striker, and three repeats of each test (two repeats for the sphere at 7.57 m/s). The test apparatus and measurement diagnostics demonstrated very good repeatability (Fig. 12.3). Wave attenuation and dispersion were more prevalent for the

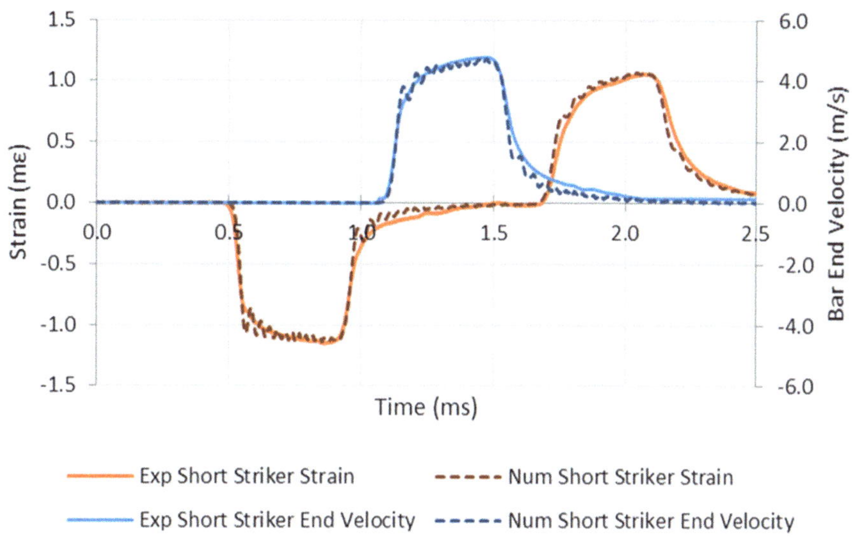

Fig. 12.5 Computational model compared to experimental test, example for long striker

spherical striker, which is why this was selected for determination of the material parameters. Only one single case (sphere at 9.168 m/s) was used for material parameter identification; however it was important to undertake initial sensitivity studies to identify an appropriate range of values for the material properties and to identify the number of parameters for the Ogden model. Additional material model parameters were considered but did not improve the goodness of fit for the constitutive model. The optimization analysis was applied to the measured strain gauge data using a multi-objective function. Initial evaluations identified that a mean squared error approach was limited in ability to address the rapid change in slope of the response on the rise and decay of the waves. The curve fitting algorithm was better suited to address the rapid changes in slope of the strain gauge signal, but required an additional criterion (peak values) to ensure convergence. Convergence was achieved within approximately eight iterations of the optimization software, and continuing the analysis for additional iterations did not significantly change the optimized material parameters. Model validation was undertaken using independent data and test conditions, and demonstrated excellent agreement with the experimental results and the PDV measurements (Fig. 12.5). Comparison was undertaken using cross-correlation to evaluate the shape and magnitude of the curves with values ranging from 0.970 to 0.997, indicating excellent agreement. The developed model can be applied to provide guidance for new configurations of the PSHPB apparatus.

References

1. Kolsky, H.: An investigation of the mechanical properties of materials at high rates of loading. Proc. Phys. Soc. B. **62**, 676–700 (1949)
2. Gray III, G.T., Blumenthal, W.: Split Hopkinson pressure bar testing of soft materials. In: ASM Handbook: Mechanical Testing and Evaluation, vol. 8, pp. 488–496 (2000)
3. Van Sligtenhorst, C., Cronin, D.S., Brodland, G.W.: High strain rate compressive properties of bovine muscle tissue found using a split Hopkinson bar apparatus. J. Biomech. **39**, 1852–1858 (2006)
4. Song, B., Chen, W.: Dynamic stress equilibration in split Hopkinson pressure bar tests on soft materials. Exp. Mech. **44**(3), 300–312 (2004)
5. Salisbury, C., Cronin, D.S., Lien, F.-S.: Deformation mechanics of a hyper-viscoelastic porous material, part I: testing and constitutive modeling of non-porous polychloroprene material. J. Dyn. Behav. Mater. **1**(3), 236–248 (2015). doi:10.1007/s40870-015-0026-2. (NSERC Discovery, Compute Canada)
6. Christman, D.R.: Dynamic Properties of Poly (Methylmethacrylate) (PMMA) (Plexiglas), Report AD743547. Materials and Structures Laboratory, Manufacturing Development, General Motors Corporation. Accessed from the National Technical Information Service. Report reproduced in Springfield, Virginia (1972)
7. Salisbury, C.P., Cronin, D.S.: Mechanical properties of ballistic gelatin at high deformation rates. Exp. Mech. **49**(6), 829–840 (2009). doi:10.1007/s11340-008-9207-4
8. Bacon, C.: An experimental method for considering dispersion and attenuation in a viscoelastic Hopkinson bar. Exp. Mech. **38**, 242–249 (1998)
9. Ahonsi, B., Harrigan, J.J., Aleyaasin, M.: On the propagation coefficient of longitudinal stress waves in viscoelastic bars. Int. J. Impact Eng. **45**, 39–51 (2012), ISSN 0734-743X. http://dx.doi.org/10.1016/j.ijimpeng.2012.01.004
10. Salisbury, C.P.: Spectral analysis of wave propagation through a polymeric Hopkinson bar. MASc Thesis, University of Waterloo, Canada (2001)
11. Thunert, C.: CORA Release 3.6 User's Manual (2012). http://www.pdb-org.com/de/information/18-cora-download.html

D.S. Cronin is a Professor in the Department of Mechanical and Mechatronics Engineering at the University of Waterloo and University Research Chair in Impact Mechanics and Material Characterization. Professor Cronin's expertise lies in the areas of biomechanics, trauma, auto crash and occupant simulation, pipeline integrity, and multi-material lightweight vehicles.

Chapter 13
Mode I Rigid Double Cantilever Beam Test and Analysis Applied to Structural Adhesives

C.-H. Liao, B. Watson, M.J. Worswick, and D.S. Cronin

Abstract Computational modeling and design using structural adhesives in dynamic events, such as vehicle collisions, requires material properties across a range of deformation rates. In this study, the analysis of the rigid double cantilever beam (RDCB) geometry [7, 8] has been evaluated to measure the critical mode I energy release rate (G_{IC}) and the traction-separation response of a structural adhesive at quasi-static rates. Standardized test geometries such as the tapered double cantilever beam (TDCB) are used to measure G_{IC}; however the large specimen size is not conducive to measuring properties at high deformation rates and requires additional tests to determine the full traction-separation curve for an adhesive. In contrast, the low mass and size of the RDCB enables use of this geometry for dynamic testing, which will be the focus of future research. Importantly, the RDCB test can provide the entire traction-separation curve of the adhesive from one test. A first series of experimental tests at quasi-static deformation rates identified high variability in the measured response. This variability was reduced by using a new fixture to improve alignment and maintain constant bond-line thickness, highlighting the sensitivity of this method to small variations in sample geometry. In addition, the surface preparation method was crucial to reducing variability in the measured response and achieving primarily cohesive failure in the bond. In this study, the RDCB analysis method was applied to a toughened epoxy structural adhesive. The average measured maximum traction and separation-to-failure of the structural adhesive were 35.3 MPa and 0.214 mm respectively. The average critical energy release rate of the tested adhesive was 3497 J/m^2. Future investigations will consider optical measurement of displacement on the sample, enhanced data analysis methods, and dynamic testing of adhesive using the RDCB geometry.

Keywords Structural adhesive • Energy release rate • Cohesive zone model • Mode I test • Material characterization

13.1 Introduction and Background

Adhesive joining of similar and dissimilar materials has become an attractive lightweighting strategy, driven by the need to reduce emissions and improve the Corporate Average Fuel Economy (CAFE) through vehicle weight reduction [1, 2]. Adhesives enable joining of lightweight multi-material structures, use of thinner gauge materials, and can improve structural stiffness compared to traditional joining techniques. Further, reduced design cycle time can be achieved through the use of accurate and validated computational vehicle models to assist in assessing vehicle performance and crashworthiness. Thus, effective implementation of adhesive joining technologies in vehicles requires characterization and material parameters for computational modeling.

Cohesive zone modelling (CZM) is a damage mechanics based approach to simulate the failure response of an adhesive joint, utilizing traction-separation response of the adhesive subjected to mode I or mode II loading [3]. Depending on the type of adhesive, the shape of the traction-separation curve can vary, but bi-linear or trapezoidal traction-separation relationships are often used [3].

Double cantilever beam (DCB) and tapered double cantilever beam (TDCB) test methods are often used in standardized mechanical tests (ASTM D3433; ISO 25217:2009) to measure the mode I critical energy release rate (G_{IC}) of adhesives [4]. This measure represents the energy required to propagate the crack over a unit area, and is an important input parameter for CZM. The DCB and TDCB tests have different data reduction methods such as the Compliance-Based Beam Method (CBBM) for DCB and Simple Beam Theory (SBT) for TDCB to determine the interface property [5] from the measured data. However, these tests do not provide adhesive peak traction and must generally be supplemented with additional testing (e.g. butt joint) to obtain peak traction values to complete the traction-separation response [6]. During the DCB tests, crack

C.-H. Liao • B. Watson • M.J. Worswick • D.S. Cronin (✉)
Department of Mechanical & Mechatronics Engineering, University of Waterloo, Waterloo, ON, Canada
e-mail: duane.cronin@uwaterloo.ca

J. Kimberley et al. (eds.), *Dynamic Behavior of Materials, Volume 1*, Conference Proceedings of the Society for Experimental Mechanics Series, DOI 10.1007/978-3-319-62956-8_13

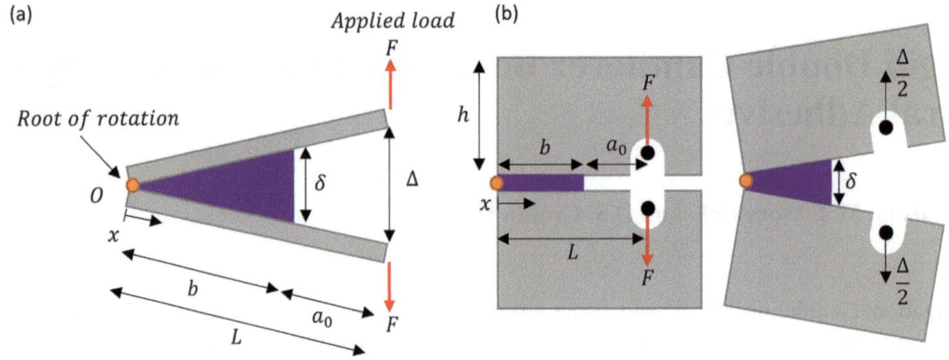

Fig. 13.1 (**a**) RDCB specimen for soft biological adhesive [7]; (**b**) RDCB specimen for industrial adhesives [8]

propagation within the adhesive is generally measured optically; however, due to the unstable crack growth, the data reduction methods that are dependent on the crack length measurement can potentially provide unreliable mode I critical energy release rates. This is often the case when testing brittle materials, such as non-toughened adhesives. Lopes et al. [5] identified the CBBM data reduction method for DCB to be more reliable than the methods that rely on measuring crack propagation. Difficulty in measuring crack growth in bonded samples poses technical challenges, especially when high deformation rate material properties are required for impact scenarios. The test sample geometries used in the standardized tests also tend to have high masses. For example, a TDCB sample constructed to the ASTM D3433 standard using A36 steel has a mass of 2045 g. This high mass can lead to large inertial effects, prohibiting testing at high rates.

More recently, Dastjerdi et al. [7] introduced a small specimen geometry (Fig. 13.1a) termed the rigid double cantilever beam (RDCB) with the benefit of measuring all mode I material properties (maximum traction, initial stiffness of the adhesive and maximum separation) in a single test. This geometry was applied to measure the critical energy release rate (G_{IC}) from the traction-separation curve for soft biological adhesives. The analysis did not require in-situ measurements of crack growth of the adhesive, but relied on the measured force and crosshead displacement measured using a linear variable differential transformer (LVDT), along with the initial sample dimensions. This method provides a full cohesive traction-separation response, so additional testing to measure the maximum traction is not required. Further, the small geometry and correspondingly small mass are well suited for testing at higher deformation rates. The original test specimen geometry (Fig. 13.1a) was updated to increase the second moment of area so that the adherend could be considered rigid (Fig. 13.1b), allowing for the testing of higher stiffness and higher strength materials (soft elastomers and thermoset polymers) [8], while following the same analysis for soft adhesive.

A comprehensive material model of an adhesive joint should predict adhesive response over a wide range of deformation rates, from quasi-static to high strain rate. An important component of this response is the mode I critical energy release rate, maximum traction, maximum separation, and initial stiffness. The focus of this study was to apply the RDCB test methodology to a commercially available toughened structural adhesive to measure the critical energy release rate and adhesive traction-separation response in mode I (tension) loading using a single mechanical test.

13.2 Method

In the current study, the dimensions of the adherend were updated to meet the rigidity assumption for the higher stiffness and higher strength structural adhesive investigated in this study. The adherends used in this study (Fig. 13.2) were machined from 6.2 mm thick structural steel with yield strength of 345 MPa and a Young's modulus of 210 GPa. The final mass of the tests samples was approximately 54 g. The adhesive investigated in this study was a 2-part toughened impact resistant structural epoxy adhesive [#07333, 3M Corporation]. The samples were cured 30 min at 80 °C [9] using a forced convection oven (Binder ED-53).

The surfaces of the adherends were grit blasted (Fig. 13.3b) to increase the surface roughness for bonding, and subsequently cleaned with methyl ethyl ketone to remove potential contaminants. Preliminary tests showed that the roughened surface reduced variability of the force-displacement response and increased the force at failure. The pre-crack was created by applying corrective fluid to mask part of the surface of the adherend so the adhesive did not bond to the metal (Fig. 13.3c), producing nominal bond length, *b*, (Fig. 13.2b) of 13.25 mm. The adhesive was applied using a single bead on

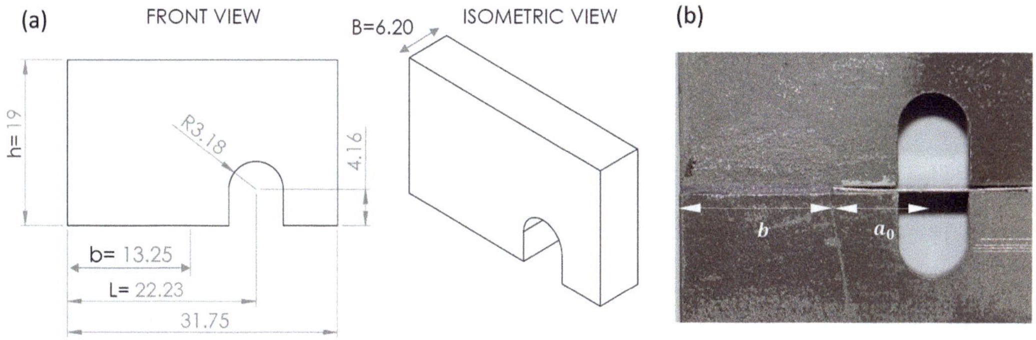

Fig. 13.2 (**a**) RDCB – specimen geometry (all dimensions are in mm); (**b**) RDCB specimen showing the bond length b and crack tip a_0

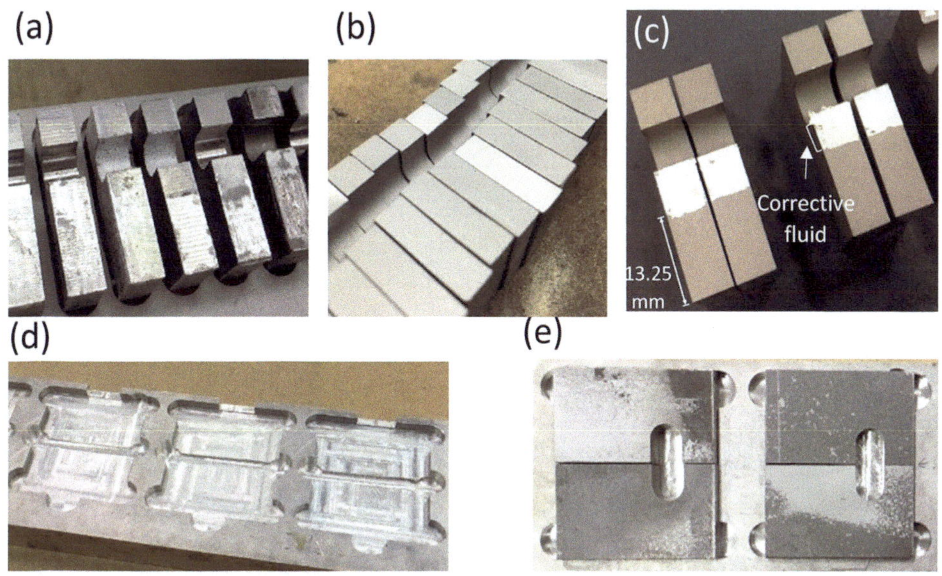

Fig. 13.3 (**a**) Adherends before surface treatment; (**b**) specimens surface after grit-blasting; (**c**) masked bonding surface; (**d**) specimen fixture; (**e**) test samples placed in fixture

each of the adherends. The test sample was then placed into the curing fixture (Fig. 13.3d) to keep the adherends in the same plane and maintain uniform bond-line thickness using shims of the desired bond-line thickness (Fig. 13.2b).

After curing and removal of excess adhesive, the bond-line thickness and bond-line length of each specimen were measured using an opto-digital microscope (VHX 5000, KEYENCE) (Fig. 13.4).

The tests were conducted using an Instron Model 4465-Standard test machine with a 5 kN capacity load cell (Fig. 13.5). The nominal cross-head speed was 0.0115 mm/sec, corresponding to the lowest crosshead speed of the test frame. This test velocity along with an average bond thickness of 0.641 mm led to a nominal strain rate of 0.018 s^{-1} in the adhesive. A total of nine specimens were tested.

In this study, the experimentally measured load cell force, F, and pin displacement, Δ, were used in the equations derived by Dastjerdi et al. [7, 8] to calculate the separation at the crack tip, δ, and traction, T, using Eqs. (13.1) and (13.2) respectively.

$$\delta = \frac{b}{L}\Delta \tag{13.1}$$

$$T(\delta) = \left(\frac{L}{B(L-a_0)^2}\right)\left(2F + \Delta\frac{dF}{d\Delta}\right) \tag{13.2}$$

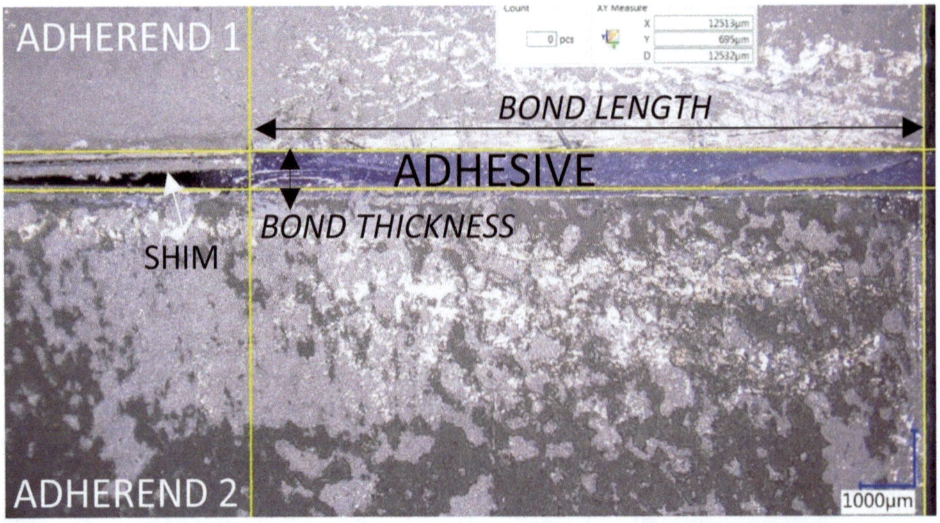

Fig. 13.4 RDCB specimen bond length and thickness

Fig. 13.5 RDCB test setup with a close-up look of a sample mounted onto the grips

Where 'b' is the bond line length, 'L' is the length between the edge of the sample and the point of loading, 'B' is the adherend thickness, and 'a_0' is the crack length. The derivative term $\frac{dF}{d\Delta}$, was calculated using a sixth order central difference scheme in this study. The critical energy release rate G_{IC}, was computed by numerically integrating the traction-separation response generated from Eqs. (13.1) and (13.2) using a first order forward difference scheme. This analysis assumes the second moment of area of the adherend is large enough to be considered rigid relative to the adhesive, which is contrary to the DCB and TDCB tests where the adherends are deformable and the adherend compliance is taken into account to measure the critical energy release rate. The rigidity of the substrate was verified using the non-dimensional ratio, α, derived by Dastjerdi et al., which is a function of the elastic moduli and geometry of the adherend and nominal loading slope measured during the tests (Eq. (13.3)).

$$\alpha = \left(1 + \frac{Bb^3 S_{int}\left((1+\upsilon)\,\alpha_s' + \alpha_b'\right)}{3L^2 E}\right)^{-1} \tag{13.3}$$

Where 'S_{int}' is the initial slope of the cohesive law, 'ν' the is Poisson's ratio of the adherend (0.3 in this study), 'E' is the Young's modulus of the adherend (210 GPa in this study) and 'α'_s' and 'α'_b' are defined by:

$$\alpha_s' = \frac{2}{Bh}\left(\frac{9}{5}L\left(\frac{L}{b} - 1\right)\right) \tag{13.4}$$

and

$$\alpha_b' = \frac{12}{Bh^3}\left(\frac{L^3}{3} + \frac{Lb^2}{10} - \frac{2L^2b}{5}\right) \tag{13.5}$$

Where 'h' is the height of the adherend (19 mm in this study). An α value of 1.0 corresponds to a rigid adherend. The geometry and material chosen for this study led to an α value of 0.992, supporting the assumption of the rigid adherend. The use of crosshead displacement, as measured using an LVDT, was assessed by measuring pin displacement optically from a video captured using a DLSR camera using an image tracking software (Tracker, Open Source Physics) [10].

13.3 Experimental Results

The bond-line length and thickness were measured for each sample (Table 13.1) to be used in sample-specific analysis for each test.

The measured LVDT displacement and pin displacement from the video tracking software were compared to assess the fixture compliance (Fig. 13.6).

The force-displacement responses for each test were recorded (Fig. 13.7a). The average maximum force measured was 592 N (standard deviation 46 N). Equations (13.1) and (13.2) were applied to the measured data to obtain the mode I traction-separation curves (Fig. 13.7b). The traction-separation response was, in turn, used to calculate the energy release rate for each test (Fig. 13.7c).

Table 13.1 RDCB specimens bond line length and thickness

Adhesive bond characteristics	Bond length, B [mm]	Bond thickness [mm]
RDCB-IRSA-3	13.46	0.637
RDCB-IRSA-5	13.04	0.697
RDCB-IRSA-6	13.52	0.681
RDCB-IRSA-7	12.67	0.604
RDCB-IRSA-8	13.05	0.655
RDCB-IRSA-9	12.51	0.730
RDCB-IRSA-10	13.27	0.584
RDCB-IRSA-11	13.72	0.586
RDCB-IRSA-12	13.12	0.590
Average	13.15	0.641
Standard deviation	0.393	0.054

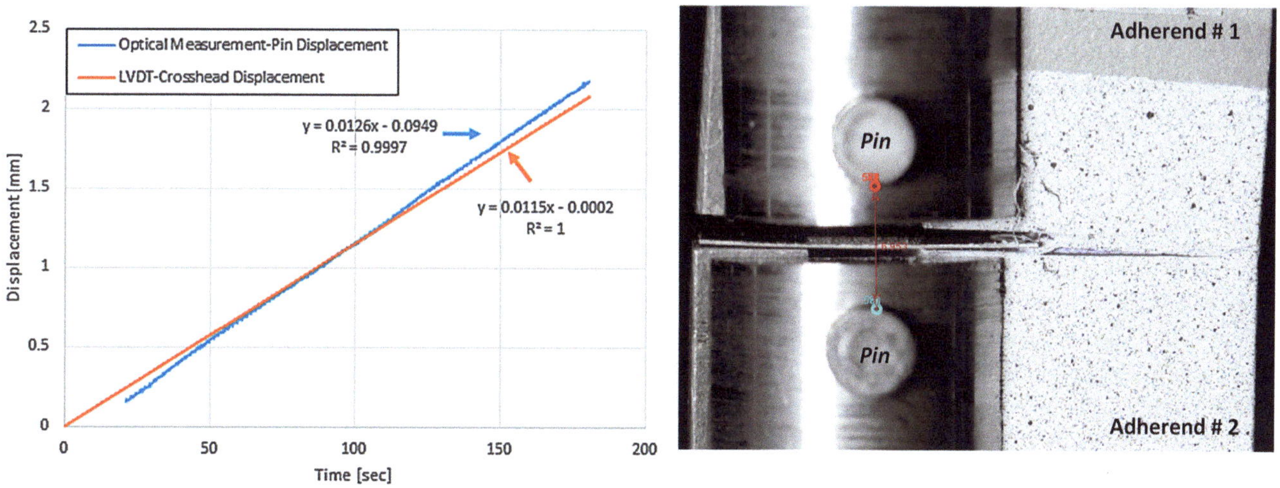

Fig. 13.6 LVDT displacement of a tested specimen compared to optically tracked pin displacement

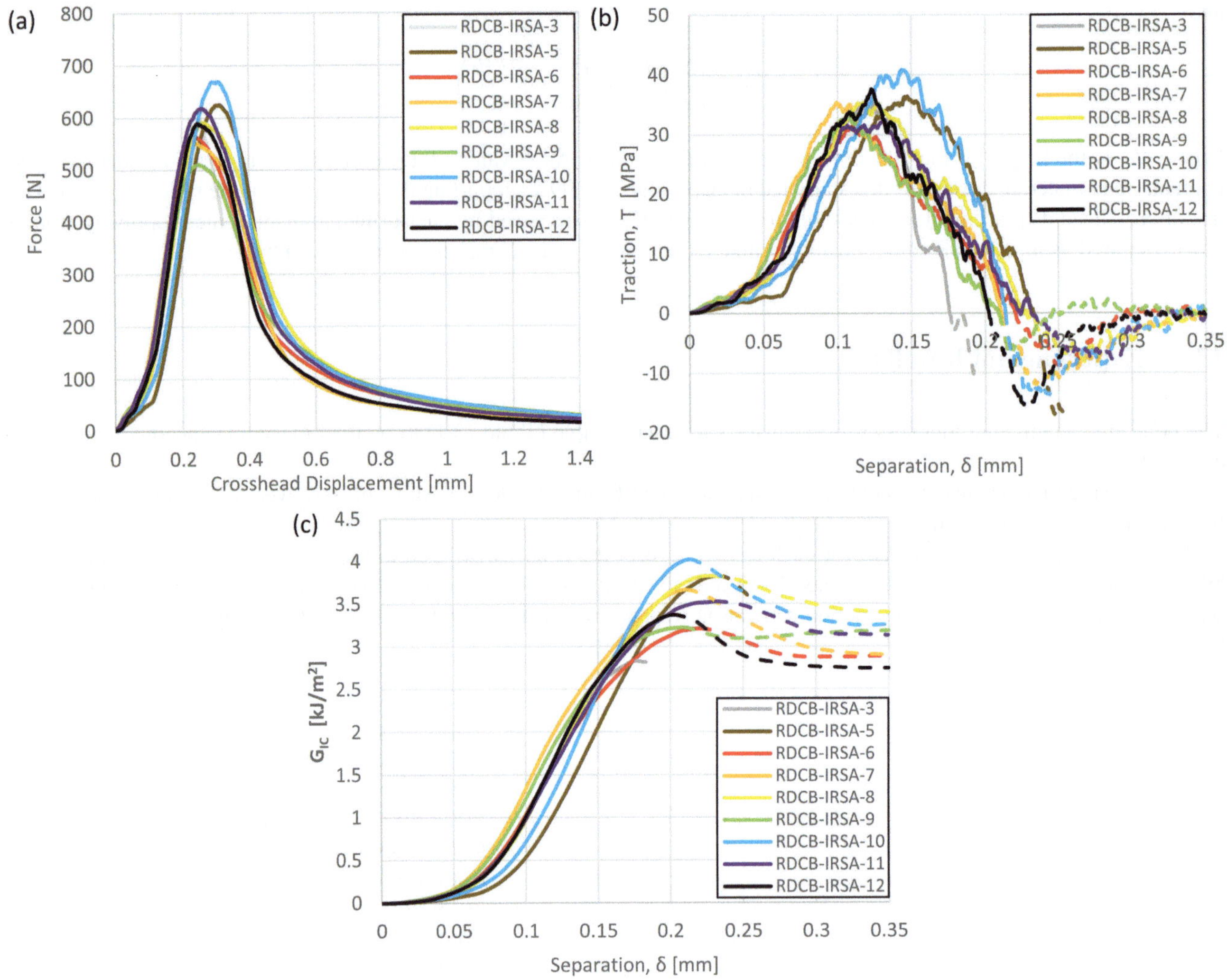

Fig. 13.7 (**a**) Force vs crosshead displacement curves; (**b**) mode I traction-separation curves; (**c**) energy release rate vs. separation

The maximum traction and critical energy release rates were calculated for each test (Fig. 13.8). The average peak traction was 35.32 MPa (standard deviation 2.84 MPa) while the average critical energy release rate was 3.497 kJ/m^2 (standard deviation 0.375 kJ/m^2). The maximum separation (i.e. the crack opening at which the traction returned to 0 MPa) was also measured for each test. The average maximum separation was 0.214 mm (standard deviation 0.018 mm).

Evaluation of the test sample fracture surfaces identified a predominantly cohesive failure with small amount of interfacial failure (Fig. 13.9).

13.4 Discussion

Early tests where grit-blasting was not performed (not included in the current study) on the adherend surface led to interfacial failure being observed and significant variability in the force to failure, highlighting the importance of surface preparation. The current surface preparation (grit-blasting) resulted in primarily cohesive failure, meaning that the failure occurred within the bond line rather than between the adhesive and adherend surface. This demonstrated that it is the properties of the adhesive that are measured, rather than the interaction of the adhesive and adherend surface. Improvements to ensure adherend alignment and consistent bond line thickness through a bonding fixture improved the consistency of the test results.

Some machine compliance was found in these tests, manifesting in the pin displacement and the cross head displacement responses showing different loading slopes (Fig. 13.6). There was also some initial slack in the test apparatus, the effect

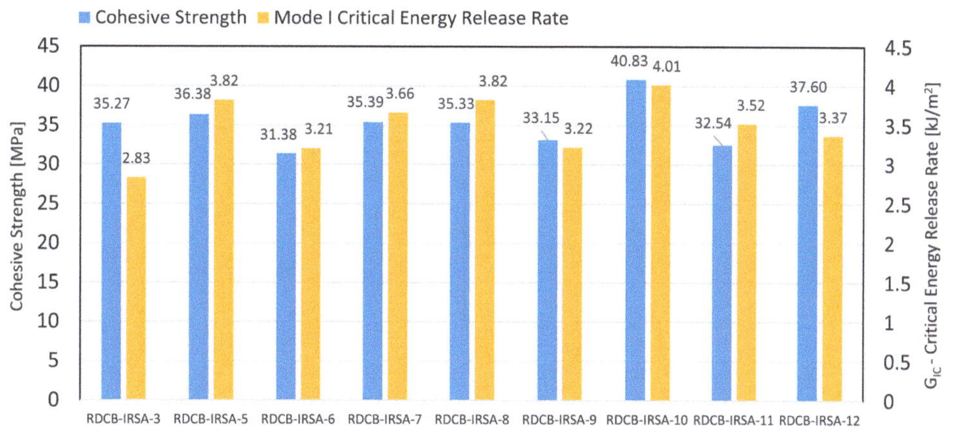

Fig. 13.8 Cohesive strength and critical energy release rate in mode I for each specimen

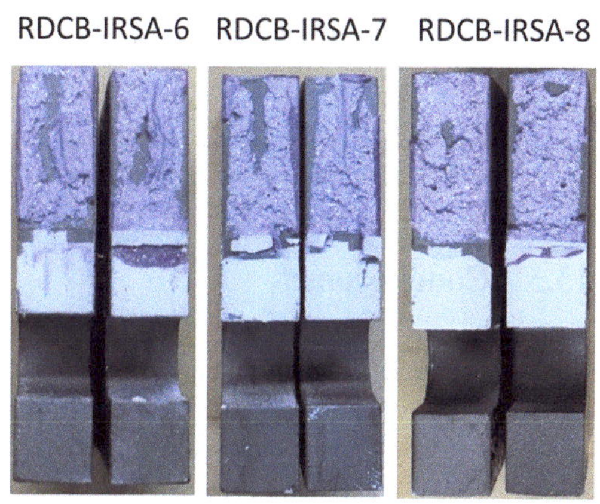

Fig. 13.9 Representative adhesive fractured surfaces of RDCB specimens

of which was visible in the initial toe-region of the force-displacement results (Fig. 13.7a). Since the traction calculation is a function of pin displacement, accurate measurement of pin displacement is critical. As the focus of this study was to assess the ability of the test methodology [8] to identify the traction-separation response of a toughened structural adhesive, modifications to the test methodology were outside the scope of the current study. However, the results from the current study suggest that future testing should include optically tracked pin displacements rather than crosshead displacement measurements.

Following the softening region of the traction-separation curves, negative traction regions were calculated. This was due to the steep decrease in force over relatively small displacements once the peak force was reached (Fig. 13.7b). These negative traction regions then carried over as a drop in energy release rate vs separation results (Fig. 13.7c). Therefore, the critical energy release rates were reported from the point of maximum separation.

Comparing the measured critical energy release rate and peak traction to the three different types of adhesive previously tested using this methodology [8], the structural adhesive investigated in this study had three and half times higher strength compared to the strongest adhesive previous tested (an epoxy based adhesive) and eight times the fracture toughness of the toughest adhesive (a polyurethane based adhesive). In previous studies the adhesive thickness was nominally 30 μm and the adherends had mirror polished surfaces compared to the 0.641 mm average bond-line thickness and grit-blasted surfaces in the current study. The bond-line thickness is known to affect fracture toughness [11] and a mirror or polished surface generally leads to poorer adhesion between the adhesive and adherend.

With stronger adhesives, the assumptions used to develop the original analysis should be reviewed since the original analysis was validated by testing softer biological adhesives and generic commercial adhesives that have lower toughness than the structural adhesive used in this study. Of particular importance, the assumption that the point of rotation lies at the edge of the sample (Fig. 13.10a) should be investigated as a compression region in the adhesive bond-line (Fig. 13.10b) may exist which was not initially taken into consideration when developing Eqs. (13.1) and (13.2).

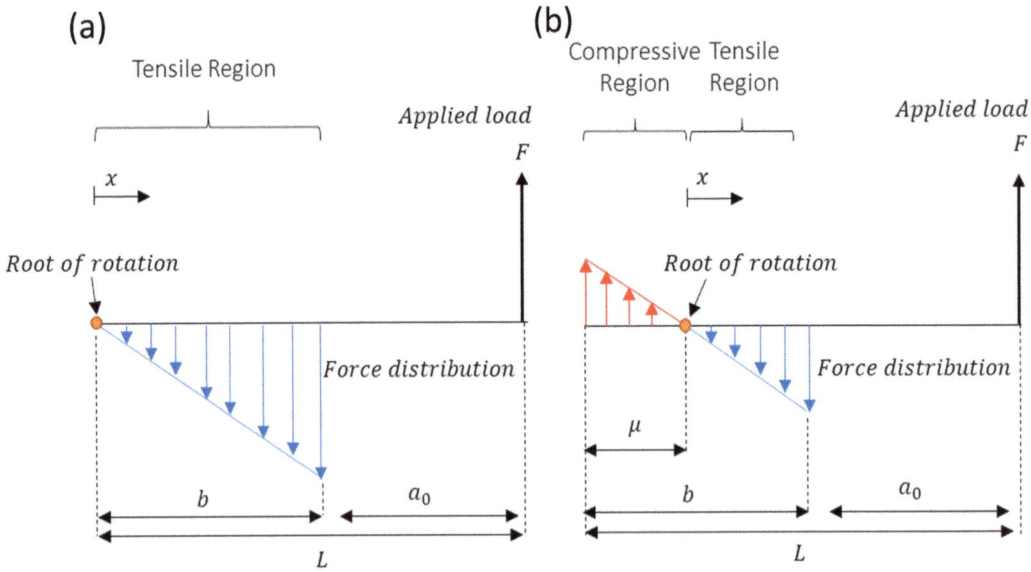

Fig. 13.10 (**a**) Original free body diagram analysis with the centre of rotation located at the edge of the adherend; (**b**) proposed free body diagram analysis introducing the offset centre of rotation, which results in compressive stresses in the bond line

13.5 Conclusions

RDCB specimens were tested under quasi-static loading to measure the adhesive stiffness, maximum traction and maximum separation of a toughened structural epoxy adhesive for mode I loading. The small geometry of the RDCB sample and relatively straightforward test setup provides an alternative to standard test methodologies to obtain the critical energy release rate along with the full-traction-separation response. The adherend geometry and material properties were selected to meet the assumption of rigid adherends as defined in the original analysis of the specimen geometry. The specimen preparation used in this study, including grit-blasting and a custom fixture to ensure good alignment of the adherends, produced repeatable experimental results. Furthermore, grit-blasting the adherend surface was found to increase the failure force and produced a predominantly cohesive failure.

Compared to the results reported by Dastjerdi, the adhesive investigated in this study exhibited a higher cohesive strength and critical energy release rate. An improved analysis for a stiffer adhesive will be part of the next steps of research that will include an enhanced analysis of the stresses in the bond line and include optically tracked pin displacement for an accurate measure of crack opening. Furthermore, the calculated G_{IC} values from RDCB tests should be compared to the values measured from standard tests such as DCB and TDCB and a validated numerical model of the RDCB test should be used to examine the updated RDCB analysis.

Acknowledgements This study would not have been possible without the adhesive material provided by 3M of Canada and the preliminary work that Luis Trimiño carried out on the geometry of the rigid double cantilever beam specimen. The authors would also like to thank the Natural Sciences and Engineering Research Council of Canada (NSERC), 3M Canada, Honda R&D Americas and ArcelorMittal Dofasco for supporting this research.

References

1. Martinsen, K., Hu, S.J., Carlson, B.E.: Joining of dissimilar materials. CIRP Ann. Manuf. Technol. **64**, 679–699 (2015)
2. Del-Colle, A.: Obama Announces 54.4 mpg CAFE Standard by 2025: Hearst Communication. Popular Mechanics. http://www.popularmechanics.com/cars/a7015/obama-announces-54-6-mpg-cafe-standard-by-2025/?click=pm_latest (2011)
3. da Silva, L.F.M., Campilho, R.D.S.G.: Advances in numerical modelling of adhesive joints. In: Springer Briefs in Computational Mechanics, p. 17. Springer, Berlin (2012)
4. ASTM Standard D3433-99: Standard Test Method for Fracture Strength in Cleavage of Adhesives in Bonded Metal Joints. ASTM International, West Conshohocken (2012)
5. Lopes, R.M., Campilho, R.D.S.G., da Silva, F.J.G., Faneco, T.M.S.: Comparative evaluation of the double-cantilever beam and tapered double-cantilever beam tests for estimation of the tensile fracture toughness of adhesive joints. Int. J. Adhes. Adhes. **67**, 103–111 (2016)

6. Marzi, S., Hesebeck, O., Brede, M., Kleiner, F.: A rate-dependent cohesive zone model for adhesively bonded joints loaded in mode I. J. Adhes. Sci. Technol. **23**, 881–898 (2009)
7. Dastjerdi, A.K., Pagano, M., Kaartinen, M.T., McKee, M.D., Barthelat, F.: Cohesive behavior of soft biological adhesives: experiments and modeling. Acta Biomater. **8**, 3349–3359 (2012)
8. Dastjerdi, A.K., Tan, E., Barthelat, F.: Direct measurement of the cohesive law of adhesives using a rigid double cantilever beam technique. Exp. Mech. **53**(9), 1763–1772 (2013)
9. 3M Company, ed.: Safety Data Sheet: Impact Resistant Structural Adhesive. St. Paul (2016)
10. Brown, D.: Tracker Version 4.95: Licensed under GNU General Public License. http://www.cabrillo.edu/~dbrown/tracker/. Aptos (2016)
11. Tvergaard, V., Hutchinson, J.W.: Toughness of an interface along a thin ductile layer joining elastic solids. Philos. Mag. A. **70**(4), 641–656 (1994)

Chi-Hsiang Liao currently pursuing a Master degree in Impact mechanics and material characterization group at University of Waterloo. The author's focus of the research is on surface treatment of aluminized ultra high strength steel for adhesive bonding and characterizing the adhesive joint through mechanical testing. Presenting author currently resides in Waterloo, Ontario, Canada.

Chapter 14
Brute Force Ceramic Constitutive Model Parameterization

Brady Aydelotte and Brian E. Schuster

Abstract We conducted simplified long rod impact experiments which were instrumented with flash X-ray and Photon Doppler Velocimetry (PDV). The time resolved experimental data was used to fit a geomaterials model using both 2D and 3D simulations. It was found that the 2D simulations could not fit the experimental data satisfactorily. The 3D simulations were much more successful with some limitations.

Keywords Impact • Ceramic • Damage • Constitutive model • Optimization

14.1 Introduction

Material model parameterization is a vital part of continuum scale modeling efforts. But it is often a challenge. Complex models often require a great deal of experimental data to parameterize or worse, there may not be a straightforward way to deduce an important material model parameter from existing data. This means that the modeler must often employ incomplete data in the search for reasonable model results. Experimental point measurement techniques such as Velocity Interferometer System for Any Reflector (VISAR) [1] or Photon Doppler Velocimetry (PDV) [2] which can be employed across a sample surface have become widely available. High speed imaging can provide full-field measurements of quantities like displacement. These techniques and the concurrent availability of tremendous computational power, naturally lead to greater interest in numerical material model calibration efforts.

Ballistic experiments range from very simple laboratory projectiles and targets to complex, realistic targets and projectiles. As such, they can expose a range of behaviors which impact experiments with cleaner boundary conditions, such as Kolsky bar and parallel plate shock experiments, may not. We have combined multiple channels of PDV measurements with full field flash X-ray measurements to provide greater insight into the process of long rod penetration into an unconfined boron carbide target.

Here we discuss our efforts to parameterize a complex geomaterial model, Kayenta [3], using the Dakota optimization toolkit [4] and the arbitrary Eulerian Lagrangian simulation package, ALEGRA [5].

14.2 Results

We conducted a series of impact experiments by forward launching 93 weight percent tungsten heavy alloy rods into hot pressed boron carbide cylinders purchased from CoorsTek. PDV probes were used to measure the rear surface velocity of the center of the boron carbide target at several different points. One hundred and fifty kilovolt flash X-ray heads were used to measure the depth of penetration of the projectiles in the targets during the penetration experiment. One particular experiment, an impact at 517 m/s, is the focus of this study.

The ALEGRA hydrocode [5] was used to construct 2D axisymmetric and 3D quarter symmetry models of this experiment using an Eulerian mesh description. We used the embedded Dakota [4] capability in ALEGRA in an effort to calibrate the Kayenta [3] geomaterials model using both depth of penetration (DOP) and rear surface velocity data. A uniform resolution

B. Aydelotte (✉) • B.E. Schuster
US Army Research Laboratory, 321 Colleran Road RDRL-WML-H, 21005, Aberdeen Proving Ground, MD, USA
e-mail: brady.b.aydelotte.civ@mail.mil

J. Kimberley et al. (eds.), *Dynamic Behavior of Materials, Volume 1*, Conference Proceedings of the Society for Experimental Mechanics Series, DOI 10.1007/978-3-319-62956-8_14

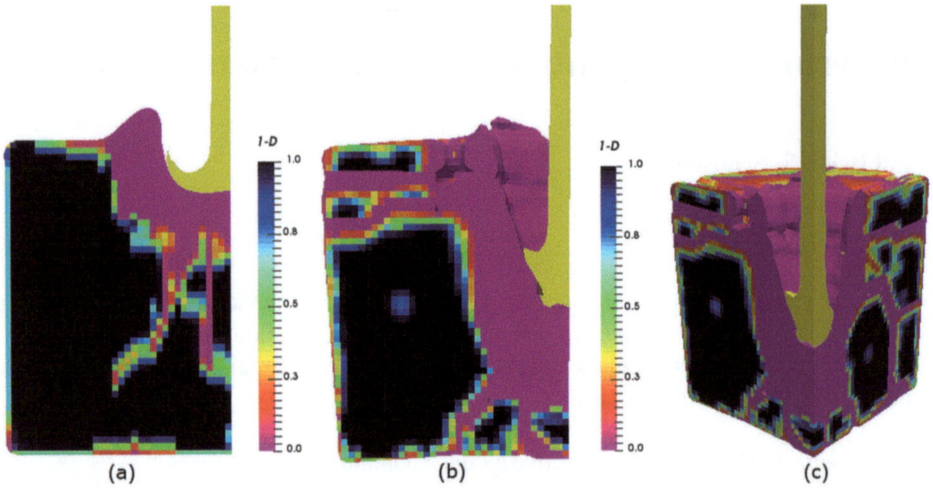

Fig. 14.1 A comparison of (**a**) 2D and (**b** and **c**) 3D impacts 70 μs post impact. Notice that the 3D impacts show significantly more rod penetration. In 3D, additional fractures are free to form, allowing greater penetration

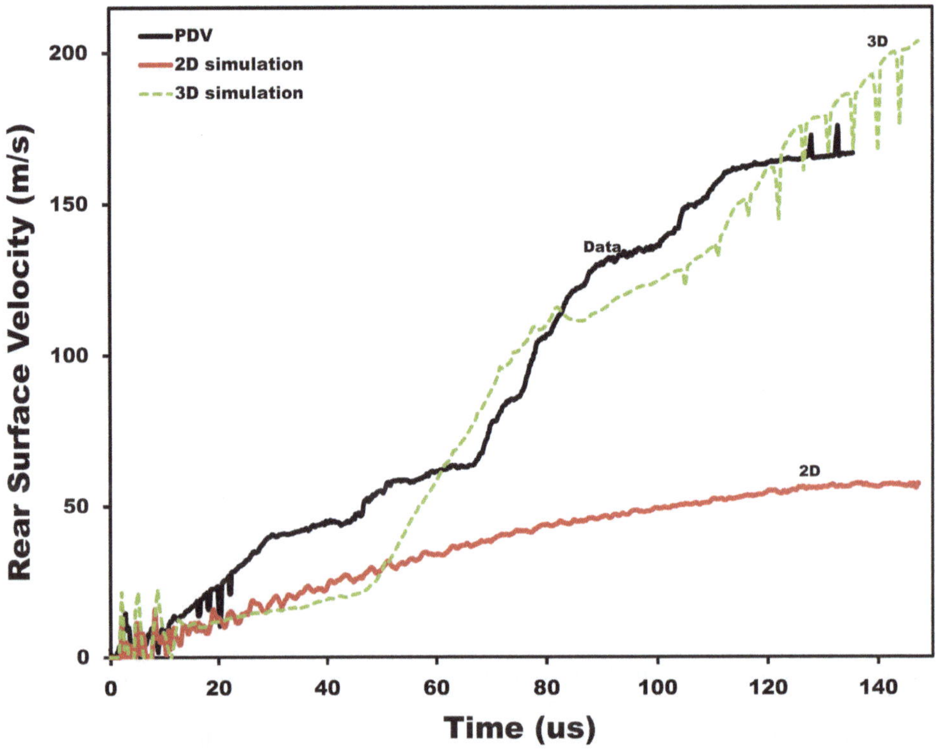

Fig. 14.2 A comparison of the actual rear surface velocity measured by PDV and the best 2D and 3D calibration simulations. The 3D simulations replicate the gross features of the experimental velocity history, but leave much to be desired. The 2D calibration effort was essentially unsuccessful

of 3 cells across the radius of the rod was used for all simulations to make the 3D simulations of tractable size and duration for running hundreds of simulations. The 2D simulations were run on 4 processors, and the 3D simulations were run on 64 processors.

As evident in Fig. 14.1, significant differences exist between the 2D and 3D simulated results. Crack patterns can change from model paramterization to parameterization, but the overall differences between 2D and 3D simulations are essentially similar. It appears that in 3D, the ceramic target can more easily break into pieces allowing the rod to penetrate. Imposing axisymmetry on the problem appears to provide extra constraints on how the damage forms, making penetration more difficult. As can be seen in Fig. 14.2, neither the 2D nor 3D calibrations to the PDV data were extremely successful, but the 2D calibration effort was far less successful at replicating the rear surface velocity history of the experiment.

Fitting the depth of penetration data was similarly challenging.

Acknowledgements We greatfully acknowledge John Niederhaus (Sandia National Laboratories) for his assistance in running ALEGRA. We also acknowledge the helpful comments of Brian Adams (Sandia National Laboratories) relative to using DAKOTA. We are also grateful for the range 309A technicians who assisted in these experiments and the High Performance Computing Modernization Program (HPCMP) for a grant of computer time on the ARL DSRC system Excalibur to run the ALEGRA models.

References

1. Barker, L.M., Hollenbach, R.E.: Laser interferometer for measuring high velocities of any reflecting surface. J. Appl. Phys. **43**, 4669–4675 (1972)
2. Strand, O.T., Goosman, D.R., Martinez, C., Whitworth, T.L., Kuhlow, W.W.: Compact system for high-speed velocimetry using heterodyne techniques. Rev. Sci. Instrum. **77**(8), 083108 (2006)
3. Brannon, R., Fossum, A., Strack, O.: Kayeta: Theory and User's Guide. No. SAND2009-2282, Mar 2009
4. Adams, B.M., Ebeida, M.S., Eldred, M.S., Jakeman, J.D., Maupin, K.A., Monschke, J.A., Swiler, L.P., Stephens, J.A., Vigil, D.M., Wildey, T.M., et al.: Dakota, A Multilevel Parallel Object-Oriented Framework for Design Optimization, Parameter Estimation, Uncertainty Quantification, and Sensitivity Analysis: Version 6.4 Reference Manual. No. SAND2014-5015, May 2016
5. Robinson, A.C., Hansen, G.A., Kramer, R., Labreche, D.A., Love, E., Luchini, C.B., Petney, S.V., Sanchez, J.J., Siefert, C., Voth, T.E., et al.: ALEGRA User Manual. No. SAND2016-11878, Nov 2016

Chapter 15
High-Strain Rate Compressive Behavior of a "Natural Soil" Under Uniaxial Strain State

Huiyang Luo, Zhenxing Hu, Tingge Xu, Lei Guo, and Hongbing Lu

Abstract Dynamic compressive behavior of a synthetic "natural soil" was investigated on an 85-ft split Hopkinson pressure bar. The "natural soil" was prepared by mixing a sorted dry clay and assorted dry Mason sand. The as-received moist clay (obtained from Boulder, Colorado) was first pulverized into powders, which were subsequently dried in oven at 105 °C for 2 days. The as-received Colorado Mason sand was dried under the same conditions. Then the clay was sieved to powers with dimensions smaller than #50 (0.3 mm) using ASTM E-11 standard sieves. The "natural soil" was obtained by mixing at different ratios of clay and sand at 5:100, 10:100, 20:100, 100:100 weight ratios. The mixed soil was capped inside a hardened steel tube, by two tungsten carbide rods at both ends in contact with the bars. This assembly was subjected to repetitive shaking to consolidate the soil to attain a desired bulk mass density; it was then placed between the incident steel bar and the transmission bar for dynamic compression. Through measurements of both axial and transverse responses of the cylindrical soil sample under confinement, both volumetric and deviatoric responses were determined. Both dry and water-saturated soil samples were characterized at high strain rates. The effects of different mixed ratio of clay/sand, and the water-saturation on the compressive behavior were investigated. The compressibility as a function of axial stress was determined. These results are compared with pure clay and sand samples.

Keywords Long split Hopkinson pressure bar • Saturated • Mason sand • Clay • Uniaxial strain

15.1 Introduction

The mechanical behaviour of soil under relatively high stress and high strain rate, important for developing constitutive models, has not been well characterized. In order to simulate soil under blast, computational modeling relies heavily on development of multiscale, multiphase hybrid Lagrangian particle-continuum computational approach. Such simulations require accurate constitutive data for soil under high strain rates encompassing a wide range of loading conditions [1–6]. Recently, a modified split Hopkinson pressure bar (SHPB) with the use of pulse shaping has been used to characterize the mechanical behaviour of soil under high strain rates, confined by a steel sleeve under moist condition at high strain rates [7–10]. A soil specimen was directly sandwiched between incident and transmission bar ends, and confined by a cylindrical tube with the same inside diameter as the bar diameter. This arrangement gives uncertain initial mass density in soil. Recently, we developed a technique to resolve this problem, to allow preparation of sand specimen with consistent initial mass density for dynamic compression [11]. We investigated the compressive behavior of Boulder clay [3], Mason sand [4], dense Eglin sand [12] and glass beads [13] at high strain rates. The constitutive behavior of soil/sand at high strain rates has been reviewed [14], and the modelling was made to simulate the dynamic response of soil under blast [15].

In this study, a "natural soil" was prepared by mixing a sorted dry clay and assorted dry sand. One type of natural sand obtained from the region around Boulder, Colorado, was chosen as the test material; it was purchased in Longmont, Colorado (Colorado Materials), and referred to as Colorado Mason sand [1, 4, 5]. A clay obtained from the region around Boulder, Colorado was chosen as the material for testing [1, 5, 6]. The clay was obtained from a stockpile of soil from a construction site on the University of Colorado Boulder campus, and is referred to as Boulder clay. The clay was processed to remove all particles greater than the #50 sieve size, and thus provided a more homogeneous and consistent material for experiments. Our technique was used to prepare soil specimens with consistent mass densities for experiments on a long SHPB. This technique was combined with other techniques developed in the recent years, including the use of strain gage attached to the outer surface of a confining cylinder tube [16, 17], to investigate the compressive behavior of soil under high strain rates.

H. Luo • Z. Hu • T. Xu • L. Guo • H. Lu (✉)
Department of Mechanical Engineering, The University of Texas at Dallas, Richardson, TX, 75080, USA
e-mail: hongbing.lu@utdallas.edu

J. Kimberley et al. (eds.), *Dynamic Behavior of Materials, Volume 1*, Conference Proceedings of the Society for Experimental Mechanics Series, DOI 10.1007/978-3-319-62956-8_15

15.2 Experimental Aspects

In this investigation, a 24.4 m long modified SHPB was used to apply dynamic compression on soil with relatively high compressive strain with a pulse shaping technique, which can reach certain long loading times (up to 3.4 ms). The SHPB, as shown schematically in Fig. 15.1a, consists of a solid 304 L stainless steel incident bars (Young's modulus 196 GPa, density 8.0 g/cm^3, and bar wave speed 4950 m/s) and a strain data acquisition system. The length of the incident bar and the transmission bar are 8.7 m and 3.7 m long, respectively, they have the same outer diameter of 19 mm.

To obtain soil samples with consistent mass densities, we used an assembly to prepare soil specimen [3, 4, 11–13]. This assembly was then inserted directly between two bar ends of the SHPB apparatus for dynamic compression. Two cemented tungsten carbide (WC) rods with diameters of 12.7 mm were used for soil assembly then were inserted into high carbon steel hollow cylinder with outside diameter 19 mm for confinement. It is noted that the WC rod has a mechanical impedance value (the product of mass density, bar wave speed, and cross sectional area) that is 97% of the corresponding values for the stainless steel bars with a diameter of 19 mm. Therefore, the dispersion due to the mismatch of the mechanical impedance between each bar end to insert rod can be neglected.

The as-received clay has moist content; the clay is in form of irregular large clumps. Steel hamper was used to crush the clay clumps into smaller chunks/powders. The clay clumps was placed in a convention temperature chamber for drying at 105 °C for 2 weeks. Subsequently the dried clay was sorted by #50 sieve (0.30 mm), to obtain dry clay samples. The as-received uncompacted Mason sand also contains moisture; it was dried at the same temperature chamber at 105 °C for 2 days. Then the dry #50 Boulder clay and dry assorted Mason sand were mixed at the given mass ratios to form synthetic "natural soil" with four weight ratios of 100:100, 20:100, 10:100, and 5:100. The soil was first poured into the hollow steel cylinder with one end closed by the WC rod. Then the soil was filled in the hollow cylinder assembly, to form a dry soil specimen. When it is not compacted, the dry mass density was 1.52 g/cm^3 with length/diameter ratio of 0.816, except the pure clay specimen which has initial density of 1.30 g/cm^3. For the saturated soil specimen, water was added into the chamber; the water was weighted by digital balance (Denver Instrument APX-300, 1 mg resolution). The open assembly was placed inside a vacuum chamber for 30 min, to remove air bubbles and make the water evenly distributed inside the soil. Then another brass rod was used to cap the testing chamber. A schematic diagram is shown in Fig. 15.1b. The net mass of the soil (2.0 g)

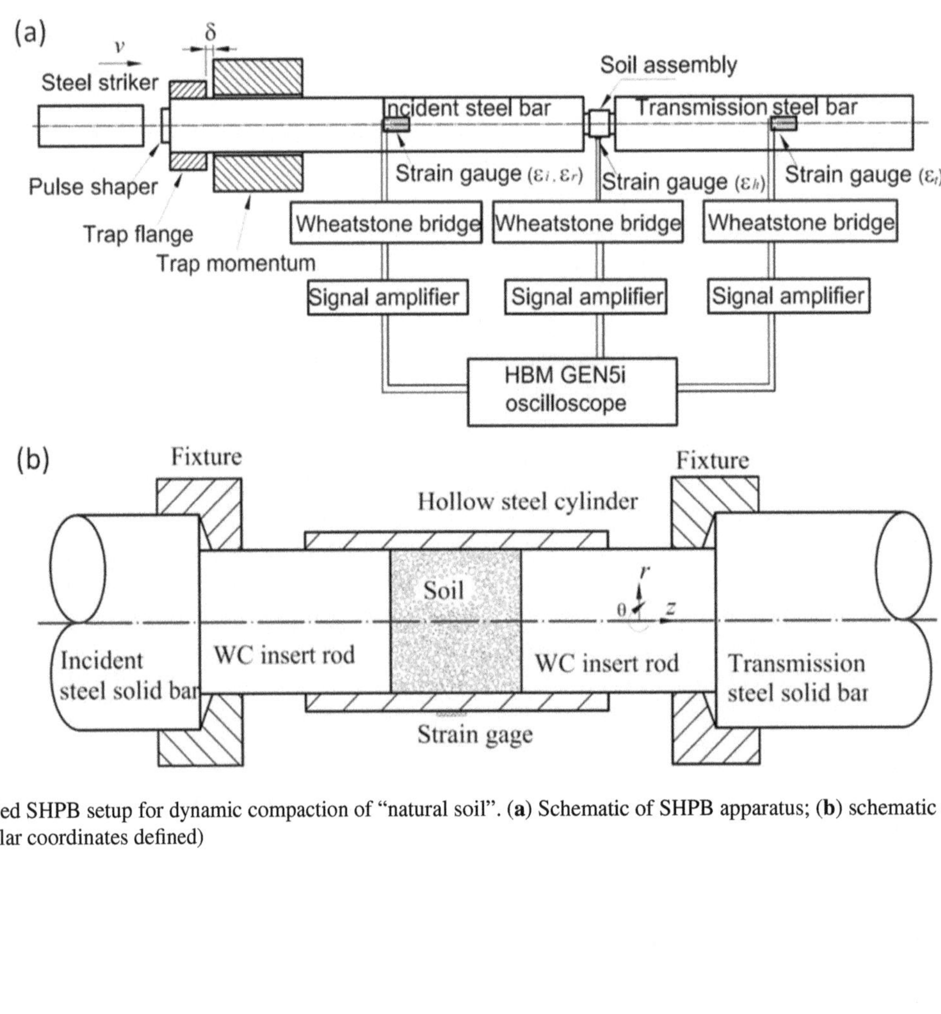

Fig. 15.1 Modified SHPB setup for dynamic compaction of "natural soil". (**a**) Schematic of SHPB apparatus; (**b**) schematic of test section of the soil assembly (polar coordinates defined)

was determined by weighing the overall assembly and then subtracting the tare (steel sleeve and inserted rods). The length of the cavity occupied by soil was measured and controlled by change in distance of the two end surfaces of the two rods. These values were then converted into bulk mass density of the soil specimen. To prepare fully saturated soil, the length of soil was controlled so that extra water can be added through the gap between WC rod and the inner surface of the sleeve.

A strain gage was attached onto the external surface of the confining steel sleeve on the circumferential direction. The lateral deformation of the soil sample was measured by the strain gage, and the radial/circumferential stress was calculated from the strain considering a linear elastic response of the confining cylinder under lateral deformation below 0.2%. This arrangement allows characterization of dynamic volumetric and deviatoric behaviours of soil. Further analysis yields other relationships, such as the void ratio - pressure relationship at high strain rates. In each experiment, the soil assembly was sandwiched between incident and transmission bars on SHPB, supported and aligned by two aluminum fixtures attached to the lateral surface of the ends of the incident and transmission bars. Compact loading was applied through WC rods in contact with incident and transmission bars.

Brass 260 disks of 7.6 mm in diameter with thicknesses varying from 1.6 to 3.1 mm were used as pulse shapers in the SHPB experiments by placing them on the impact surface of the incident bar. A thin layer of Dow Corning High-vacuum grease (McMaster #2966 k52) was used at the interface between the pulse shaper and the incident end of the bar. A 2.13 m long steel striker, with the same diameter as that of both the incident and transmission bars, launched by compressed air in a gas gun, impacts the pulse shaper and generates a desired compressive wave traveling in the incident bar to induce a nearly constant strain rate in the soil specimen. Trap momentum was used to apply a single loading. A HBM Genesis digital oscilloscope (15-bit resolution, 25 Ms/s sampling rate) was used to acquire strain signals through Wheatstone bridges and a Vishay 2310B signal conditioning amplifier (Gain 1–11,000 continuously varying; 180 kHz frequency response, 180 kHz bandwidth and ±0.02% linearity) connected to the strain gages.

A hollow cylinder was used to confine soil specimen passively. For a specimen confined by a hollow cylinder within the elastic range, the calculation for the confining radial and circumferential pressures, as well as circumferential strain has been well documented [16, 17]. The polar coordinates with r and θ are defined in Fig. 15.1b. In this study, we used the confinement method for dynamic compression of sand on SHPB to characterize the hydrostatic and deviatoric behaviors of soil at high strain rates. Since the incident bar is the same as the transmission bar, the calculation follows the previous approach [18]. With the known axial, confining stress and strain histories, the mean stress (e.g., hydrostatic pressure, isotropic stress) component σ_m and the volumetric dilatation ε_m are well documented, as well as the octahedral shear stress τ_{oct} [16, 17]. With the known values for hydrostatic and shear stresses, the hydrostatic and deviatoric behaviors of soil under multi-axial stress states can be determined at high strain rates. Thus, the known stresses can be used as input for the modeling in the future. In this investigation, all tests were conducted at room temperature (23 °C) and near 26% relative humidity environment. Under each testing condition, eight or more specimens were tested to ensure reproducibility of experimental results, and the average values are reported.

15.3 Results and Discussion

The physical properties of Boulder clay and Mason sand have been measured according to ASTM soil testing standards [5, 6]. For Boulder clay, the relevant grain size distribution parameters are determined as $D_{10} < 1.7 \times 10^{-4}$ mm, $D_{30} < 0.001$ mm, $D_{50} = 0.001$ mm, and soil index properties are used to determine the liquid limit LL = 43 and plasticity index PI = 22. Boulder clay is classified as a low plasticity clay (CL) according to the Unified Soil Classification System (USCS). The specific gravity G_s was measured as 2.70. From the standard Proctor compaction test, an optimal water content of 18% corresponds to the maximum dry unit weight of 17.5 kN/m^3. The clay has an apparent preconsolidation stress of 0.30 MPa, a compression index c_c of 0.23, and a recompression index c_r of 0.041. For the Colorado Mason sand, the relevant grain size distribution parameters are determined as $D_{10} = 0.20$ mm, $D_{30} = 0.44$ mm, $D_{60} = 0.90$ mm. The sand is classified as poorly graded sand (SP) according to the USCS. The specific gravity G_s was measured to be 2.62 g/cm^3. The minimum void ratio of 0.50 corresponds to a maximum dry density of 1.74 kg/m^3, while the maximum void ratio of 0.78 corresponds to minimum dry density of 1.47 kg/m^3.

Typical SHPB results are shown in Fig. 15.2 for dry soil under a strain rate in the range of 400–600 s^{-1}. The mean stress-volumetric strain curve (Fig. 15.2a) and compressibility e-log p curves are shown in Fig. 15.2b, reported as the average strain rate and average stress-strain curves. In Fig. 15.2a, the stress is plotted in logarithmic scale while the strain is plotted in linear scale. In these semi-logarithmic plots, the stress follows approximately a two-segment linear relationship with the corresponding strains. The first linear segment occurs within the initial 2% compressive strain, most likely due to elastic deformation of soil grains. It is referred as soil grain-level elastic behaviour, similar to the situation under quasi-static

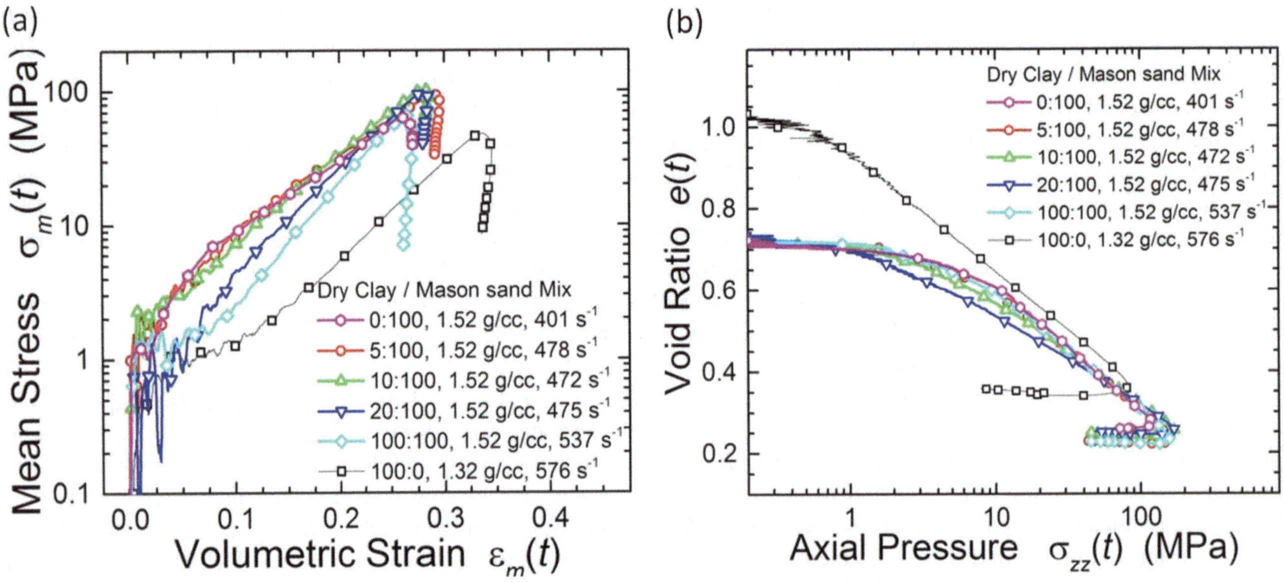

Fig. 15.2 SHPB results of dry soil with different mixture levels. (**a**) Mean stress-volumetric strain curves; (**b**) compressibility *e*-log *p* curves

compression [19]. The second linear segment on a semi-logarithmic scale occurs at compressive strains larger than 4%. It is perhaps associated with sand grain compaction through grain crushing, rearrangement of crushed fragments and intact grains under higher pressure. This stage can be referred to as grain-level plastic behaviour. At compressive strains between 2% to 4%, there is a gradual transition region. Since the dry soil is not compacted in its natural stacking stage with minimum disturbance during the preparation before sandwiched between two bar ends, the mean stress-volumetric strain curve of soil has a small oscillated stage in the elastic and transition regions. However, the grain-level plastic stage shows that it deformed in smooth curves. It is noted that the pure clay has an initial density of 1.32 g/cm^3, lower than 1.52 g/cm^3 for the soil, the curve of pure clay has weak mechanical properties than that of soil contained sand. Under lower ratios of clay/sand in soil (5:100 mix soil and 10:100 mix soil), the curves are very close to that of pure Mason sand. For other two ratio of soil mixtures (20:100 and 100:100), the first two stages are similar to the pure clay, while it converges to the curves of pure sand and the low ratio of clay/sand in soil during grain-level plastic stage.

Figure 15.2b shows the compressibility of dry soils as a function of pressure. The relationship is given in terms of the void ratio, *e*, and the applied axial stress, σ_z, often termed as *p* used in the soil mechanics. The measurement of both radial and circumferential stresses and strains in the soil sample in this work made it possible to report the relationship under high strain rates. A linear segment clearly appears in the *e*-log *p* curves for dry soils, representing an exponent relationship between *e* and *p*. The *e*-log *p* curve during unloading also follows nearly a linear relationship. At these high strain rates, four soils with the clay/sand ratios of 0:100, 5:100, 10:100 and 100:100 has the close *e*-log *p* curves; while the pure clay have higher void ratio and easier compressibility, and the 20:100 soil mixture is associated with less compressibility.

Figure 15.3 shows the average SHPB results for fully saturated soil under strain rate of 400–600 s^{-1}. These are mean stress-volumetric strain curve (Fig. 15.3a) and compressibility *e*-log *p* curves (Fig. 15.3b). In Fig. 15.3a, the stress is plotted in logarithmic scale while the strain is plotted in linear scale. In these semi-logarithmic plots, the stress-strain curves are curved. They do not follow two-segment linear stage as shown Fig. 15.2a, due to the fact that the water has carried load. The soils mixed at four lower ratios of clay/sand have higher mechanical properties than those of pure clay and pure sand under saturated state. At high strain level, in larger than 12% and high stress level (200–300 MPa), all curves converge and are close to each other. It indicated that water carried more load in soil than situations for pure clay or sand.

Figure 15.3b shows the compressibility of saturated soils as a function of pressure in terms of the *e*-log *p* curves. Figure 15.3b shows a linear segment before preconsolidation, representing an exponential relationship between *e* and *p*; it becomes curved for saturated soils. The *e*-log *p* curve during unloading also follows nearly a linear relationship. At these high strain rates, four soils with the clay/sand ratios of 5:100, 10:100, 20:100 and 100:100 have *e*-log *p* curves close to each other, indicating that the clay/sand ratios did not have significant effect when water in the saturated soil is capable to carry load. The pure clay has higher void ratio and easier compressibility, and the pure sand has the lower void ratio and less compressibility. As compared with dry soils, water plays more significant role, for load-bearing capability than the sand grains. The soil with lower clay/sand ratios has stiffened the mechanical properties under water saturated state, and pure clay and pure sand have reduced mechanical properties.

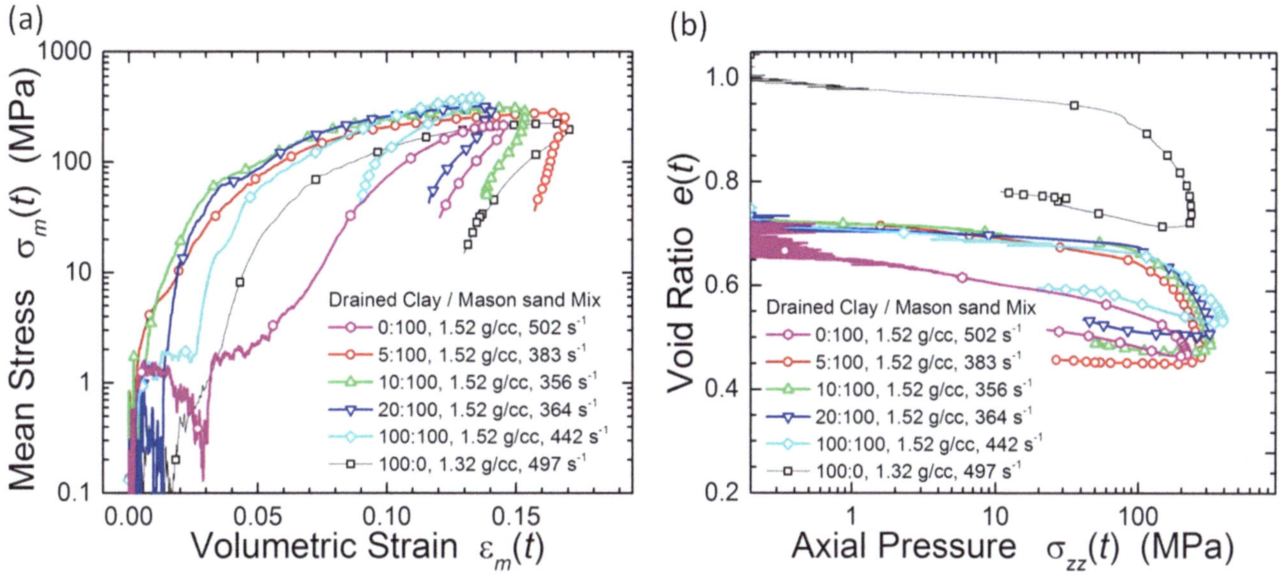

Fig. 15.3 SHPB results of saturated soil with different mixture levels. (**a**) Mean stress-volumetric strain curves; (**b**) compressibility *e*-log *p* curves

15.4 Conclusions

The dynamic compressive behaviour of synthetic natural soil under stiff confinement was investigated using a long split Hopkinson pressure bar (SHPB) at high strain rates (400–600 s^{-1}) up to 350 MPa compressive stress. Soil samples under six mixing ratios (Boulder clay to Mason sand) sand at 0:100 (pure clay), 5:100, 10:100, 20:100, 100:100 and 100:0 (pure sand) at dry initial mass densities 1.52 g/cm^3 were compressed at high strain rates. Fully water-saturated soil samples were also characterized at high strain rates. Using a strain gage attached on the external surface of the confining tubes, both the bulk and the deviatoric responses were determined. The curves of mean stress-volumetric strain and compressibility *e*-log *p* curves under 3D stress state of soils are determined up to 32% compressive strain. Stronger saturation effects have been identified for soil than clay and sand. The stress-strain relationships follow an exponential power relationship for dry soil. At high level of strain and stress, water has strengthened the soil mechanical properties. These results can be analysed further for constitutive modelling and for mesoscale simulations to understand the dynamics of soil under high pressure and high rate deformation.

Acknowledgments We acknowledge the support of ONR MURI grant N0. 0014-11-1-0691, NSF ECCS-1307997 and NSF CMMI-1636306. Lu Also acknowledges the Louis A. Beecherl Jr. Chair for additional support.

References

1. Regueiro, R., Pak, R., McCartney, J., Sture, S., Yan, B., Duan, Z., Svoboda, J., Mun, W., Vasilyev, O., Kasimov, N., Brown-Dymkoski, E., Hansen, C., Li, S., Ren, B., Alshibli, K., Druckrey, A., Lu, H., Luo, H., Brannon, R., Bonifasi-Lista, C., Yarahmadi, A., Ghodrati, E., Colovos, J.: ONR MURI project on soil blast modeling and simulation. In: Dynamic Behavior of Materials 1: Conf. Proc. Society for Experimental Mechanics Series, pp. 341–353. Springer, Cham (2014)
2. Luo, H., Xu, T., Wang, X., Lu, H.: Interaction of shock wave with granular materials. In: Antoun, B. (ed.) Challenges in Mechanics of Time Dependent Materials, 2: Proc. 2015 Ann. Conf. Exp. Appl. Mech, pp. 35–43. Springer, Cham (2016)
3. Luo, H., Hu, Z., Xu, T., Lu, H.: High-strain rate compressive behavior of a clay under uniaxial strain state. In: Dynamic Behavior of Materials 1: Proc. 2016 Ann. Conf. on Exp. Appl. Mech, pp. 117–122. Springer, Cham (2017)
4. Luo, H., Du, Y., Hu, Z., Lu, H.: High-strain rate compressive behavior of dry Mason sand under confinement. In: Song, B., et al. (eds.) Dynamic Behavior of Materials, 1: Proc. 2014 Ann. Conf. Exp. Appl. Mech, pp. 325–333. Springer, Cham (2015)
5. Svoboda, J.S., McCartney, J.S.: Shearing rate effects on dense sand and compacted clay. In: Dynamic Behavior of Materials 1: Conf. Proc. Society for Experimental Mechanics Series, pp. 389–395. Springer, Cham (2014)
6. Mun, W., McCartney, J.S.: Compression mechanisms of unsaturated clay under high stresses. Can. Geotech. J. **52**(12), 2099–2112 (2015)
7. Martin, B.E., Chen, W., Song, B., Akers, S.A.: Moisture effects on the high strain-rate behavior of sand. Mech. Mater. **41**(6), 786–798 (2009)
8. Song, B., Chen, W., Luk, V.: Impact compressive response of dry sand. Mech. Mater. **41**(6), 777–785 (2009)

9. Huang, J., Xu, S., Hu, S.: Effects of grain size and gradation on the dynamic responses of quartz sands. Int. J. Impact Eng. **59**, 1–10 (2013)
10. Martin, B.E., Kabir, E., Chen, W.: Undrained high-pressure and high strain-rate response of dry sand under triaxial loading. Int. J. Impact Eng. **54**, 51–63 (2013)
11. Luo, H., Lu, H., Cooper, W.L., Komanduri, R.: Effect of mass density on the compressive behavior of dry sand under confinement at high strain rates. Exp. Mech. **51**(9), 1499–1510 (2011)
12. Luo, H., Cooper, W.L., Lu, H.: Effects of particle size and moisture on the compressive behavior of dense Eglin sand under confinement at high strain rates. Int. J. Impact Eng. **65**, 40–55 (2014)
13. Luo, H., Du, Y., Hu, Z., Cooper, W.L., Lu, H.: High-strain rate compressive behavior of glass beads under confinement. Exp. Mech. **55**(5), 935–950 (2015)
14. Omidvar, M., Iskander, M., Bless, S.: Stress-strain behavior of sand at high strain rates. Int. J. Impact Eng. **49**, 192–213 (2012)
15. Higgins, W., Chakraborty, T., Basu, D.: A high strain-rate constitutive model for sand and its application in finite-element analysis of tunnels subjected to blast. Int. J. Numer. Anal. Methods Geomech. **37**(15), 2590–2610 (2013)
16. Ravi-Chandar, K., Ma, Z.: Inelastic deformation in polymers under multiaxial compression. Mech. Time-Depend. Mater. **4**, 333–357 (2000)
17. Knauss, W.G., Emri, I., Lu, H.: Mechanics of polymers: viscoelasticity. In: Sharpe, W.N. (ed.) Handbook of Experimental Solid Mechanics, pp. 49–95. Springer, New York (2008)
18. Gray III, G.T.: Classic split-Hopkinson pressure bar testing. In: Mechanical Testing and Evaluation 8, Metals Handbook, pp. 462–476. ASM, Materials Park (2000)
19. Hagerty, M.M., Hite, D.R., Ullrich, C.R., Hagerty, D.J.: One-dimensional high-pressure compression of granular material. J. Geotech. Eng. **119**(1), 1–18 (1993)

Hongbing Lu is full Professor, Louis Beecherl Jr. Chair, Associate Department Head of Mechanical Engineering at UT Dallas. He received his PhD in Aeronautics at California Institute of Technology in 1997. His research interests focus on experimental mechanics, nanomechanics, mechanics of time-dependent materials, mechanical behavior of nanomaterial, and dynamic behavior of material.

Chapter 16
Latest Results for Elasto-Plastic Identification at High Rates Using Inertial Impact

Sarah Dreuilhe, Frances Davis, Clive Sivour, and Fabrice Pierron

16.1 Introduction

The behavior and failure mechanisms of materials often change at high strain rates when compared with their quasi-static response. In automotive manufacturing the use of aluminum alloys has doubled in the last decade as the introduction of aluminum frames can reduce the body weight. During crash events these aluminum alloys can reach strain rates of $10^4\ s^{-1}$ [1] and understanding their dynamic response is necessary to design safe vehicles. In this paper, inertial impact tests coupled with ultra-high speed imaging and the virtual fields method were used to identify the high strain rate elasto-plastic properties of Al6082-T6.

16.2 Materials and Methods

A series of inertial impact tests were performed using the gas gun facilities at the Solid Mechanics and Material Engineering Group at Oxford. The gas gun was used to fire cylindrical projectiles at a nominal speed of 50 m/s. Each aluminum specimen was L = 48 mm by w = 31 mm with a thickness of 4 mm (Fig. 16.1). A grid with a pitch of 0.6 mm was adhered to the surface of the sample using an epoxy glue. An ultra-high speed camera (HPV-X, Shimadzu) was used to capture images of the sample during the impact test. Light gates inside the gas gun barrel were used to trigger the flashes and a custom make trigger between the waveguide and the projectile was used to trigger the camera. A total of 128 images were recorded using the HPV-X at a rate of 5 Mfps.

The grid method [2] was used to obtain time-resolved maps of displacement from the collected images. The resulting displacement was smoothed temporally using a 3rd order low pass Butterworth filter with a normalized cut-off frequency of 0.09. The smoothed displacement was double differentiated to obtain maps of the acceleration. Before calculating the strain, the displacement was spatially smoothed using a robust filter. The calculated strains and acceleration fields were used as inputs to the virtual fields method.

S. Dreuilhe • F. Davis • F. Pierron (✉)
Faculty of Engineering and the Environment, University of Southampton, SO17 1BJ, Highfield, UK
e-mail: F.Pierron@soton.ac.uk;
e-mail: hongbing.lu@utdallas.edu

C. Sivour
Department of Engineering Science, University of Oxford, OX1 3PJ, Oxford, UK

J. Kimberley et al. (eds.), *Dynamic Behavior of Materials, Volume 1*, Conference Proceedings of the Society for Experimental Mechanics Series, DOI 10.1007/978-3-319-62956-8_16

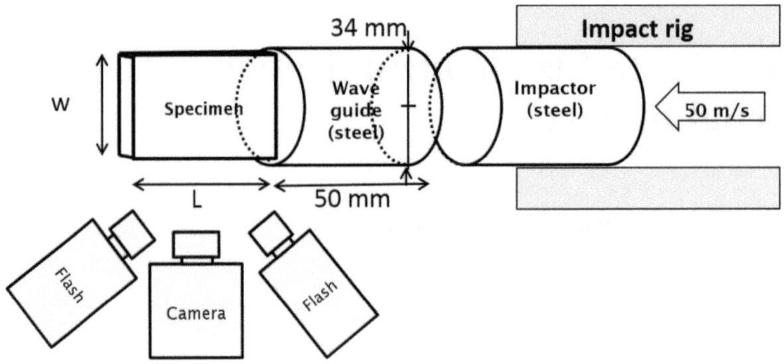

Fig. 16.1 Schematic of the impact rig

16.3 Virtual Fields Method

The principle of virtual work is derived from an integral form of equilibrium equations [3]. Ignoring body forces the following equation is obtained:

$$\overbrace{\int_{S_f} \boldsymbol{T}\cdot\boldsymbol{u}^{*}\,\mathrm{d}S}^{W^*_{ext}} - \overbrace{\int_V \boldsymbol{\sigma} : \boldsymbol{\epsilon}^*\,\mathrm{d}V}^{W^*_{int}} = \overbrace{\int_V \rho\,\boldsymbol{a}\cdot\boldsymbol{u}^{*}\,\mathrm{d}V}^{W^*_{acc}} \tag{16.1}$$

where $\boldsymbol{T}$ is the traction vector, $\boldsymbol{\sigma}$ is the stress tensor, $\boldsymbol{a}$ is the acceleration vector, ρ is the density of the material, and $\boldsymbol{u}^*$ and $\boldsymbol{\epsilon}^*$ are the virtual displacement vector and associated virtual strain tensor, respectively. Assuming a state of plane stress and imposing a virtual displacement of $u_1^* = 1$ and $u_2^* = 0$, the first integral in Eq. 16.1 will equal the average stress in the x_1 direction. This set of virtual fields also cancels out the contribution of internal virtual work, W_{int}, leaving only the inertial virtual work, W_{acc}. Replacing the integrals with discrete sums the following equation was found:

$$\overline{\sigma_1(x_1,t)}^{x_2} = \rho\, x_1\, \overline{a_1(x_1,t)}^{S}. \tag{16.2}$$

In Eq. 16.2, the overbars indicate averages over the line or area specified in the superscript. By varying the value of x_1, the average stress for any transverse section can be identified from the acceleration. To identify the elasto-plastic material parameters, a set of non-linear equations is minimized. The cost function to be minimized is sum of the squared difference between the average stress in each section calculated from the acceleration (Eq. 16.2) and the average stress in each section calculated using the constitutive model. The identification is iterative; the cost function is evaluated for several different combinations of the material parameters (through the constitutive model) searching for the set of parameters that minimize the cost function. In this study, a linear elastic model with linear hardening was used to describe the elasto-plastic response requiring two parameters to be identified, the yield stress, σ_0, and hardening modulus, H.

16.4 Results and Discussion

A total of nine aluminium specimens were successfully tested at two impact different speeds. The values of the yield stress and hardening modulus identified can be found in Table 16.1. Previous studies in the literature show that Al6082-T6 shows limited strain rate sensitivity. This is confirmed by the present results, at least for the yield stress. The hardening modulus for low speed impact presents very high scatter, which was expected as the value is very small and therefore, will be hard to identify. At higher speed, the values are more stable as more plastic strain is introduced in the specimen. This can be seen also in Fig. 16.2. At lower speeds, the maximum strains are around 3% and the strain rate maps show that the large strain rate values only last for a few microseconds. Looking at the faster impact, plasticity builds up longer leading to much larger strains.

In the future, optimized virtual fields which minimize the influence of experimental noise will be explored to determine if the hardening modulus can be more accurately identified. The presentation will also report data on 316L stainless steel.

Table 16.1 Identified plastic properties of Al 6082-T6

Low speed				
Specimen name	σ_0(MPa)	H (MPa)	Max$\|\epsilon_1\|$(%)	Impact speed (m.s^{-1})
AL30	280	50	4.2	51
AL26	252	1173	3.5	56
AL23	264	1352	3.7	57
AL14	270	946	5.7	56
AL07	259	2095	3.3	57
Mean ± std	266 ± 9.7	1123 ± 661	4.1 ± 1.0	
Quasi-static ref. ± std	282 ± 1.3	1002 ± 78.7	1.5 ± 0.03	
High speed				
Specimen name	σ(MPa)	H (MPa)	Max $\|\epsilon_1\|$ (%)	Impact speed (m.s^{-1})
AL28	265	2081	7.1	132
AL24	276	1913	7.1	128
AL12	281	1613	7.4	130
AL11	280	1326	7.9	131
Mean ± std	275 ± 6.5	1734 ± 289	7.4 ± 0.4	
Quasi-static ref. ± std	282 ± 1.3	1002 ± 78.7	1.5 ± 0.03	

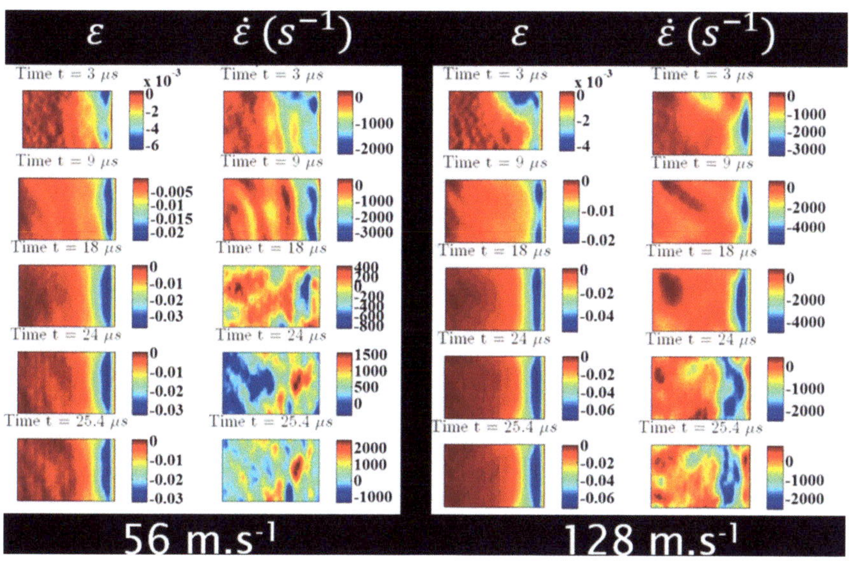

Fig. 16.2 Longitudinal strain and strain rate maps at some particular times

Acknowledgements This material is based on research sponsored by the Air Force Research Laboratory, under agreement number FA8655-13-1-3041. The U.S. Government is authorized to reproduce and distribute reprints for Governmental purposes notwithstanding any copyright notation thereon. The views and conclusions contained herein are those of the authors and should not be interpreted as necessarily representing the official policies or endorsements, either expressed or implied, of the Air Force Research Laboratory or the U.S. Government. Professor Fabrice Pierron is grateful to the Wolfson Foundation for a Royal Society Research Merit Award. Prof. Pierron and Dr Davis acknowledge funding through EPSRC grant EP/L026910/1.

References

1. Moćko, W., et al.: Compressive viscoplastic response of 6082-T6 and 7075-T6 aluminium alloys under wide range of strain rate at room temperature: experiments and modelling. Strain **48**(6), 498–509 (2012)
2. Badulescu, C., Grédiac, M., Mathias, J.D.: Investigation of the grid method for accurate in-plane strain measurement. Meas. Sci. Technol. **20**(9), 095102 (2009)
3. Pierron, F., Grédiac, M.: The Virtual Fields Method: Extracting Constitutive Mechanical Parameters from Full-Field Deformation Measurements. Springer, New York (2012)

Chapter 17
Mesoscale Modeling of Porous Materials Using New Methodology for Fracture and Frictional Contact in the Material Point Method

M.A. Homel and E.B. Herbold

Abstract Using a new damage-field gradient partitioning approach to fracture and frictional contact in the material point method (MPM) we have demonstrated a capability to produce mesh-independent predictions of particle size distribution in simulations of comminution of brittle materials. Here, we apply this new methodology to simulate damage and dynamic loading in mesoscale models of porous, heterogeneous materials. Simulations are validated against experimental data, and we discuss the advantages and limitations of this new methodology.

Track: Dynamic Behavior of Materials
Session: Dynamic Behavior of Geomaterials

Keywords Material point method • Fracture • Comminution • Fragmentation • Contact • Brittle • Porous • Heterogeneous

17.1 Background

Porous heterogeneous materials have a complex constitutive response, where continuum properties that describe the stiffness, strength, softening, shear bulking, and compaction are governed by microscale mechanisms. For example, concrete has a hierarchical structure with heterogeneity that spans many length scales; there are sub-micron particulates within the mortar, coarse aggregates in the 100 micron to centimeter range, porosity with 10–100 micron pore sizes, and the possibility for micro fractures and interface failures spanning these scales. Similarly, natural geomaterials such as sandstone and granite may have multiple levels of porosity, aggregates of varying strength, and a distribution of initial flaws, all of which must be accounted for to predict the material response. Advances in imaging and experimental methods continue to enhance our understanding of the microstructure of these materials and our ability to estimate the physical properties of the discrete phases that comprise the microstructural heterogeneity. However, even with a perfect description of the microstructure and the properties of the individual phases, there are many technical challenges to predicting the continuum response of the bulk material in complex loading paths.

Our focus is on mesoscale modeling of the mechanical response of these materials, whereby we explicitly resolve the microstructure features with length scales 1-2 orders of magnitude below that of the representative volume element (RVE), and apply a continuum description to capture the effect of features below the resolvable scale. For example, mesoscale simulation of a 1mm^3 volume of a high-performance concrete might resolve pores down to 10 microns, but the nanometer "gel porosity" within the calcium-silicate-hydrate cement paste would be accounted for by including a compaction model in the continuum description of the paste.

To simulate continuum compaction, damage, or softening for the bulk material it is necessary to account for fracture, contact, large displacements, and large deformation while accurately representing the complex geometry of the mesoscale model. As failure progresses in these materials, fracture and fragmentation cause an initially solid material to evolve into a (locally) granular state that undergoes comminution with successive fracture and division of particulates. This complicates the numerical solution by extending the range of length scales in the problem since the dimensions of a representative

This work was performed under the auspices of the U.S. Department of Energy by Lawrence Livermore National Laboratory under contract DE-AC52-07NA27344. Lawrence Livermore National Security, LLC. LLNL-CONF-725442.

M.A. Homel (✉) • E.B. Herbold
Lawrence Livermore National Laboratory, Atmospheric, Earth, and Energy Division, Livermore, CA, 94550, USA
e-mail: homel1@llnl.gov

J. Kimberley et al. (eds.), *Dynamic Behavior of Materials, Volume 1*, Conference Proceedings of the Society for Experimental Mechanics Series, DOI 10.1007/978-3-319-62956-8_17

material element may be many orders of magnitude greater than that of the finest particles produced. Inevitably, there are computational limitations to the range of particle sizes that can be resolved, and it is therefore necessary to use a continuum approximation to describe the material response of a collection of the finest particles. As these particles flow, the continuum deformation can be quite large, introducing an additional challenge to numerical simulation. Many common numerical simulation methods are inadequate to this task, which has motivated the development of new approaches.

One promising approach is the material point method (MPM), a mixed Eulerian-Lagrangian approach in which the material state is defined by Lagrangian particles and the equation of motion is solved on a fixed computational mesh [1–4]. This approach has four key advantages, that make it well suited for mesoscale modeling of porous heterogeneous materials.

(i) The problem geometry is initialized through particle creation, avoiding the need for mesh generation, allowing for direct use of voxelized microCT data in defining material microstructure (cf. [5]).
(ii) The method avoids the advection errors that occur in Eulerian and ALE methods, allowing for the use of complex constitutive models to describe the response of each phase in the material (cf. [6]).
(iii) The fixed computational mesh eliminates mesh entanglement that occurs in Lagrangian methods, allowing for robust simulation of the large deformation that occurs when during material failure at the grain and pore scales (c.f. [2]).
(iv) The method automatically prevents interpenetration at contact surfaces, and frictional contact can be computed efficiently—an important consideration in modeling pore collapse and compaction [7].

The traditional MPM method has two important limitations that can limit its ability to simulate the mesoscale response of porous heterogenous materials. The first is the *velocity field smearing error*, which arises when particles in difference damage states occupy the same grid cell (sharing a common velocity gradient) and which can manifest as a dependence of the softening response on both the grid resolution and the relative velocity of the particles to the background mesh [8]. The second limitation is that frictional contact requires an *a priori* partitioning of material into contact groups, which adds significant cost to the simulation of granular materials and it does not allow for frictional "self contact" at internal pore surfaces or at newly formed fracture surfaces. There are modifications to the MPM that address these limitations through the introduction of explicit fracture and contact surfaces [9, 10], but such approaches have not yet been demonstrated to describe the complex 3-D crack topologies and large displacement contacts that occur in simulations of comminution of porous heterogeneous materials. To address these limitations in the MPM, we have developed the damage-field gradient (DFG) partitioning method, which was presented along with full algorithmic detail and verification examples in [11]. Here we will provide only a conceptual summary of the method, and give examples to illustrate its value for mesoscale simulation.

17.2 Methods

In the material point method, strong discontinuities (i.e. fracture and contact interfaces) are accounted for by defining two coincident computational grids, each describing a continuous velocity field, so that material on either side of an interface can map to its own grid. The DFG-partitioning method is a way to dynamically repartition each particle into one of these two velocity fields based on the relative location of the particle to a fracture surface. The key to the effectiveness of the method in complex problems is that it is not actually necessary to construct or track fracture surfaces, rather the partitioning is defined based on a smooth kernel field constructed from a scalar damage measure at each particle. Figure 17.1a shows a notional damage field for a notched tensile test specimen, where we observe that the gradient of the damage field is normal to the ideal fracture plane.

Figure 17.1b shows the particle partitioning around a damage field that describes a curved crack. To partition the material into contact pairs it is necessary only to identify regions where adjacent particles have opposing damage field gradients. This is done as part of the MPM solution algorithm where nodal masses velocities and forces are computed from the state of the particles surrounding the node. First, the damage-field gradient is computed at each node based on the infinite norm of the gradient field surrounding the node. Then, when particles are mapped to that node, the sign of the dot product of the particle and grid-node gradient vectors is used to determine whether the particle maps to the A or B field at that node. In this way, contact pairs are automatically constructed from the damage field. A separability criterion is also evaluated to ensure that contact surfaces only form where the damage state is sufficient to represent a fracture surface.

The damage-field gradient partitioning approach can be applied to an arbitrary scalar field, and can solve the self-contact class of problems that are challenging with an *a priori* material partitioning. To do this, a scalar contact flag is defined at each particle at problem initialization, either by computing the density gradient or by "painting in" a surface flag in algorithmically generated problem geometries (Fig. 17.2a, b). With the surface flag so-defined, the damage-field gradient partitioning algorithm is modified slightly so that surface particles are treated as fully damaged (this does not affect the

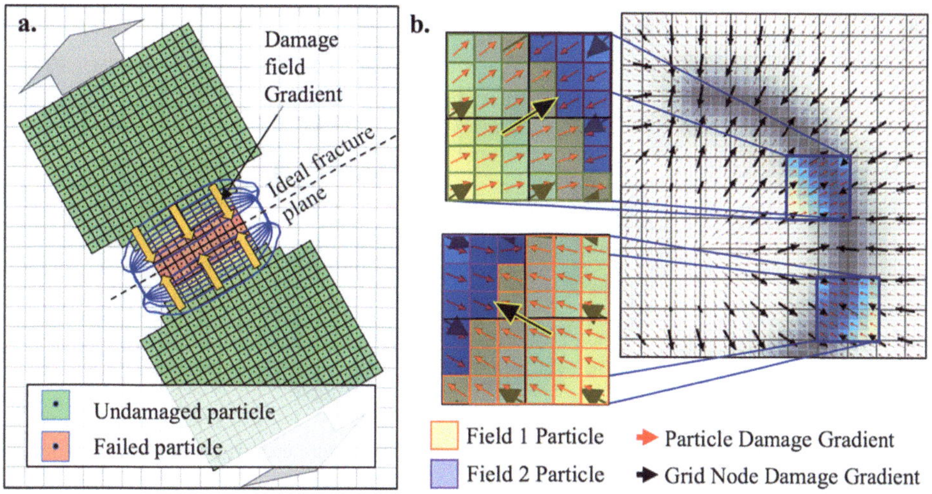

Fig. 17.1 Concept of the damage-field gradient partitioning showing (**a**) damage-field contours in a tensile bar, and (**b**) particle partitioning around a curved crack (Adapted from [11])

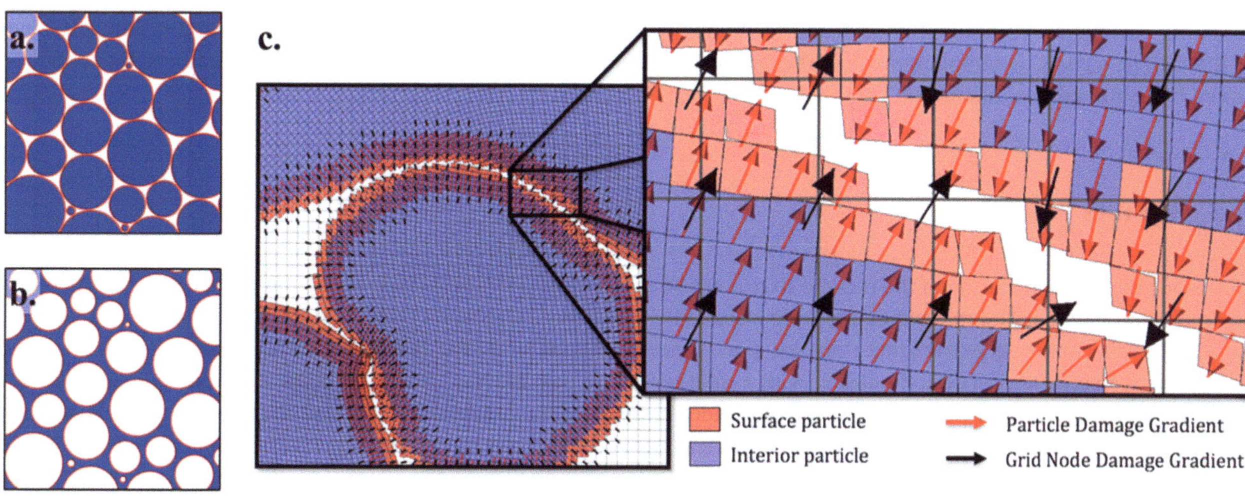

Fig. 17.2 Self contact with field-gradient partitioning. (**a**) Surface flags for a granular material, (**b**) surface flags for a porous material, (**c**) effective damage gradient and field partitioning at contact surface (Adapted from [11])

constitutive response of the surface particle), which naturally leads to separation or frictional contact at the surface, as shown in Fig. 17.2c. Additionally, this approach accommodates frictional contact between these prescribed surface particles and the damaged particles at fracture surfaces. In this way, it becomes possible to efficiently model fracture and subsequent granular flow that may significantly change the contact topology relative to the initial configuration.

The performance of the DFG partitioning method was evaluated with a series of quantified verification tests, as reported in [11]. Snapshots of the key results are shown in Fig. 17.3, which demonstrated (a) the DFG-partitioning method produces a fracture energy consistent with the constitutive model parameters in a notched Charpy impact even as the mesh resolution is varied, (b) the self-contact approach produces the correct response for frictional contact with large deformation and displacement, (c) the method is efficiently implemented in a 3-D parallel code, and (d) the method produces dynamic fracture response in agreement with experimental results. These verification results give confidence in the method when it is applied to complex mesoscale models where the correct material response is unknown.

A key capability of the DFG partitioning method is its ability to robustly simulate comminution in brittle materials with mesh-independent predictions of fragment velocities and size distributions compared to other methods. This was demonstrated in [11] by simulating the crushing of a brittle disk using both a traditional single-field MPM and the DFG-partitioning method, producing the results shown in Fig. 17.4. Here we see that the traditional "single-field" MPM solution has a strong dependence of fracture energy on mesh resolution, manifesting in an increase in velocity and fracture surface

Fig. 17.3 Test problems for the DFG-partitioning method from [11] to verify (**a**) fracture energy, (**b**) large-deformation frictional self-contact, (**c**) 3-D fragmentation, and (**d**) dynamic crack propagation and branching

area (i.e. decreased fragment size) as the mesh is refined. In contrast, the DFG partitioning method gives more consistent results at all resolutions. To our knowledge, the MPM/DFG-partitioning approach is the only method that has demonstrated the ability to simulate comminution of brittle materials with this degree of mesh independence, making it uniquely suited for the task of mesoscale simulation where the spatial resolution that can be used to model a single grain will generally be limited by computational resources.

17.3 Results

To demonstrate the application of the method to mesoscale simulation, we model the compaction of a porous heterogeneous material, defining the initial microstructure from a scanning-electron micrograph of a high-strength concrete. A thresholding is applied to the image to extract three phases, pore space, a mortar, and a hard aggregate. The material model for each solid phase is linear elastic-perfectly plastic with a time-to-failure fracture model and a Weibull distributed tensile strength. The material is compressed between steel platens under a distributed compressive load, and there is a gap to allow lateral expansion between the sample and the lateral boundaries. Figure 17.5 shows the simulation results for the compression of the porous heterogeneous material. Initially, the failure propagates from the stress concentrations at pore surfaces and interface failures between the mortar and aggregate. We see the fracture surfaces slide and open, producing the shear bulking effect. As the loading continues the mortar material fragments and continues to flow, increasing the load on the hard aggregates. Fracture in the aggregates initiates from pore surfaces, and transgranular fractures start to form. Finally the material flows to fill the space, with force chains forming in the damaged material as it is recompressed.

This example illustrates the challenging numerical aspects in modeling failure in porous and heterogeneous materials, and shows the capabilities of the new DFG partitioning method for this class of problems.

17.4 Conclusion

The DFG partitioning method enhances our capability for multi-scale simulation of porous brittle materials, including geomaterials, concrete, sand, as well as the post-failure response of ceramic materials. This capability will lead to deeper understanding of the relationship between microscale mechanisms and bulk material properties, supporting the development of enhanced constitutive models and improving the performance of engineered materials.

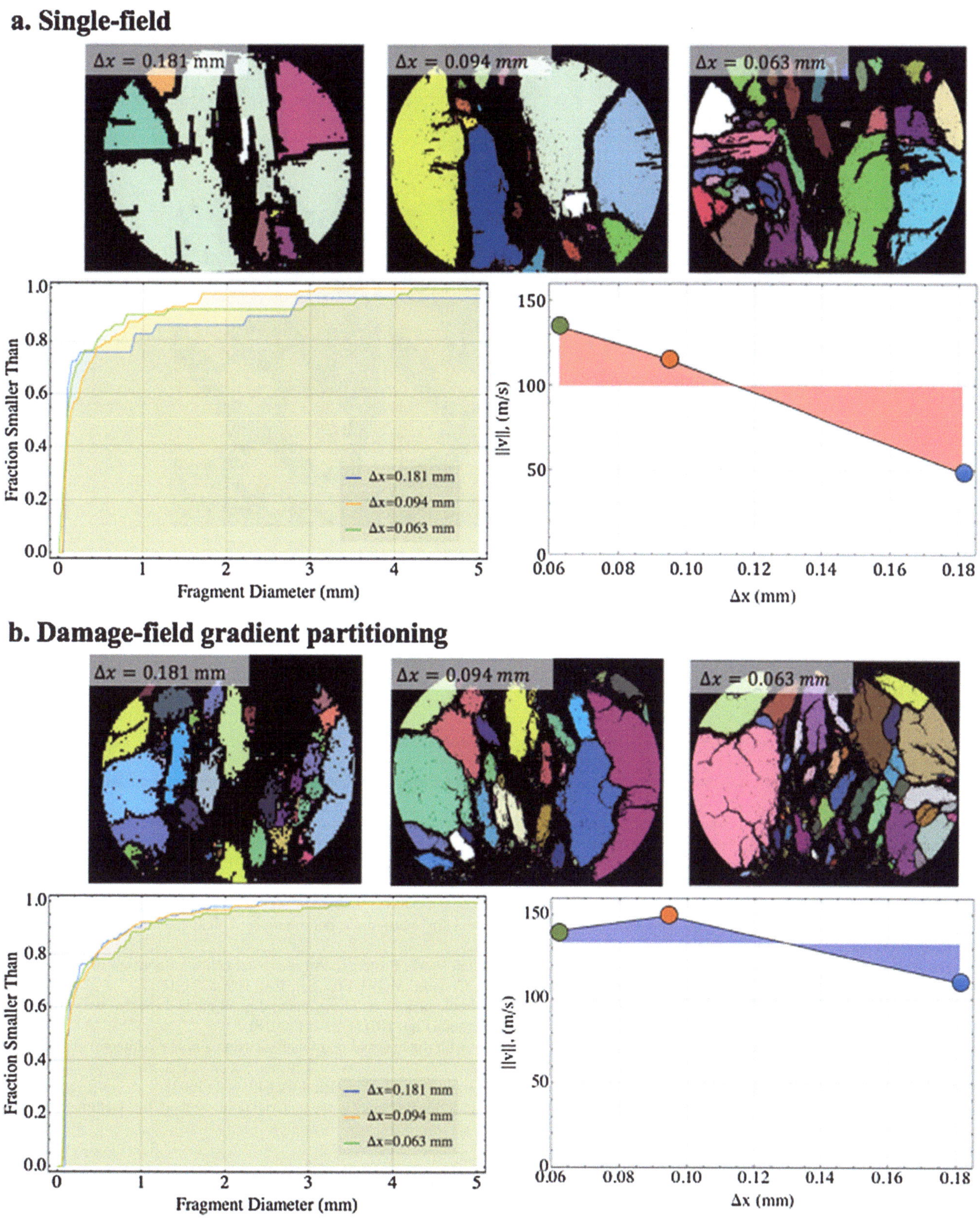

Fig. 17.4 Comminution of a brittle disk simulated using (**a**) the traditional single-field MPM, and (**b**) damage-field gradient partitioning. The DFG-partitioning method gives more consistent results for the fragment size distribution and velocity as the mesh is refined (Adapted from [11])

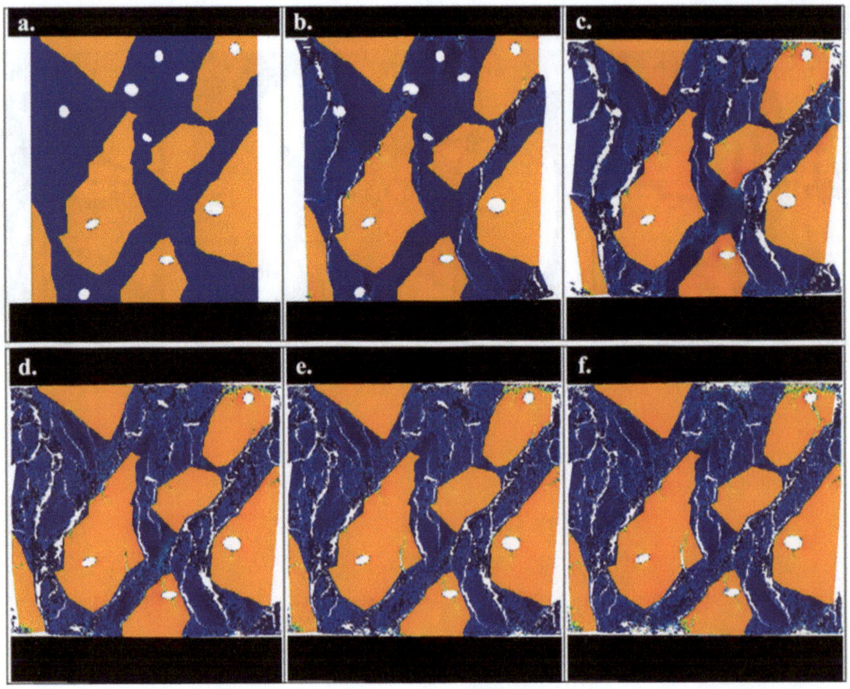

Fig. 17.5 Application of the MPM with damage-field gradient partitioning to mesoscale simulation. (**a**) Initial state of the two-phase porous material, (**b**) fractures propagate through the mortar from pore surfaces and aggregate interfaces, (**c**) large displacement and sliding at fracture surfaces, (**d**) comminution of the mortar material, and initiation of fracture in the aggregate phase, (**e**) propagation of transgranular fractures, (**f**) stable final state under applied compressive load

References

1. Sulsky, D., Chen, A., Schreyer, H.L.: A particle method for history-dependent materials. Comput. Methods Appl. Mech. Eng. **118**, 179–196 (1994)
2. Bardenhagen, S.G., Brackbill, J.U., Sulsky, D.: The material-point method for granular materials. Comput. Methods Appl. Mech. Eng. **187**(3), 529–541 (2000)
3. Bardenhagen, S.G., Kober, E.M.: The generalized interpolation material point method 2. Comput. Model. Eng. Sci. **5**, 477–495 (2004)
4. Sadeghirad, A., Brannon, R.M., Burghardt, J.: A convected particle domain interpolation technique to extend applicability of the material point method for problems involving massive deformations. Int. J. Numer. Methods Eng. **86**(12), 1435–1456 (2011)
5. Homel, M.A., Guilkey, J., Brannon, R.M.: Mesoscale validation of simplifying assumptions for modeling the plastic deformation of fluid-saturated porous material. J. Dyn. Behav. Mater. **3**(1), 23–44 (2017)
6. Homel, M. A., Brannon, R. M.: Semi-empirical effective stress model for quasi-static deformation of fluid-saturated porous materials, (Submitted not yet reviewed). Int. Symp. Plast. Its Curr. Appl., Jan. 2012
7. Bardenhagen, S.G., Guilkey, J.E., Roessig, K.M., Brackbill, J.U., Witzel, W.M., Foster, J.: An improved contact algorithm for the material point method and application to stress propagation in granular material. Comput. Model. Eng. Sci. **2**(4), 509–522 (2001)
8. Homel, M.: Controlling the onset of numerical fracture in parallelized implementations of the material point method (MPM) with convective particle domain interpolation (CPDI) domain scaling. Int. J. Numer. Methods Eng. **107**(1), 31–48 (2016)
9. Bardenhagen, S.G., Nairn, J.A., Lu, H.: Simulation of dynamic fracture with the material point method using a mixed J-integral and cohesive law approach. Int. J. Fract. **170**, 49–66 (2011)
10. Nairn, J.A.: Material point method calculations with explicit cracks. Comput. Model. Eng. Sci. **4**(6), 649–663 (2003)
11. Homel, M.A., Herbold, E.B.: Field gradient partitioning for fracture and frictional contact in the material point method. Int. J. Numer. Methods Eng. **109**(7), 1013–1044 (2017)

M.A. Homel received a Ph.D. in mechanical engineering from the University of Utah and is now a computational geoscientist at Lawrence Livermore National Laboratory. His research focus is on methods for mesoscale simulations of geomaterials to validate and inform full-stress constitutive models of compaction and comminution processes, and improved constitutive models.

Chapter 18
Underwater Blast Response of Weathered Carbon Composite Plates

Helio Matos, Carlos Javier, James LeBlanc, and Arun Shukla

Abstract An experimental study was conducted to evaluate the response of weathered unidirectional composite plates subjected to near-field blast loading. The composite materials in this study are carbon-epoxy composite plates with a [45, −45]s layup and was subjected to simulated marine aging through submersion in seawater baths for 70 days at 65 degrees Celsius in order to simulate approximately 20 years of real life operating conditions. Experiments were performed by fully clamping the specimen plates to an air-backed enclosure in a water tank. An RP-503 explosive was placed underwater behind the composite structure to be loaded. During the experiments, transducers measured the pressure emitted by the explosive, and three high-speed cameras captured the entire event. Two of the cameras were placed apart facing the specimen to measure full-field displacements through 3-D Digital Image Correlation analysis. Results show that the diffusion of water into the composite material leads to degradation of the blast response behavior as well as a loss of flexural strength and modulus.

Keywords Weathered composites • Accelerated aging • Property degradation • Blast loading • Digital image correlation

18.1 Introduction

In this study, an experimental investigation was conducted to evaluate the response of weathered unidirectional composite plates subjected to near-field explosive/blast loading. This research arises from the concern of damage to naval and marine composite structures such as ships, submarines, and underwater vehicles [1, 2]. During the service life of these structures, their mechanical properties degrade due to the continuous exposure to an aggressive environment [3]. In undesirable circumstances, marine structures can be further subjected to shock and/or blast loadings. If the degradation of mechanical properties is not accounted for under these highly dynamic conditions, the damages and losses could be fatal.

The leading cause of property degradation in composites in a marine environment is the diffusion of water into the matrix material [3]. The diffusion process is relatively well established and can be described by a diffusion coefficient that is a function of many parameters such as temperature, the composition of resin and curing agent, fillers, and so on. The value for diffusion coefficient, and the theoretical models used to describe the diffusion, varies in previous studies of diffusion in composites [4–10]. The aim of this study is to better understand how a composite plate's blast performance is affected by long time exposure to sea water. This work experimentally analyses the dynamic response of weathered composite plates subjected to near-field underwater blasts from explosives.

H. Matos • C. Javier • A. Shukla (✉)
Dynamic Photo Mechanics Laboratory, Department of Mechanical, Industrial and Systems Engineering,
University of Rhode Island, Kingston, RI, 02881, USA
e-mail: shuklaa@uri.edu

J. LeBlanc
Naval Undersea Warfare Center (Division Newport), Newport, RI, 02841, USA

J. Kimberley et al. (eds.), *Dynamic Behavior of Materials, Volume 1*, Conference Proceedings
of the Society for Experimental Mechanics Series, DOI 10.1007/978-3-319-62956-8_18

18.2 Experimental Procedures

18.2.1 Composite Material

The composite materials used consists of four unidirectional carbon fiber sheets with a [45, −45]s layup. These materials were manufactured by the University of Rhode Island students at TPI Composites Inc. in Warren, RI. The composites were made from two layers of +/− 45° biaxial carbon fabric and an epoxy resin/hardener mixture. Also, the resin/hardener is a 100/30 weight mixture of the RIMR135/RIMH137 epoxy from Momentive Performance Materials Inc. in Waterford, NY. The epoxy mixture was drawn into the fabric by Vacuum Infusion at a constant pressure of 730 mmHg. After hardening, curing was performed by placing the composite plate in an oven at 70 °C for 10 h. The final product was a 1.26 mm (0.050 in) thick composite plate with 1% void content (measured in accordance to ASTM Standard D2734 [11]) and 60% fiber volume content.

18.2.2 Weathering Facility

The composite materials are placed in a 3.5% NaCl solution (prepared in accordance to ASTM Standard D1141 [12]). Four water heaters (Model LXC from PolyScience in Niles, IL) are used to maintain a temperature of 65 °C. The composite materials in this study are exposed to the salt water for consecutive 70 days. Experiments are initiated immediately after the specimens are removed from the salt water exposure to avoid moisture loss as suggested by ASTM Standard D5229 [13].

18.2.3 Blast Facility

To perform the blast experiments, the 1.26 mm (0.050 in) thick carbon-epoxy composite plate is fully clamped inside an 1800 L (475 gallons) water tank. The water tank has an inner 45 L (12 gallons) air chamber. The composite specimen is clamped between the water and air chambers with a 25.4 (1 in) all-around clamping width; leaving a 254 × 254 mm^2 (10 × 10 in^2) exposed area as shown in Fig. 18.1.

An RP-503 explosive was used to load the composite structure; it is submerged in the water, centered to the specimen, and placed at a 152 mm (6 in) standoff distance. Two dynamic pressure transducers (PCB 138A05, PCB Piezotronics Inc. in Depew, NY) are located next to the specimen and explosive at 152 mm (6 in) and 203 mm (8 in) distances from the explosive. During the experiments, a Dash 8HF data acquisition system (from AstroNova Inc. in Warwick, RI) captured the pressure data at 2 mega samples per second. Furthermore, two Photron SA1 high-speed cameras (from Photron USA Inc. in San Diego, CA) are placed 14° apart outside the blast facility and used to capture high-speed images of the specimen at

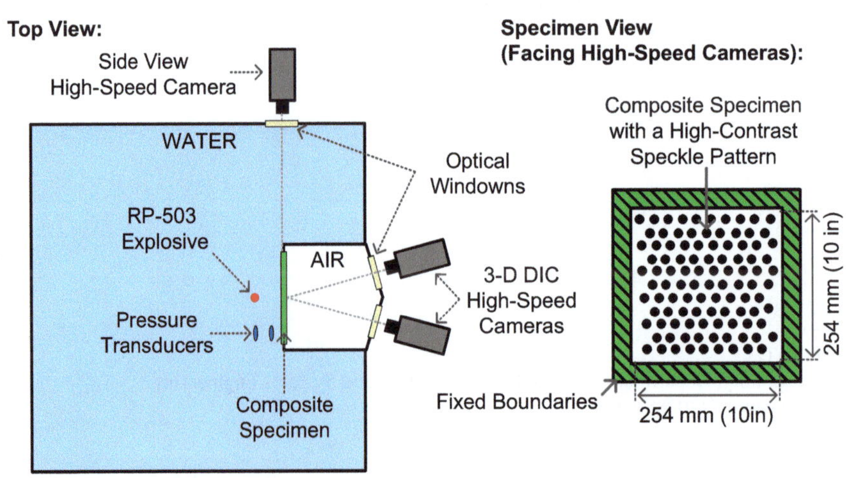

Fig. 18.1 Blast facility experimental setup

Table 18.1 Experimental cases details

Cases	Layup	Standoff distance, mm (in)	Weathering exposure, days
45s_0WD	[45,−45]s	152 (6)	0
45s_70WD	[45,-45]s	152 (6)	70

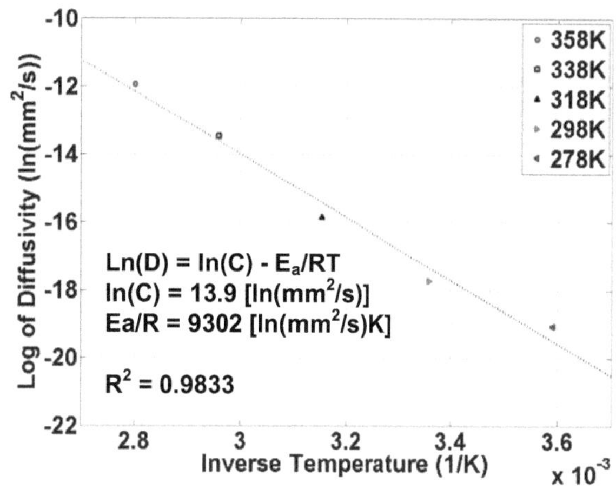

Fig. 18.2 Logarithmic relationship between diffusivity and temperature

10,000 frames per second. The photographs from the high-speed cameras are captured through optically clear windows in the tank. These images are later used for the Digital Image Correlation (DIC) analysis. The composite specimen is coated with high-contrast speckle pattern for the DIC analysis. The experimental cases and its details are summarized in Table 18.1. Each experimental case has been repeated three times to validate the results.

The high-speed images are analyzed using commercially available DIC software (VIC3D 7 from Correlated Solutions, Inc., Columbia, SC) to measure full-field displacements across the viewable surface of the specimen. Previous work [14] outlines the calibration procedures that validate the accuracy of the DIC results when capturing images through an optical window. For this study, the in-plane displacement errors are $\leq 1\%$, and the out-of-plane errors are $\leq 2\%$.

18.3 Results and Discussions

18.3.1 Weathering

Since the activation energy (E_a) for a material is constant, a mass diffusion study can be performed at various temperatures to obtain the acceleration factor (AF) of submersion at for the material a specific temperature. For this study, moisture absorption was measured for composites submerged in 3.5% NaCl solutions at 5, 25, 45, 65, and 85 °C in accordance to ASTM Standard D5229 [13] and the logarithmic relationship between diffusivity and temperature was obtained as shown in Fig. 18.2.

After obtaining the activation energy for the composite material, AF can be found as the ratio of working over experimental diffusion rates as shown in Eq. (18.1) [15–17]. Additionally, the submersion experiments are performed at a constant temperature ($T_1 = 338$ K), but the service temperature (T_2) can vary depending on application; hence, AF is application dependent. Assuming an average ocean temperature of 16 °C, 70 days of submersion approximates to 20 years of service.

$$AF = \frac{Ce^{-\frac{E_a}{RT_2}}}{Ce^{-\frac{E_a}{RT_1}}} = e^{\left(\frac{E_a}{R}\right)\left(\frac{T_2 - T_1}{T_1 T_2}\right)} \tag{18.1}$$

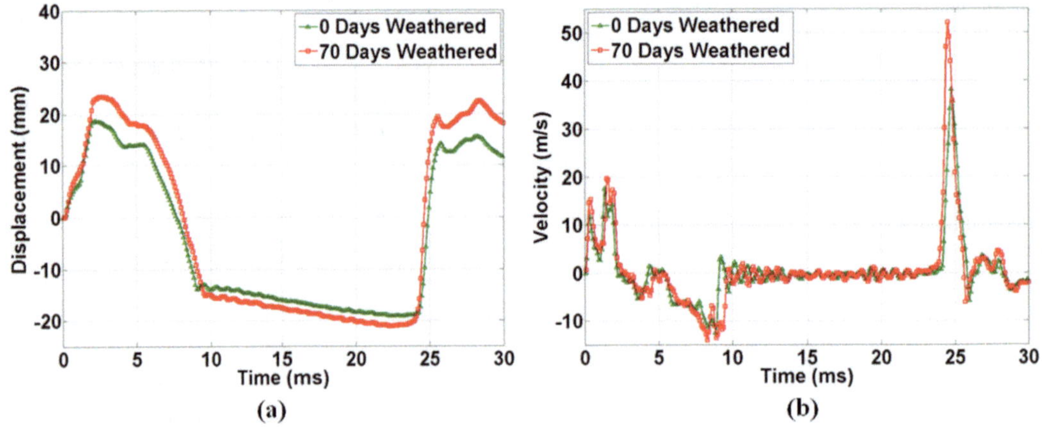

Fig. 18.3 Composite's center point (**a**) displacement and (**b**) velocity measured by DIC

18.3.2 Blast Response

The blast performance of the non-weathered and 70 days weathered composite plate is illustrated in Fig. 18.3 in terms of center point displacement and velocity. At $t = 0$, the underwater explosive (UNDEX) is ignited. The specimen deforms towards the camera between $t = 0$ and 5 ms. Soon after detonation, a cavitation bubble forms underwater and it engulfs the specimen's surface. Since the cavitation bubble is a vacuum, the specimen deforms out of plane towards the bubble between $t = 5$ and 10 ms. At about 25 ms the bubble collapses onto the specimen's surface causing a rapid deformation towards the camera. Throughout the loading event, some differences are observed between the non-weathered and weathered composite. The weathered composite reached higher displacements (in turn, higher strains) as shown in Fig. 18.3a. Also, the weathered composite reached higher velocities, especially during bubble collapse (inferring loss of stiffness) as shown in Fig. 18.3b.

18.4 Conclusions

This work experimentally analyzed the dynamic response of weathered composite plates subjected to near-field underwater blasts from an RP-503 explosive. This work shows that weathered composites have degraded dynamic properties; leading to higher deformation and velocities when subjected to explosive loading. Moreover the higher velocities infer a loss of stiffness that could not be accomplished through saturation alone; hence, there are further degradation mechanisms in play.

Acknowledgements The authors kindly acknowledge the financial support provided by Kirk Jenne from the Naval Engineering Education Consortium (NEEC) Grant No. N00174-16-C-0012; the composite expertise of Tim Fallon, and the manufacturing assistance given by Mike Trapela from TPI Composites in Warren, RI.

References

1. Mouritz, A., Gellert, E., Burchill, P., Challis, K.: Review of advanced composite structures for naval ships and submarines. Compos. Struct. **53**(1), 21–42 (2001). doi:10.1016/s0263-8223(00)00175-6
2. Graham-Jones, J., Summerscales, J.: Marine Applications of Advanced Fibre-Reinforced Composites. Woodhead Publishing (2016). doi:10.1016/c2013-0-16504-x
3. Davies, P.: Environmental degradation of composites for marine structures: new materials and new applications. Phil. Trans. R. Soc. Philos Trans Roy Soc: Math, Phys Eng Sci. **374**(2071), 20150272 (2016). doi:10.1098/rsta.2015.0272
4. Browning, C., Husman, G., Whitney, J.: Moisture effects in epoxy matrix composites. In: Davis, J. (ed.) Composite Materials: Testing and Design (Fourth Conference), STP26961S, pp. 481–496. ASTM International, West Conshohocken (1977). doi:10.1520/STP26961S
5. Fichera, M., Totten, K., Carlsson, L.A.: Seawater effects on transverse tensile strength of carbon/vinylester as determined from single-fiber and macroscopic specimens. J. Mater. Sci. **50**(22), 7248–7261 (2015). doi:10.1007/s10853-015-9279-3
6. Popineau, S., Rondeau-Mouro, C., Sulpice-Gaillet, C., Shanahan, M.E.: Free/bound water absorption in an epoxy adhesive. Polymer. **46**(24), 10733–10740 (2005). doi:10.1016/j.polymer.2005.09.008

7. Choqueuse, D., Davies, P., Mazéas, F., Baizeau, R.: Aging of Composites in Water: Comparison of Five Materials in Terms of Absorption Kinetics and Evolution of Mechanical Properties. High Temperature and Environmental Effects on Polymeric Composites: 2nd Volume (1997). doi:10.1520/stp11369s
8. Faguaga, E., Pérez, C., Villarreal, N., Rodriguez, E., Alvarez, V.: Effect of water absorption on the dynamic mechanical properties of composites used for windmill blades. Mater Des (1980–2015). **36**, 609–616 (2012). doi:10.1016/j.matdes.2011.11.059
9. Tual, N., Carrere, N., Davies, P., Bonnemains, T., Lolive, E.: Characterization of sea water ageing effects on mechanical properties of carbon/epoxy composites for tidal turbine blades. Compos. A: Appl. Sci. Manuf. **78**, 380–389 (2015). doi:10.1016/j.compositesa.2015.08.035
10. Gunti, R., Prasad, A.R., Gupta, A.: Mechanical and degradation properties of natural fiber reinforced PLA composites: jute, sisal, and elephant grass. Polym. Compos. (2016). doi:10.1002/pc.24041
11. ASTM Standard D2734-16: Standard Test Methods for Void Content of Reinforced Plastics. ASTM International, West Conshohocken (2016). doi:10.1520/D2734-16
12. ASTM Standard D1141-98: Standard Practice for the Preparation of Substitute Ocean Water. ASTM International, West Conshohocken (2013). doi:10.1520/D1141-98R13
13. ASTM Standard D5229-14: Standard Test Method for Moisture Absorption Properties and Equilibrium Conditioning of Polymer Matrix Composite Materials. ASTM International, West Conshohocken (2014). doi:10.1520/D5229_D5229M-14
14. Gupta, S., Parameswaran, V., Sutton, M.A., Shukla, A.: Study of dynamic underwater implosion mechanics using digital image correlation. Proc Roy Soc: Math, Phys Eng Sci. **470**(2172), 20140576–20140576 (2014). doi:10.1098/rspa.2014.0576
15. Davies, P., Rajapakse, Y.: Durability of Composites in a Marine Environment. Springer, Dordrecht (2014)
16. Crank, J.: The Mathematics of Diffusion, 2nd edn. Oxford University Press UK, Oxford (1975)
17. Rice, M.: Activation Energy Calculation for the Diffusion of Water into PR-1590 and Pellethane 2103-80AW Polyurethanes. NUWC-NPT Technical Memo 11-062

Chapter 19
Characterization of a Visco-Hyperelastic Synthetic Gel for Ballistic Impacts Assessment

A. Bracq, G. Haugou, B. Bourel, R. Delille, C. Maréchal, F. Lauro, S. Roth, and O. Mauzac

Abstract Understanding the human body response during impact has become a primary concern, especially in the field of ballistic impacts. Among various materials employed as human body substitutes, synthetic gel named SEBS is used in the present study. Mechanical characterization from quasi-static to very high strain rates are performed allowing the constitutive modeling of the gel material and its implementation in an explicit finite element code. Tensile testing are completed using appropriate tensile fixtures and specific strain measurement and point out a hyperelastic behavior (strain over 300%) with strain rate sensitivity. Compression tests are performed from static strain rates to dynamic strain rates using for very high strain rates polymeric split Hopkinson bars and high-speed imaging. Mooney-Rivlin hyperelastic material model with specific strain rate dependence is implemented using optimization process by inverse method of dynamic compression testing results for material parameters identification. The material law is employed during modeling of direct impact of less lethal kinetic energy projectiles over various velocities and gives satisfactory results compared to the experimental impact analysis.

Keywords Soft tissue simulant • Mechanical characterization • High strain rate • Constitutive modeling • Impact modeling

19.1 Introduction

The past decade had seen the rapid increase in ballistic impact studies, especially to understand the human body response, as well as to improve protective equipment. Among various materials employed as human body substitutes such as ballistic gelatin, clay and rubbers, a synthetic gel named SEBS (styrene-ethylene-butylene-styrene) has shown many benefits such as its transparency, mechanical consistency and environmental stability [1, 2]. Then it leads the French Ministry of the Interior to adopt this soft material as tissue simulant for blunt ballistic impact analysis. Although gel transparency allows direct impact analysis using high-speed camera and provides information on the back face deformation as wall displacement, volume of deformation, these macroscopic information do not allow a direct evaluation of a blunt ballistic trauma. Thence the authors focus their research on the use of numerical tools as finite element method for impact modeling to obtain more accurate data and extend information as strain, strain rate, pressure. Indeed, recent studies have studied ballistic blunt trauma and more precisely behind armor blunt trauma (BABT) through finite element modeling [3, 4]. However to obtain proper data, a material model must be determined to correctly describe its mechanical behavior over a large range of strain rates with experimental tensile and compressive testing.

The compliant nature of the gel and its low mechanical impedance require the use of adequate device. A previous research on SEBS gel depicts an interesting fixture design as well as strain measuring method to obtain valid and expendable tensile testing. Moreover, authors have dedicated their study on the mechanical characterization of the SEBS gel in compression at

A. Bracq (✉) • G. Haugou • B. Bourel • R. Delille • C. Maréchal • F. Lauro
Laboratory LAMIH UMR 8201 CNRS, University of Valenciennes and Hainaut Cambrésis, 59313, Valenciennes, France
e-mail: anthony.bracq@univ-valenciennes.fr

S. Roth
University of Bourgogne Franche-Comté, University of Technology of Belfort Montbéliard, 90010, Belfort, France

O. Mauzac
French Ministry of the Interior, CREL/SAELSI, Place Beauveau, 75800, Paris Cedex 08, France

J. Kimberley et al. (eds.), *Dynamic Behavior of Materials, Volume 1*, Conference Proceedings of the Society for Experimental Mechanics Series, DOI 10.1007/978-3-319-62956-8_19

high strain rates [5]. They use the well-known Hopkinson bars or Kolsky bars to obtain the material response [6]. As several authors, polymeric split Hopkinson bars are adopted to correctly characterize the compressive behavior of such soft materials at very high strain rates [7, 8]. However, literature review is not sufficient to obtain usable true stress-strain curves over a wide range of strain rate for the studied material. Actually, SEBS gel is obtained by mixing SEBS powder and mineral oil producing 30 wt% SEBS with a styrene/elastomer ratio of about 30/70. Consequently tensile and compressive testing are performed at very high strain and strain rates using dedicated machine and measuring system.

19.2 Experimental Study

Based on literature, tensile testing are conducted using aluminum fixture with a specific design including the tensile sample [2]. Indeed, fixture shape allows to the sample to be only maintained by its shoulders. Black spots are drawn on the specimen, their local deformation leads to calculate the true stress-strain response. Actually, through the use of high-speed camera, measuring the black spot deformation, it results in the longitudinal and transverse strain measurements. It conducts to the validation of incompressible material assumption, quasi-constant engineering strain rate as well as homogeneity of deformation. Several tests are performed from 0.07 to 59.5 s^{-1} and true stress-strain curves at different strain rates are presented in Fig. 19.1 (left). It can be highlighted that the material has a hyperelastic behavior (strain over 300%) with a strong strain rate dependence. Conventional compressive tests are also completed over a range of strain rate. It is important to note that, despite the use of lubricant, only engineering stress-strain curves can be presented until 40% strain in Fig. 19.1 (right). Likewise, a strain rate sensitivity is pointed out, where the stress increases with the strain rate.

During ballistic impacts, the gel block used as soft tissue simulant is subjected to very large compressive at dynamic strain rates. Therefore, extensive effort is made to characterize the material at very high strain rates. Polymeric split Hopkinson pressure bars (SHPB) of 20 mm in diameter are employed along with high-speed camera to observe the specimen deformation during loading. Dedicated data processing software is adopted to determine the material response through the analysis of strain gauges recording data placed on the input and output bars.

Experimental apparatus and material low impedance result in the gel mechanical response from 440 to 1520 s^{-1}. Figure 19.2 presents typical strain gauges raw signals allowing the material characterization. However, even though multitude of precautions are undertaken, high-speed imaging reveals non-homogeneous deformation during loading which prevents the use of stress-strain curves determined at high strain rates for direct material constitutive modeling. Nevertheless, modeling compressive testing, especially the Hopkinson bars can be employed to identify material model parameters and will be depicted afterwards.

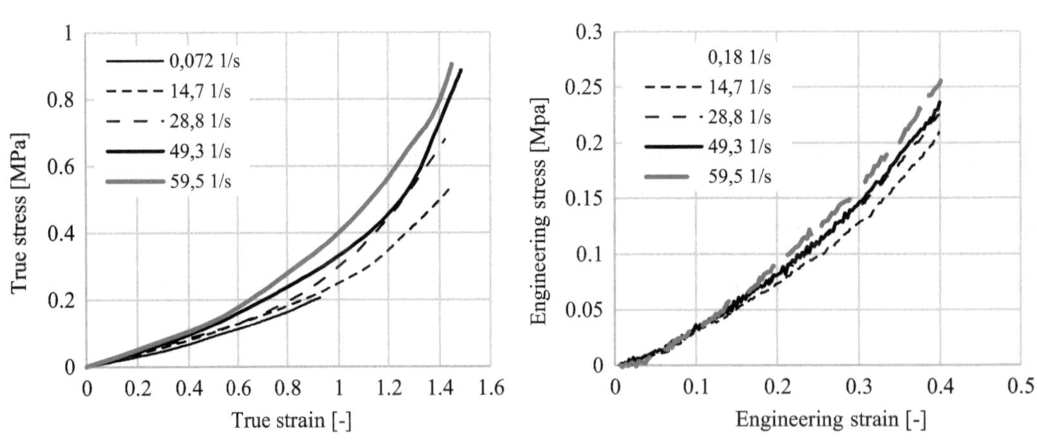

Fig. 19.1 True stress-strain curves from tensile testing (*left*) and engineering stress-strain curves from compression testing (*right*) at various strain rates

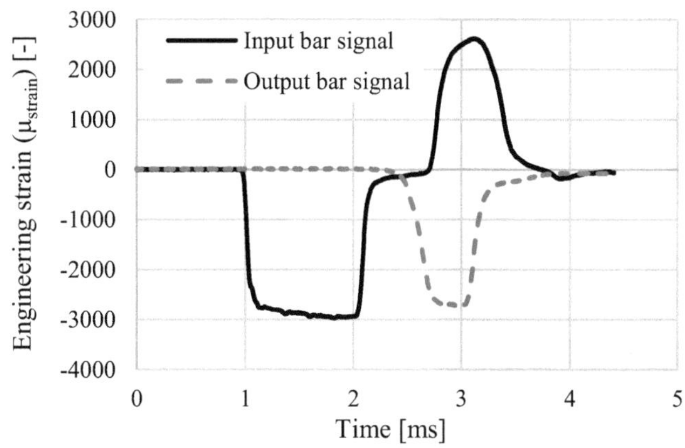

Fig. 19.2 Recording from strain gauges for dynamic compression testing at 1520 s^{-1}

19.3 Numerical Study

As experimental testing highlights a hyperelastic behavior, the authors take an active interest in hyperelastic material models. Ogden material model is one of the most use to describe non-linear material response until 600–700% strain [9]. This model can be described with a strain energy density function Eq. (19.1) based on principal stretch.

$$W(\lambda_1, \lambda_2, \lambda_3) = \sum_{k=1}^{N} \frac{\mu_k}{\alpha_k} \left(\lambda_1^{a_k} + \lambda_2^{a_k} + \lambda_3^{a_k} - 3 \right) \quad (19.1)$$

For particular values of material constants ($N = 2, \alpha_1 = 2, \alpha_2 = -2$), Ogden model can be reduced to the Mooney-Rivlin material model [10]. This latter model is chosen for the few parameters to identify. Each mechanical testing is assumed to be at constant strain rate. In such case, model parameters can be identified for each strain rate. μ_1 parameter is constant and is identified from tensile testing. On the other side, μ_2 is function of strain rate and its identification is based on compression testing. Each μ_2 parameter considered at constant strain rate is determined through the use of multi-objective optimization using global response surface method. Load measurements and strain gauge recording data are used to find a suitable parameter respectively for each quasi-static and high strain rate. The implementation of such constitutive model with special care for strain rate filtering is conducted in the explicit finite element code Radioss (Altair Hyperworks). In a concern of blunt trauma assessment, the authors decide to replicate on gel block the impact conditions of Bir reference cases, which are performed on postmortem human subjects (PMHS) [11]. These tests consist in the impact of round rigid projectile on the mid-sternum of PMHS. They represent important information in terms of autopsy results and quantitative data (load, sternum deflection). Model validity is evaluated by comparing experimental and numerical impact testing on gel block at different impact velocities. Figure 19.3 illustrates qualitative comparison of experimental impact (left) of 140 g projectile with 37 mm in diameter on gel block of size 25 × 25 × 25 cm, at 20 m/s and numerical modeling (right). It can be affirmed that numerical modeling and consequently constitutive material law is correctly described with an excellent representation of gel deformation.

To go further on the model effectiveness, experimental and numerical impact testing are carried out at several velocities: 12, 20 and 30 m/s. Gel wall displacement in function of time is chosen to correlate experimental and numerical testing (see Fig. 19.4). This comparison highlights satisfactory model abilities to describe blunt ballistic impacts by considering finite element modeling limits at very high deformation.

19.4 Conclusion

With an aim of assessing blunt ballistic trauma through the use of soft tissue simulant, mechanical characterization testing are performed on SEBS gel sample at various strain rates. Hyperelastic behavior with strain rate sensitivity is pointed out through specific tensile and compressive testing. Very high strain rates (until 1520 s^{-1}) are achieved using polymeric SHPB system. Numerical tools are adopted and a material constitutive law is implemented in the explicit code Radioss. Mooney-

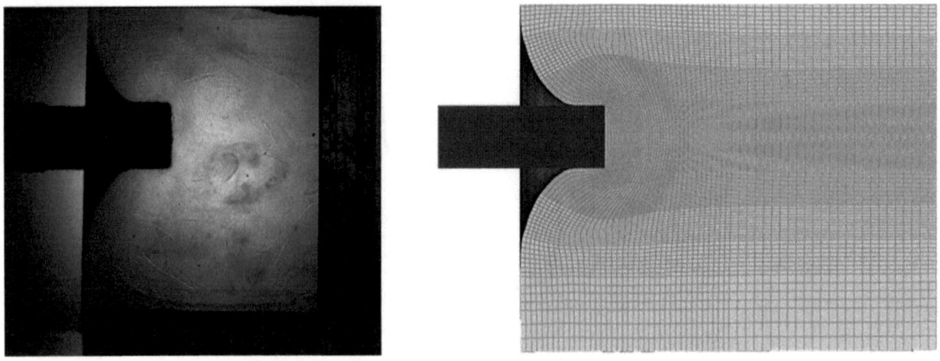

Fig. 19.3 Experimental (*left*) and numerical (*right*) impact testing on gel block at maximal deflection

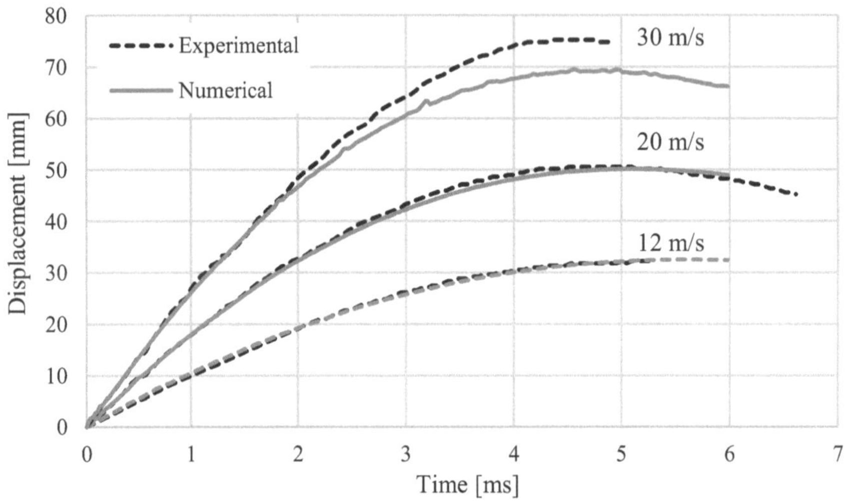

Fig. 19.4 Experimental and numerical gel wall displacement in function of time at impact velocities: 12, 20 and 30 m/s

Rivlin hyperelastic model is chosen and parameters are identified through the use of optimization by inverse method. Each mechanical testing and its numerical modeling are used to identify each parameter associated to a constant strain rate. Considering strain rate dependence in the constitutive model, experimental and numerical replications of Bir reference cases on gel block are performed at various impact velocities. Qualitative and quantitative data prove the model abilities to reproduce accurately blunt ballistic impacts. Therefore, extensive study on the evaluation of injury risk will be the focus of future investigations.

References

1. Mauzac, O., Paquier, C., Debord, E., Jacquet, J.F.: A substitute of gelatin for the measurement of dynamic back face deformation, presented at the Personal armour systems symposium, Canada (2010)
2. Mrozek, R.A., et al.: The relationship between mechanical properties and ballistic penetration depth in a viscoelastic gel. J. Mech. Behav. Biomed. Mater. **44**, 109–120 (2015)
3. Liu, L., Fan, Y., Li, W.: Viscoelastic shock wave in ballistic gelatin behind soft body armor. J. Mech. Behav. Biomed. Mater. **34**, 199–207 (2014)
4. Wen, Y., Xu, C., Wang, S., Batra, R.C.: Analysis of behind the armor ballistic trauma. J. Mech. Behav. Biomed. Mater. **45**, 11–21 (2015)
5. Casem, D.T., Dwivedi, A.K., Mrozek, R.A., Lenhart, J.L.: Compression response of a thermoplastic elastomer gel tissue surrogate over a range of strain-rates. Int. J. Solids Struct. **51**(11–12), 2037–2046 (2014)
6. Kolsky, H.: An investigation of the mechanical properties of materials at very high rates of loading. Proc. Phys. Soc. Sect. B. **62**(11), 676 (1949)
7. Salisbury, C.P., Cronin, D.S.: Mechanical properties of ballistic gelatin at high deformation rates. Exp. Mech. **49**(6), 829–840 (2009)
8. Van Sligtenhorst, C., Cronin, D.S., Wayne Brodland, G.: High strain rate compressive properties of bovine muscle tissue determined using a split Hopkinson bar apparatus. J. Biomech. **39**(10), 1852–1858 (2006)

9. Ogden, R.W.: Large deformation isotropic elasticity – on the correlation of theory and experiment for incompressible rubberlike solids. Proc. R. Soc. Lond. Math. Phys. Sci. **326**(1567), 565 (1972)
10. Rivlin, R.S.: Large elastic deformations of isotropic materials. IV. Further developments of the general theory. Philos. Trans. R. Soc. Math. Phys. Eng. Sci. **241**(835), 379–397 (1948)
11. Bir, C., Viano, D., King, A.: Development of biomechanical response corridors of the thorax to blunt ballistic impacts. J. Biomech. **37**(1), 73–79 (2004)

Chapter 20
On the Microstructural Aspects of Shock Induced Failure in Magnesium Alloys

C.L. Williams, J. Ligda, L. Farbaniec, and N. Krywopusk

Abstract The microstructural aspects of shock induced plasticity and consequent failure in metals were studied using AZ31B and AMX602 magnesium alloys which were mechanically processed via Equal Channel Angular Extrusion (ECAE) and Spinning Water Atomization Process (SWAP) respectively. Results show that microstructure and microstructure evolution can strongly influence the mechanical response and consequent failure of these materials. Failure in both ECAE processed AZ31B-4E and AZ31B-$4_{B/C}$ magnesium alloy was dominated by debonding of the matrix magnesium from large Al-Mn-rich intermetallic inclusions and/or cracking of inclusions after the passage of shock stresses ranging from 1.5 to 4.5 GPa. However, the failure characteristic in the SWAP AMX602 magnesium alloy was distinctly different from that of the AZ31B magnesium alloy for the similar shock stress range. Numerous isolated cracks were observed around the spall plane possibly emanating from nucleation sites other than the Al_2Ca intermetallic inclusions in AMX602. The spall surfaces of the AMX602 samples were striated possibly due to corrosion of the SWAP powder prior to green-compaction. Mixed-mode failure was observed in all three materials possibly due to homogeneous and heterogeneous nucleation, growth, and coalescence of nanovoids and microvoids respectively.

Keywords Microstructure • Magnesium • Failure • Spall • Intermetallic

20.1 Introduction

The relative high strength-to-weight ratio of magnesium alloys has drawn considerable attention in high strain-rate applications such as those experienced in shock compression. However, the response of magnesium (Mg) and magnesium alloys under shock compression has not been extensively explored when compared to other lightweight metals such as aluminum. Magnesium has a hexagonal closed pack (HCP) lattice structure and therefore, has limited slip systems compared to their Face Centered Cubic (FCC) and Body Centered Cubic (BCC) counterparts. Due to their limited slip systems, they are relatively difficult to form mechanically at room temperature, which makes their use in practical applications very challenging.

Deformation slip in magnesium occurs on the basal, prism, and pyramidal planes. The Critical Resolve Shear Stress (CRSS) for basal slip in Mg at room temperature is quite low, approximately 1 MPa and therefore, basal slip is the dominant deformation mode [1]. In addition to basal slip $\{0001\}\langle-12\text{–}10\rangle$, other secondary slip systems such as prismatic $\{10\text{–}10\}\langle11\text{–}20\rangle$ and pyramidal $\{10\text{–}11\}\langle11\text{–}20\rangle$ slip can be activated at slightly elevated temperatures. Similar to basal slip, the Burgers vector for slip on the prismatic and pyramidal planes is in the $\langle a\rangle$ direction. Therefore, basal, prismatic and pyramidal slip systems represent only four independent slip systems and hence cannot accommodate deformation along the $\langle c\rangle$ direction and do not satisfy the five independent slip systems required for arbitrary homogeneous deformation [2]. In order to satisfy the von Mises criterion [2], $\langle c + a\rangle$ slip or deformation twinning may be activated.

Shown in Fig. 20.1 are the $\{10\text{–}12\}\langle10\text{–}11\rangle$ extension and $\{10\text{–}11\}\langle10\text{–}12\rangle$ contraction twinning systems commonly observed in HCP materials such as Mg alloys with $\{10\text{–}12\}$ extension twins being the dominant deformation mode at

C.L. Williams (✉) • J. Ligda
Army Research Laboratory, Aberdeen Proving Ground, MD, 21005, USA
e-mail: cyril.l.williams.civ@mail.mil

L. Farbaniec
Institute of Shock Physics, Imperial College London, London SW7 2AZ, UK

N. Krywopusk
The Johns Hopkins University, Baltimore, MD 21218, USA

J. Kimberley et al. (eds.), *Dynamic Behavior of Materials, Volume 1*, Conference Proceedings of the Society for Experimental Mechanics Series, DOI 10.1007/978-3-319-62956-8_20

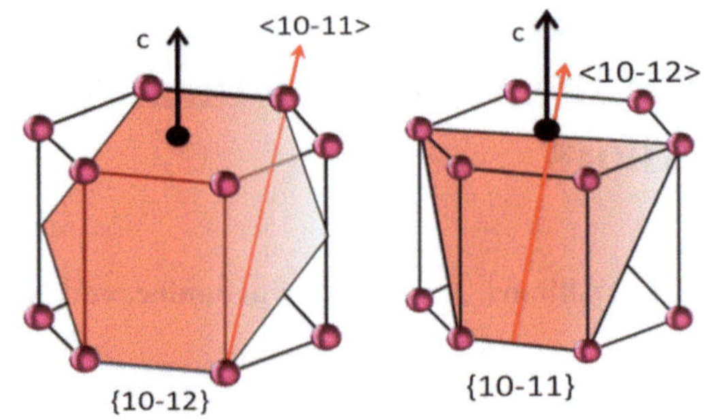

Fig. 20.1 {10–12}⟨10–11⟩ extension and {10–11}⟨10–12⟩ contraction twinning systems commonly observed the HCP materials

low temperatures. In addition to the extension and contraction twinning systems, double twinning can also occur in HCP materials. For double twinning, the contraction twin of the {10–11} type is nucleated first then an extension twin of the {10–12} type is nucleated within the contraction twin. Fracture has previously been observed to originate largely from compression twins or shear bands that develop from compression twins [3–5].

The {10–12} extension twin plays an extremely important role on the shock response of magnesium alloys. When the HCP lattice is compressed along the basal plane, extension twins are nucleated and then reorients with respect to the parent grain by 86.3°. The reoriented lattice of the twinned region is favorable to de-twinning during tensile release along the c- axis of the twinned region. Recent shock studies by Renganathan et al. [6] on single crystal magnesium show distinctly different shock response between the a-axis and c-axis. They found that the single crystal Mg exhibited strong anisotropy under shock compression and release. The release portion of the velocity profile was featureless for a-axis loading which is in contrast to the structured feature observed for c-axis.

The shock response of magnesium and magnesium alloys has not been well characterized when compared to other metals such as aluminum, copper, or titanium. However, some limited work on magnesium and magnesium alloys exist in the open literature. For instance, Schmidt and his colleagues [7] studied the spall response of AZ31B-H24 at room temperature conditions and reported a spall strength value of approximately 1.5 GPa at the onset of incipient spall. Furthermore, McQueen et al. [8] estimated the spall strength of AZ31B from the free surface velocity profiles obtained by Marsh [9] to be approximately 0.8 GPa. Kanel et al. [10] studied the effects of initial temperature on the spall strength of Mg95 (99.95% Mg by weight) and found that as the initial temperature approaches the melting point of Mg95, the spall strength drops due to the spontaneous nucleation of point defects at elevated temperatures. Also, the effect of microstructure on the shock response of Elektron 675 was studied by Hazell et al. [11] and their research show that the spall strengths in the extrusion direction were higher than those obtained in the normal direction due to the presence of striations of small grains. Recent work by Farbaniec et al. [12] showed that the strengthening of AZ31B-4E magnesium alloy by ECAE-processing resulted in adverse effects on its microstructure and spall behavior because of the process-induced cracking of intermetallic inclusions and their weak interface strengths. This ongoing research will improve the current knowledge on the shock response of two magnesium alloys processed via ECAE and SWAP.

20.2 Materials and Methods

The 152.4 mm × 152.4 mm × 12.7 mm ECAE plate was obtained by extruding AZ31B-H24 through a die using a hybrid 4E route. The 4E route implies that the plate was rotated about the plate normal by 180° after the first pass, 90° after the second pass, and 180° after the third pass (a total of four passes [13]). The plate processing temperature was varied from 498 K during the first two passes then was dropped to 473 K on the last two passes. The extrusions were performed with a backpressure of ranging from 4.14 to 6.89 kPa and the extrusion speed was 4.6 mm/min. The nominal grain size resulting from the ECAE process was approximately 3 μm. A similar extrusion procedure was employed for producing the AZ31B-$4_{B/C}$ square bar using the $4_{B/C}$ route. Recovery specimens with dimensions of 19 mm diameter and 3 mm thick were fabricated using wire electro-discharge machining (EDM) from the through-thickness direction of the plate and bar. Listed in Table 20.1 are the measured density and wave speeds from which the elastic constants were calculated. The values listed in Table 20.1 are consistent with those reported in the open literature.

On the other hand, the AMX602 plate was obtained from noncombustible AMX602 magnesium alloy ingot using the SWAP process [14]. The resulting AMX602 powder was then green compacted and hot extruded to produce a 25 mm × 102 mm plate. The nominal grain size resulting from the SWAP process was less than 0.8 μm. Similarly, recovery

Table 20.1 Measured wave speeds, density, and calculated elastic constants for AZ31B-4E

	Long. wave speed	Shear wave speed	Density	Bulk modulus	Lame's constant	Elastic modulus	Shear modulus	Poisson ratio
	C_L (km/s)	C_S (km/s)	ρ (kg/m^3)	K (GPa)	λ (GPa)	E (GPa)	μ (GPa)	v
Average	5.820	3.065	1773.75	37.867	26.758	43.596	16.664	0.308
Std. dev.	0.020	0.009	5.355	0.218	0.211	0.136	0.054	0.001

Table 20.2 Measured wave speeds, density, and calculated elastic constants for AMX602

	Long. wave speed	Shear wave speed	Density	Bulk modulus	Lame's constant	Elastic modulus	Shear modulus	Poisson ratio
	C_L (km/s)	C_S (km/s)	ρ (kg/m^3)	K (GPa)	λ (GPa)	E (GPa)	μ (GPa)	v
Average	5.701	3.220	1806.35	33.737	21.251	47.413	18.729	0.266
Std. dev.	0.020	0.008	1.163	0.022	0.014	0.031	0.012	–

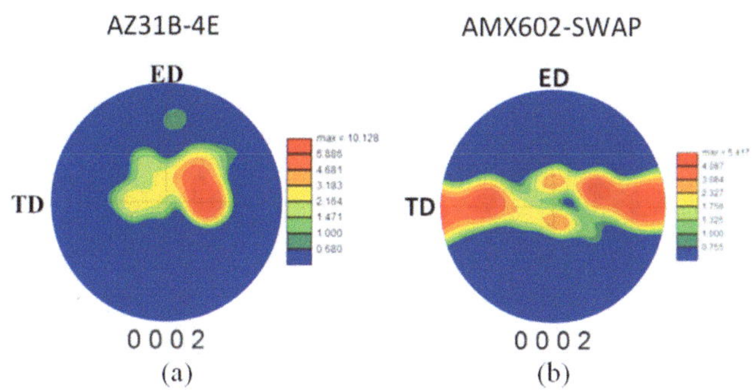

Fig. 20.2 Pole figures for (**a**) AZ31B-4E magnesium alloy and (**b**) AMX602-SWAP magnesium alloy

specimens with dimensions of 19 mm diameter and 3 mm thick were fabricated using wire electro-discharge machining (EDM) from the through-thickness direction of the plate. Listed in Table 20.2 are the measured density and wave speeds from which the elastic constants were calculated.

20.3 Experimental Techniques

Ex-situ (recovery) shock experiments were conducted using a single stage 102 mm (slotted bore) diameter gas gun. All recovery experiments were executed using the standard plate impact technique, which was previously described by Williams et al. [15, 16]. All plate impact experiments for this investigation were symmetric, that is, both flyer and target materials were identical. A series of charged pins were used to determine the flyer velocity and the uncertainty associated with the pin positions was determined to be less than 1.0×10^{-4} mm, corresponding to an accuracy associated with the final velocity of the flyer of less than 2%. Laser alignment was used to determine the tilt, which was normal to within 0.5 mrad.

20.4 Results

The nominal Hugoniot Elastic Limit (HEL) for the AZ31B-4E and AMX602 were previously determined from in-situ shock experiments to be approximately 0.181 ± 0.003 GPa [12] and 0.187 ± 0.012 GPa [17] respectively. The corrected spall strengths for AZ31B-4E and AMX602 were estimated at various shock stresses ranging from approximately 1.5 GPa to 5.0 GPa in accordance with [18] to account for elastic-plastic effects. The corrected spall strength for the AZ31B-4E samples exhibit approximately 5% reduction between 1.7 GPa and 4.6 GPa shock stress [12]. However, the AMX602 samples show an 8% increase in the spall strength as a function of shock stress [17] similar shock stress range. The spall strength as a function of shock stress of the AZ31B-$4_{B/C}$ material is yet to be determined.

The pole figures for as-received AZ31B-4E and AMX602-SWAP are shown in Fig. 20.2. The c-axis in Fig. 20.2a is oriented almost parallel to the normal direction (normal to the plane of the pole figure), which in this case is the shock

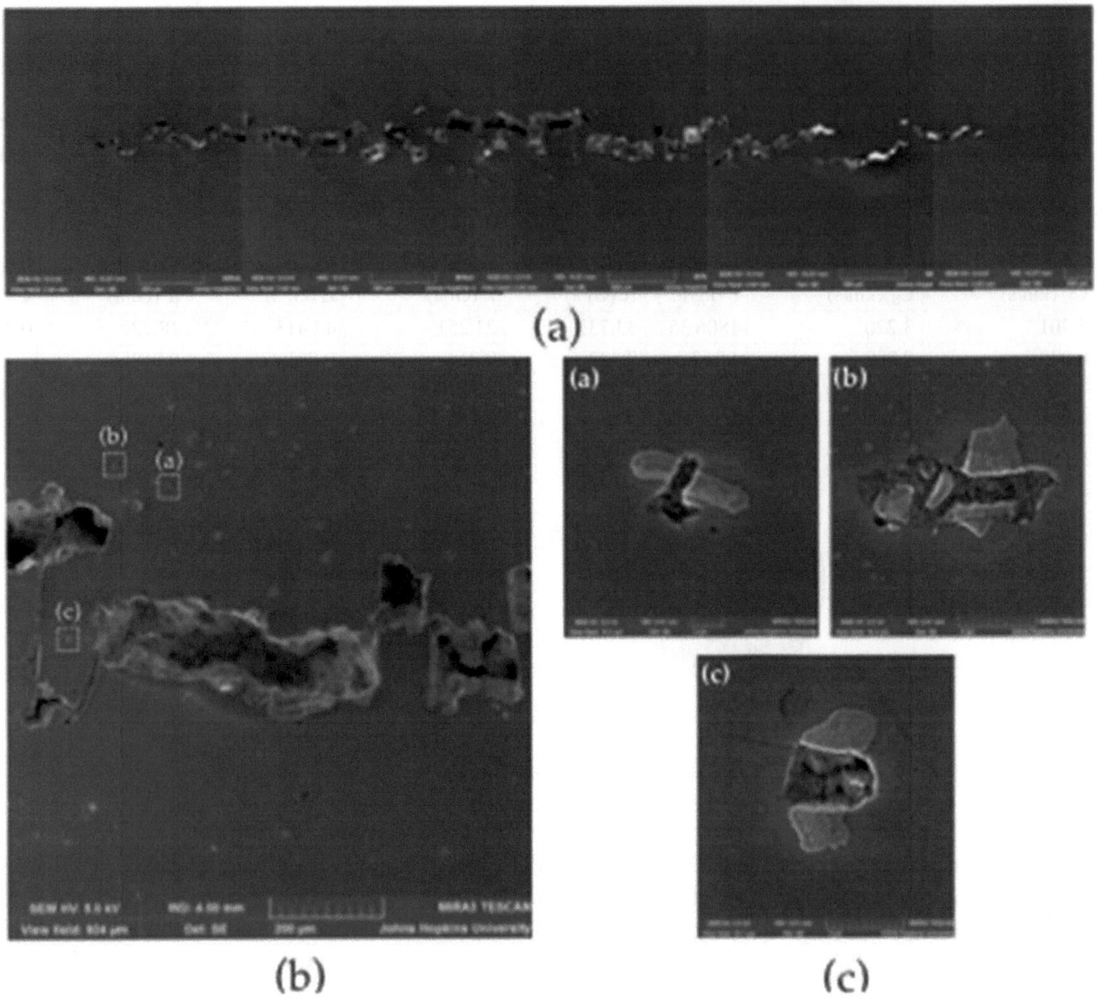

Fig. 20.3 SEM micrographs of the residual microstructure of AZ31B-4E shock loaded to ~1.3 GPa (~300 m/s). Spallation in the interior of the sample is shown in Fig. 20.3a. The areas labeled (**a**), (**b**), and (**c**) in Fig. 20.3b are shown in higher magnifications in Fig. 20.3c.

loading direction for the AZ31B-4E. However, the AMX602-SWAP exhibits a fiber texture with the c-axis oriented about the extrusion direction; the shock direction in this case is also the plate normal direction. Pole figure for the AZ31B-$4_{B/C}$ material is yet to be determined. Ex-situ recovery experiments were conducted on all three materials in accordance with the techniques employed by Williams et al. [16] to study spall failure characteristics and identify failure initiation sites in all three magnesium alloys.

The Scanning Electron Microscopy (SEM) micrographs in Fig. 20.3 show the residual microstructures of AZ31B-4E shock loaded to approximately 1.3 GPa (~300 m/s). The AZ31B-4E sample failed internally by spallation but did not separate into two halves as shown in Fig. 20.3a. Large aluminum-manganese-rich (Al-Mn) intermetallic inclusions are widely distributed throughout the AZ31B-4E microstructure (Fig. 20.3b). These intermetallic inclusions crack and de-bond from the matrix magnesium under shock compression as shown in Fig. 20.3b, c and are potential failure initiation sites for spallation. The spalled sample shock loaded to approximately 1.7 GPa also suggests that these intermetallic inclusions act as initiation sites for spall failure (Fig. 20.4). It is clearly evident from Fig. 20.4 that the spall surface is populated with these intermetallic particles. As a consequence, they are perhaps responsible for the reduction in the spall strength as a function of shock stress between 1.7 and 4.6 GPa reported by Farbaniec et al. [12].

Recovery experiments were also conducted on AMX602 magnesium samples in order to study the spall failure characteristics. As shown in Fig. 20.5a, the sample shock loaded to 0.8 GPa (~200 m/s) exhibited numerous isolated cracks in the vicinity of the spall plane; spanning approximately 1500 μm. The cracks are normal to the shock loading direction. Some of these isolated cracks are connected by smaller cracks, which are almost parallel to the shock direction as revealed in Fig. 20.5b. A magnified view of the crack interior is shown in Fig. 20.5c and it reveals a mixed-mode failure consisting

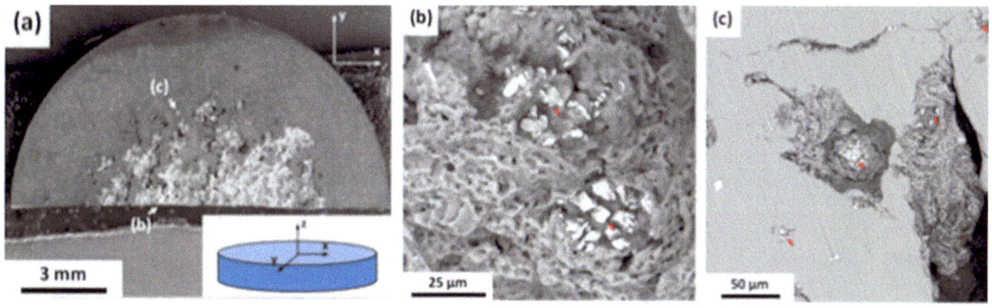

Fig. 20.4 SEM micrographs of the spall surface for the recovered sample shocked at approximately 1.7 GPa (~400 m/s). (**a**) Location of the spall surface under interrogation, (**b**) an enlargement of the area labeled (**b**) in Fig. 20.4a, c an enlargement of the area labeled (**c**) in Fig. 20.4a revealing Al-Mn-rich intermetallics

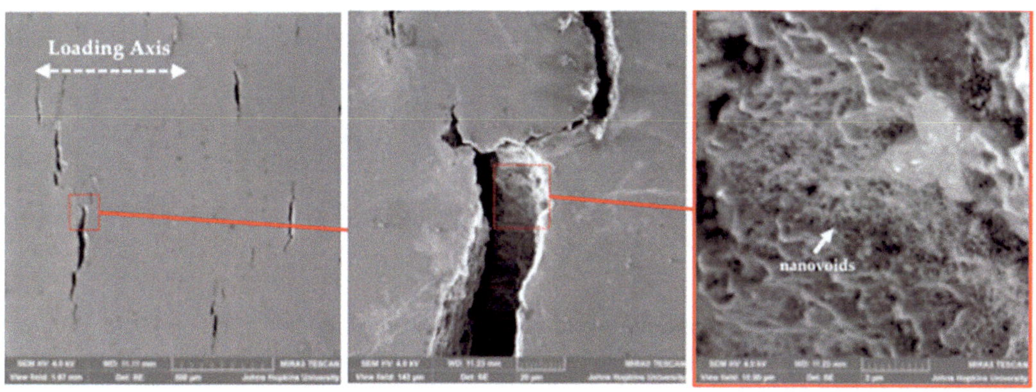

(a) (b) (c)

Fig. 20.5 SEM micrographs of the (**a**) residual microstructure of the failed AMX602 magnesium shock loaded to approximately 0.8 GPa (200 m/s). (**b**) An enlarged view of two vertical cracks in (**a**) connected by a horizontal crack. (**c**) A magnified view within the crack shown in (**b**)

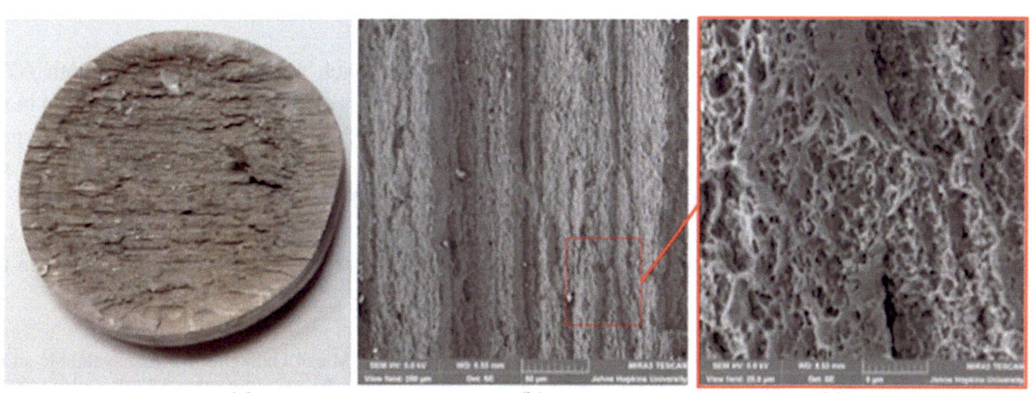

(a) (b) (c)

Fig. 20.6 SEM micrographs of the failed AMX602. (**a**) Recovered sample shock loaded to approximately 1.7 GPa (400 m/s). (**b**) A magnified view within the striated region shown in (**a**). (**c**) A magnified view of (**b**) showing voids within the striated region

of microvoids and nanovoids. The cause of the nanovoids is yet to be explored but may have resulted from homogeneous nucleation from the nano size intermetallic Al_2Ca inclusions or vacancies. Furthermore, at 1.7 GPa (~400 m/s) shock stress, the sample totally separated by spallation and the failed surface is striated as shown in Fig. 20.6a. A closer look at the striated surface (Fig. 20.6b) reveals nanovoids as shown in Fig. 20.6c.

The AZ31B-$4_{B/C}$ magnesium samples were shock loaded to approximately 0.8 GPa (~200 m/s) and 1.7 GPa (~400 m/s) in order to study the failure characteristics. At 0.8 GPa shock stress, there were no visible sign of spallation. However, the failure character was similar to that observed in the AZ31B-4E magnesium. Cracking and debonding of the Al-Mn rich

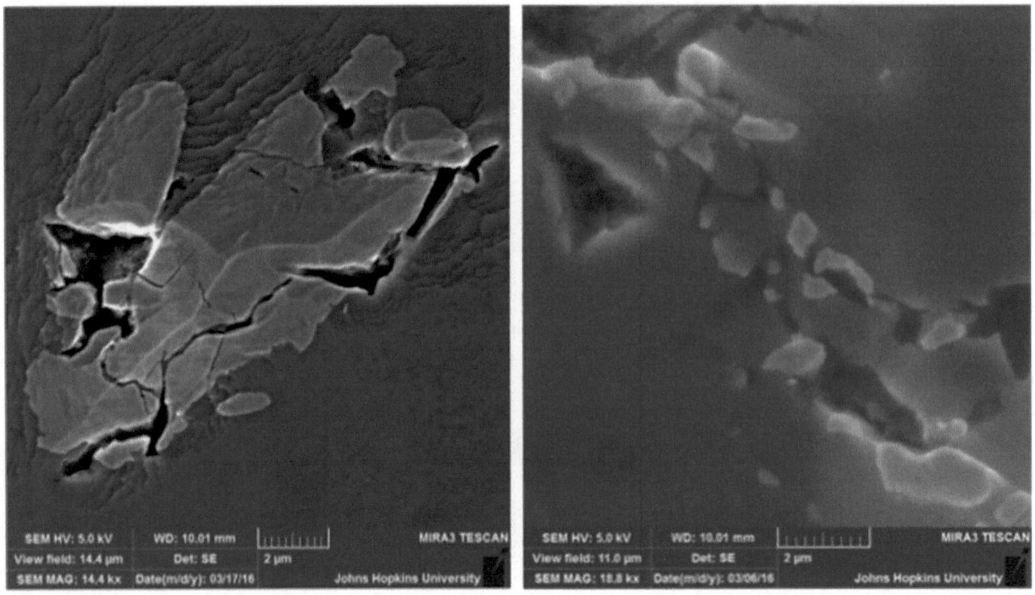

Fig. 20.7 SEM micrographs of the failed AZ31B-$4_{B/C}$. (**a**) Recovered sample shock loaded to approximately 0.8 GPa (200 m/s) and recovered sample shock loaded to approximately 1.7 GPa (400 m/s)

intermetallic inclusions from the matrix magnesium were generally observed as shown in Fig. 20.7a. Similar observations were made for the sample shock loaded to approximately 1.7 GPa but in this case spallation was evident in the interior of the material. Also, Cracking and debonding of the Al-Mn rich intermetallic inclusions from the matrix magnesium were quite severe (Fig. 20.7b).

20.5 Summary

Recovery experiments and SEM analyses show that the large manganese-rich particles present in both AZ31B-4E and AZ31B-$4_{B/C}$ magnesium alloys acted as failure initiation sites, which perhaps is responsible for the reduction in spall strength of the AZ31B-4E. Also, the spalled surface of the AZ31B-4E magnesium exhibited mixed-mode failure by nanovoids and microvoids. In contrast, the AMX602 magnesium shock loaded to 0.8 GPa exhibited multiple cracks within a wide region and mixed mode failure was observed within the cracks. However, the spalled surface created at approximately 1.7 GPa was striated with nanovoids within the striated surface. The aforementioned results warrant further in-depth studies to determine the reasons for nucleation of nanovoids in both materials and also the reasons for the striated failed surface in the AMX602 magnesium alloy.

Acknowledgements The authors gratefully acknowledge Mr. Jermaine Bradley and Mr. Micah Gallagher for their valuable support in sample fabrication and recovery experiments. Thanks to Dr. Laszlo Kecskes and Mr. Tyrone Jones for providing the AZ31B and AMX602 plates from which plate impact samples were fabricated.

References

1. Tenckhoff, E.: Review of deformation mechanisms, texture, and mechanical anisotropy in zirconium and Zirconium Base alloys. Astm Special Tech- Nical Publication. **2**(4), 1–26 (2006)
2. Mises, R.V.: Mechanik der plastischen formnderung von kristallen. ZAMM (Journal of Applied Mathematics and Mechanics/Zeitschrift fr Angewandte Mathematik und Mechanik). **8**(3), 161–185 (1928)
3. Klimanek, P., Poetzsch, A.: Mater. Sci. Eng. A. **324**, 145–150 (2002)
4. Kelley, E.W., Hosford Jr., W.F.: Trans. AIME. **242**, 5–13 (1968)
5. Wonsiewicz, B.C., Backofen, W.A.: Trans. AIME. **239**, 1422–1431 (1967)
6. Renganathan, P., Winey, J.M., Gupta, Y.M.: J. Appl. Phys. **121**, 035901 (2017)

7. Schmidt, R.M., Davies, F.W., Lempriere, B.M., Holsapple, K.A.: J. Phys. Chem. Solids. **39**, 375 (1978)
8. McQueen, R.G., Marsh, S.P., Taylor, J.W., Fritz, J.N., Carter, W.J.: In: Kinslow, R. (ed.) High Velocity Impact Phenomena, pp. 293–417. Academic, New York (1970)
9. Marsh, S.P.: LASL Shock Hugoniot Data. University of California Press, Berkeley (1980)
10. Kanel, G.I., Razorenov, S.V., Bogatch, A., Utkin, A.V., Grady, D.E.: Int J Impact Eng. **20**, 467 (1997)
11. Hazell, P.J., Appleby-Thomas, G.J., Wielewski, E., Stennett, C., Siviour, C.: Acta Mater. **60**, 6042–6050 (2012)
12. Farbaniec, L., Williams, C.L., Kecskes, L., Ramesh, K.T., Becker, R.: Int. J. Impact Eng. **98**, 34–41 (2016)
13. Barber, R.E., Dudo, T., Yasskin, P.B., Hartwig, K.T.: Scr. Mater. **51**(5), 373–377 (2004)
14. Kondoh, K., El-Sayed, A.H., Imai, H., Umeda, J.: ARL-CR-647 US Army Research Laboratory (2010)
15. Williams, C.L., Chen, C.Q., Ramesh, K.T., Dandekar, D.P.: J. Appl. Phys. **114**, 093502 (2013)
16. Williams, C.L., Chen, C.Q., Ramesh, K.T., Dandekar, D.P.: J. MSEA. **618**, 596–604 (2014)
17. Farbaniec, L., Williams, C.L., Kecskes, L., Becker, R., Ramesh, K.T.: Unpublished (2017)
18. Kanel, G.I.: J. Appl. Mech. Tech. Phys. **42**(2), 358 (2001)

Chapter 21
High Speed Imaging Techniques to Study Effects of Pressure Waves from Detonating Explosive Charges on Biological Materials

Thuvan Piehler, Nicole Zander, Rohan Banton, Richard Benjamin, Ray Sparks, Kimberly Byrnes, Josh Duckworth, and Ben A. Bahr

Abstract Threshold shock-impulse levels required to induce cellular injury and cumulative effects upon single and/or multiple exposures are not well-characterized. There are very few *in vitro* experimental model with blast pressure waves generated by using explosives in the laboratory for investigating the effects of primary blast induced traumatic brain injury. Hence, it is critical to develop a reliable experimental research tool in the laboratory to study how functional and structural alteration or damage at the cellular and tissue levels following explosive blast impacts for future improvement of detection and subsequent treatment and protection. US Army Research Laboratory (ARL) developed the unique *in vitro* indoor experimental platform that uses military explosive charges to accurately represent battle-field blast exposure and to probe the effects of primary explosive blast on dissociated neurons and tissue slices. We believe this is one of the controlled experimental methods to analyze, characterize primary explosive blast induced cellular injury, and understand threshold injury phenomenon.

Keywords High speed imaging • Pressure waves • Explosive charges • Neurons • Tissues

21.1 Introduction

There are very few in vitro experimental models with blast pressure waves generated by using explosives in the laboratory environment for investigating the effects of primary blast-induced traumatic brain injury. Hence, it is critical to develop a novel in vitro indoor experimental model that uses military explosive charges to more accurately represent battlefield blast explosives and to study how functional and structural alteration or damage at the cellular and tissue levels following real explosive blast impacts cognitive performance for future improvement of detection and subsequent treatment of mild traumatic brain injury, as well as improving personal blast protection equipment for our soldiers.

21.2 Experimental Setup

All explosive tests were performed in a 6- × 10- × 6-m blast room certified for testing up to 25 g of explosive. Figure 21.1 shows a schematic of the overall experimental setup. Neuronal cell cultured samples or tissue slices were contained in a water-filled aquarium (30.5 × 34.5 × 65 cm, poly methyl methacrylate (PMMA) construction aquarium) (Fig. 21.2a). The thickness of the aquarium walls was 0.2 cm. The tank was filled with water maintained at 37 °C by a heating element coupled to a small stirring device. The clear walls of the aquarium enable a visual record of the water shock passage and its interaction with the cell cultured samples.

T. Piehler (✉) • N. Zander • R. Banton • R. Benjamin • R. Sparks
US Army Research Laboratory, Aberdeen Proving Ground, Adelphi, MD, USA
e-mail: thuvan.n.piehler.civ@mail.mil

K. Byrnes • J. Duckworth
The Uniformed Services University of the Health Sciences, Bethesda, MD, USA

B.A. Bahr
University of North Carolina – Pembroke, Pembroke, NC, USA

J. Kimberley et al. (eds.), *Dynamic Behavior of Materials, Volume 1*, Conference Proceedings of the Society for Experimental Mechanics Series, DOI 10.1007/978-3-319-62956-8_21

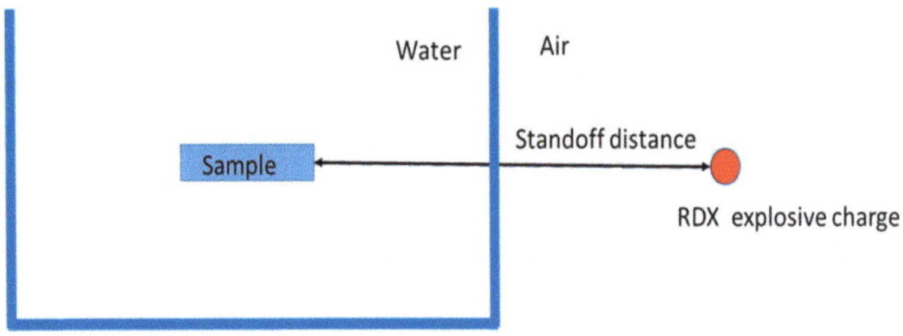

Fig. 21.1 Schematic of the experimental setup showing the positions of cyclotrimethylenetrinitramine (RDX) explosive charge, aquarium tank, and sample

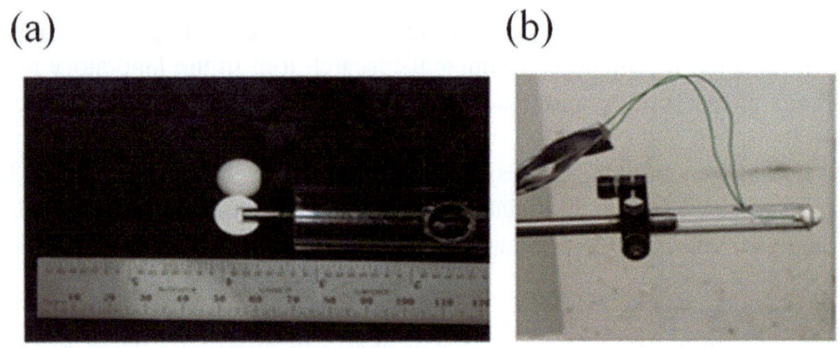

Fig. 21.2 Spherical RDX charge: (**a**) matching hemispherical charges (**b**) fully assembled 1.7-g spherical charges

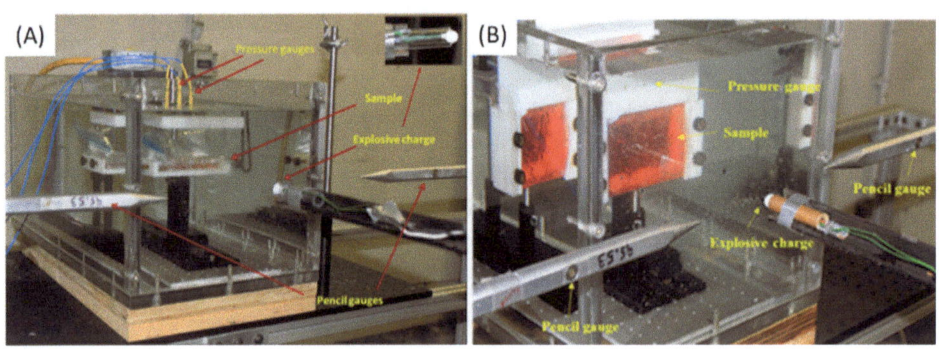

Fig. 21.3 Blast-induced injury of cells: (**a**) Side–on pressure measurement setup, and (**b**) Face-on pressure measurement setup. Both (**a**) and (**b**) show the location of pressure sensors with samples submerged in aquarium tank before each test

Spherical cyclotrimethylene trinitramine Class 5 ($C_3H_6N_6O_6$; RDX Class V, 1.7 g weight) charges are composed of matching hemispherical charge and used to generate the primary blast. Each hemispherical charge has a nominal mass of 0.85 g and is pressed to a density of 1.77 g/cm^3 (98.4% theoretical maximum density, TMD). Figure 21.2a shows matching hemispherical charges and Fig. 21.2b shows an assembled spherical charge of RDX. The spherical RDX explosive charge detonates 116 cm above the ground with RP-87 detonators.

Three piezoelectric high-frequency dynamic pressure sensors (ICP model 102A, PCB Piezotronics Inc., Depew, NY) were modified for underwater use and used to measure the shock wave pressure histories at a position of 2 cm above the cultured samples using a custom-designed pressure gauge holder. For cellular blast impact study, all pressure gauges were mounted on top of the cell culture plate with a custom-designed lid, submerged approximately 10.2 cm underwater, and positioned side-on to the blast wave direction. The caps were designed such that the pressures in different rows or columns of the well plate could be measured by moving the pressure sensors to the desired locations (Fig. 21.3a). For tissue slice blast impact study, all pressure gauges were mounted on top of the tissue sample bag, submerged in the aquarium approximately 10.2 cm under water, and positioned face-on to the blast wave direction (Fig. 21.3b). Two free-field air-blast pencil probes (model 137A23, PCB Piezotronics Inc., Depew, NY) were positioned in front of the aquarium to measure the free-field air-shock pressure before the shock wave entered the water medium Pressure-time history traces, peak overpressure, high speed images were recorded for each blast.

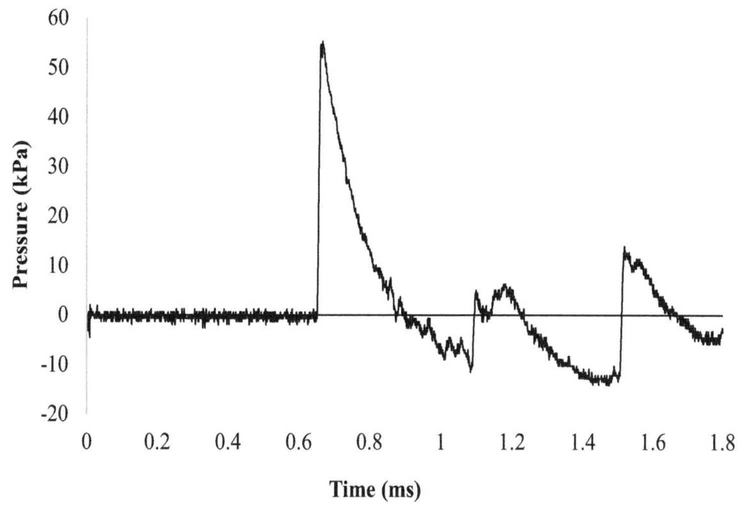

Fig. 21.4 Typical side-on pressure-time history signature recorded in air at 45 cm standoff distance

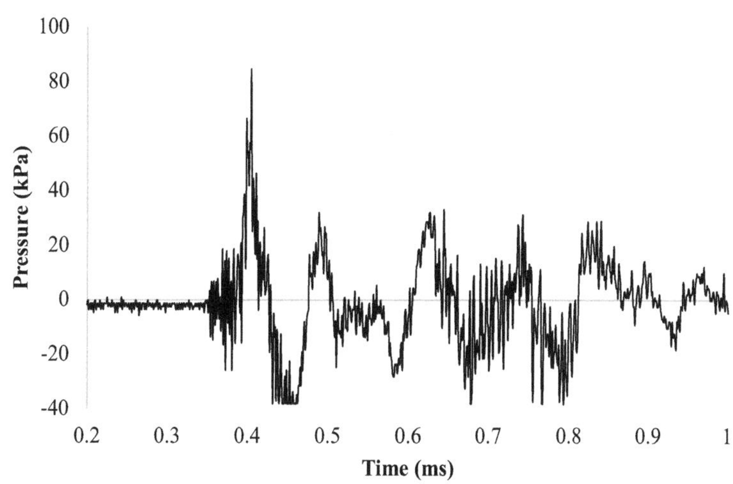

Fig. 21.5 Typical side-on pressure history signature in water at 45 cm standoff distance from RDX explosive

21.3 Results

Details of the blast-induced cellular injury experimental procedures and results have been previously reported [1–4]. Figure 21.4 illustrates a side-on pressure-time history (molecules in the leading shock are not reflected by the gauge element) observed in air at a 45 cm standoff distance from the point of initiation. The peak pressure was measured at 55 kPa. Figure 21.5 shows a typical side-on pressure history in water measured above the sample well plate for the same standoff distance of 45 cm. The peak pressure was measured at 84 kPa. The time duration of the pressure excursion was approximately 100 μs.

In work with neurons, physical insults from explosive blast created axonal beadings and caused an influx of sodium ions. Figure 21.6 shows images of the neuron morphology of sham (Fig. 21.6a), and injured cells (Fig. 21.6b) following multiple blast exposures of ca. 10 psi. There is evidence of a couple of axonal beads in the control and sham cells but significantly more beading is present in the injured cells.

21.4 Conclusions

A comprehensive experimental approach to investigate the primary blast effects on brain cells is reported. The work involved an accurate modeling of the RDX explosive for blast wave propagation and interactions with the 24-well plate. More real-time or near real-time points are needed to better elucidate the mechanical injury mechanisms, particularly in the case of

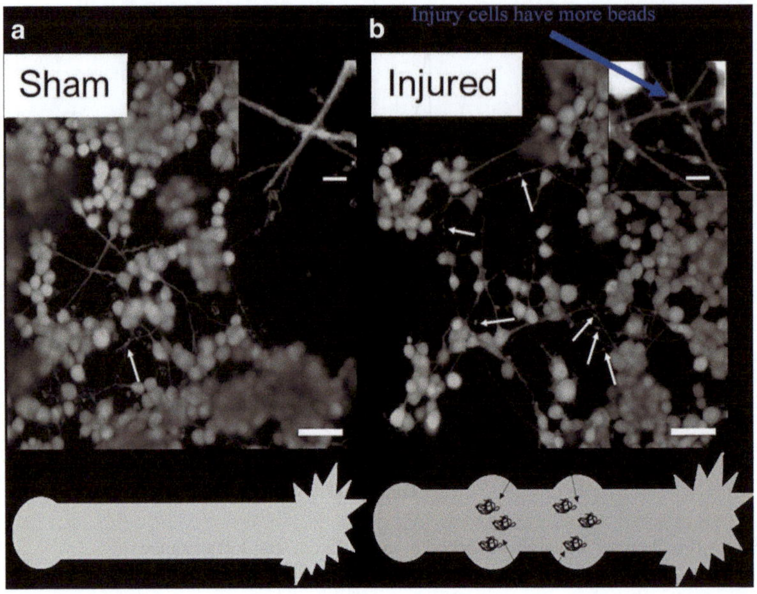

Fig. 21.6 Morphology of cells 24 h after exposure to multiple explosive blast of ca. 10 psi. (**a**) Sham cells that were transported to blast site but were not injured, and (**b**) injured cells subjected to multiple blast. Arrows denote locations of selected axonal beads. Scale bars represent 50 μm. Inset displays magnified image of beads. Inset scale bars represent 10 μm

repeated blast injury. Finally, the development of threshold parameters for cell injury and death are needed. Together, this information can help aid in the treatment and diagnosis of brain injuries, as well as in the design of better head protective equipment.

References

1. Zander, N.E., Piehler, T., Boggs, M.E., Banton, R., Benjamin, R.: In vitro studies of primary explosive blast loading on neurons. J. Neurosci. Res. **93**, 1353–1363 (2015)
2. Zander, N.E., Piehler, T., Banton, R., Boggs, M.: The effect of explosive blast loading on human neuroblastoma cells. Anal. Biochem. **504**, 4–6 (2016)
3. Wisniewski, M.L., Hwang, J., Bahr, B.A.: Submicromolar Aβ42 reduces hippocampal glutamate receptors and presynaptic markers in an aggregation-dependent manner. Biochim. Biophys. Acta. **1812**, 1664–1674 (2011)
4. Smith, M., Piehler, T., Benjamin, R., Farizatto, K.L., Pait, M.C., Almeida, M.F., Ghukasyan, V.V., Bahr, B.A.: Blast waves from detonated RDX explosive reduce GluR1 and synaptophysin levels in hippocampal slice cultures. Exp. Neurol. **286**, 107–115 (2016)

Chapter 22
Computational Study on the Driver Section Design of an Explosively Driven Conical Shock Tube

Joel B. Stewart

Abstract The work presented in this paper investigates the use of a shock tube designed to propagate a shock wave through the air (commonly denoted as "air blast") resulting from the detonation of an explosive charge. This explosively driven shock tube has a conical geometry, which allows for the use of relatively small charges to approximate the Friedlander waveform due to the free-field detonation of much more massive explosive charges. Both previous experimental data and new computational work are presented to illustrate the influence of various design parameters of the shock tube's explosive charge and driver section—the portion of the tube that confines the explosive charge and transitions to the main tube section confining the flow—on the resulting air blast.

Keywords Air blast • Shock tube • Hydrocode • Explosively driven shock tube • Explosive

22.1 Introduction

Shock tubes have been used for centuries to provide controlled environments for the propagation of shock waves through fluids and the study of the interaction of these waves with structures of interest [1]. Shock tubes generally fall under the category of either compression-driven shock tubes, which generate the air blast using high pressure gases combined with burst discs, or explosively driven shock tubes, which generate the air blast by detonating explosive charges. Each category of shock tube has its own set of advantages and disadvantages, which have been discussed elsewhere (e.g., [2]) and will not be touched on in this paper.

This paper is concerned with the design of explosively driven shock tubes.[1] There are various parts that comprise an explosively driven shock tube and, in this paper, some basic components will be denoted as follows:

1. Explosive driver charge – the explosive material used to generate the air blast
2. Driver section – the portion of the tube that confines the explosive charge and transitions to the channel section
   (a) Driver housing section – the part of the driver section that houses the explosive charge and includes the entry of the shock tube
   (b) Driver transition section – the part of the driver section that connects the driver housing section to the channel section of the shock tube
3. Channel Section – the portion of the shock tube designed to confine the air blast

Stewart and Pecora [4] investigated the use of a 17° (full angle) conical shock tube with an explosive driver charge to approximate the free-field shock wave moving through air due to a large (relative to the explosive driver charge) explosive detonation. The computational study presented in this paper seeks to build upon the previous research by looking in more detail at various modifications to the explosive charge and driver section in an explosively driven 10° (full angle) conical shock tube. Specifically, the influence of these modifications on the air blast downstream of the shock tube's driver section will be investigated.

[1]The work presented in this paper is drawn from a larger effort documented by the author in an internal ARL technical report [3].

J.B. Stewart (✉)
US Army Research Laboratory, Aberdeen Proving Ground, MD, 21005, USA
e-mail: joel.b.stewart2.civ@mail.mil

J. Kimberley et al. (eds.), *Dynamic Behavior of Materials, Volume 1*, Conference Proceedings of the Society for Experimental Mechanics Series, DOI 10.1007/978-3-319-62956-8_22

22.2 Computational Approach

The Lawrence Livermore National Laboratory shock physics code ALE3D [5] was used for all computations shown in this paper. All computations were performed in 2D axisymmetry with 0.5 mm per square computational zone. This mesh resolution resulted in problems sizes of approximately 10 million elements. The Arbitrary Lagrangian Eulerian (ALE) formulation was used in all calculations to weight the mesh toward the steel shock tube walls as well as toward the shock front by assigning a higher weight to the artificial viscosity in the air. All pressures listed in this paper are overpressures and all reported impulses are obtained by integrating these overpressures over time (i.e., impulse per unit area).

Pressure time histories due to air blast have a distinct shape in the mid and far fields (i.e., the regions beyond the fireball where all loading is due to the air shock, rather than some combination of the air shock and product gases). Friedlander [6] suggested an equation to describe the shock profile that agrees well with experimental pressure data in the far-field region of a free-field blast environment. In the mid-field region (i.e., beyond the fireball but where the peak incident overpressures are still large–greater than 14.7 psi), Dewey [7] recommended the following modified form of the Friedlander waveform:

$$P(t) = P_S e^{-\alpha t}\left(1 - \frac{t}{t^+}\right), \qquad (22.1)$$

where P is the time-dependent overpressure at the gauge location, P_S is the peak overpressure, t is the time after the air blast arrival, t^+ is the positive phase duration, and α is a fitting parameter (the original Friedlander equation may be recovered by setting $\alpha = 1/t^+$).

The Friedlander waveform is an idealized representation of a free-field air blast, without any influence of shock reflections (either from the ground or from shock tube surfaces). Since the shock tube is a tool designed to approximate the free-field air blast, it is instructive to compare the shock tube data in the mid-field region back to a modified Friedlander waveform. Therefore, in some of the plots presented in the following pages, shock tube experimental and computational data are compared against the modified Friedlander waveform of Eq. (22.1). For the comparison, the peak overpressure, P_S, and the distance from the charge to the gauge location are obtained from the experiment or computation while the corresponding values of t^+ and α are obtained from ConWep [8]. The air blast time of arrival, t_0, is adjusted to line up with the data being compared against (typically t_0 needs to be adjusted on the order of 10% to line up with the shock tube data).

Prior to conducting a computational study on geometries for which no experimental data exist, an attempt was made to ensure that the experimental data presented in reference [4] for a 17° (full angle) conical shock tube could be replicated computationally with reasonable accuracy. The shock tube in that article was 1.933 m long and used a 14.5 g cylindrical composition C4 (C4) driver charge, 28.2 mm in diameter and 15.2 mm long, placed with its rear surface flush with the entry of a conical steel driver section. The minimum diameter at the entry of the driver section was 32.0 mm, with a 101.6 mm outer diameter, which transitions after 233.2 mm to the constant 4.8-mm-thick channel section. Both ends of the shock tube were open to the air. The ALE3D calculations neglected both the blue foam centering device for the explosive charge and the Teledyne RISI RP-83 detonator. For computational purposes, a Jones-Wilkes-Lee (JWL) model was used to describe the C4 product gases and detonation was described using programmed burn with point initiation at the center of the back surface. Air was described using a tabular equation of state and the steel was modeled as 4340 steel using a Steinberg-Guinan rate-independent strength model with a Mie-Grüneisen equation of state.

A comparison of the computational results with experimental results, taken from shot seven of reference [4], is shown in Fig. 22.1 along with a modified Friedlander waveform taken from ConWep and corresponding to the free-field detonation of an 861.8 g (1.9 lb) sphere of C4 at the specified gauge distance from the charge. All data are taken at the exit of the 1.933 m-long shock tube (i.e., 1.918 m from the closest surface of the C4 charge), where the inner diameter of the shock tube is 0.61 m. The data shown in Fig. 22.1 were taken 0.15 m off-axis (referred to as "OA" in the plots); data were also collected on the shock tube axis but are not presented here. Both the computational pressures (Fig. 22.1a) and impulse (Fig. 22.1b) under-predict the experimental results. The arrival of release waves from the shock tube exit, resulting in a pronounced negative pressure phase, is captured in the ALE3D calculation; obviously, the modified Friedlander waveform does not exhibit this behavior since release waves due to exiting the shock tube would not be present in a free field detonation.

In general, the ALE3D computational results shown in Fig. 22.1 were deemed to qualitatively replicate the experimental data from reference [4] sufficiently well in terms of the peak pressure, time of arrival, and impulse (per unit area) to warrant proceeding with the parametric study detailed in the next section. Therefore, the same computational approach (i.e., mesh resolution, ALE weighting, air and steel material models, etc.) will be used in the next section to investigate the influence that various design choices for an explosively driven 10° conical shock tube have on the resultant loading profile downstream of the explosive charge. All explosive charges considered for this parametric study are 50/50 pentolite (modeled using a JWL EOS with programmed burn) as opposed to the C4 used previously. This change was made due to experimental considerations

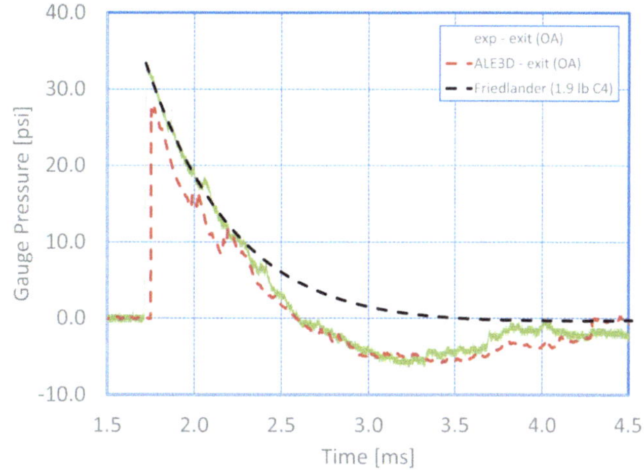

(a)

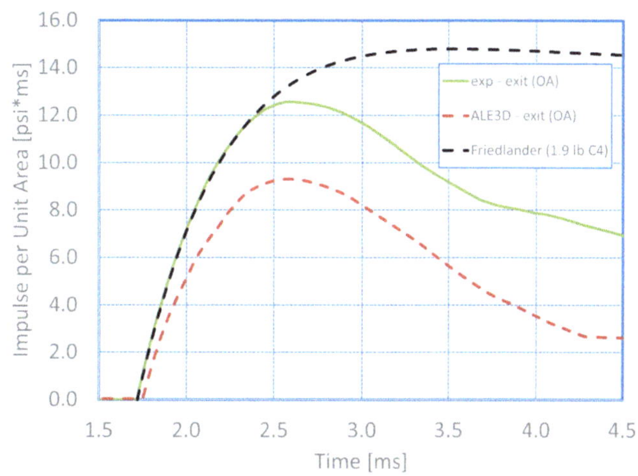

(b)

Fig. 22.1 Comparison of experimental (taken from reference [4]) versus computational data 15 cm off the axis of symmetry and at the exit (192 cm from the nearest surface of the C4 explosive driver charge) of a 17° shock tube (**a**) overpressure and (**b**) impulse per unit area

(pentolite is castable and should lead to increased repeatability compared to the putty-like C4). The explosive charge has the detonation initiated at a point located at the center of the back surface.

For all following computations, a 10° (full angle) conical channel section 2.5 m long with 4.8-mm-thick steel walls is used. The exterior geometry of the driver section is a cylinder with diameter 3.5 times that of the explosive charge (for the special case of the conical frustum, "diameter" refers to the smallest diameter). In this paper, the driver section of the shock tube will frequently be referred to as two separate sections: (1) the piece housing the explosive charge, which is denoted as the driver housing section; and (2) the part connecting the driver housing section with the channel section of the shock tube, which is denoted as the driver transition section. The interior geometry of the driver housing section corresponds to that of the explosive charge (i.e., cylindrical in all but the frustum case) before immediately transitioning to a conical geometry in the driver transition section for mating with the channel section of the shock tube. The length of the driver transition section is always taken to be 2 times the charge diameter and the channel section of the shock tube is always 2.5 m long. All computational data that follow are computed at a fixed gauge axially 2.0 m downstream of the nearest charge surface and radially halfway between the tube axis and the shock tube wall (off-axis data are considered due to the data containing numerical noise on the axis of symmetry). The length of the tube is taken to extend well beyond the gauge location so that release waves from the exit do not complicate the analysis.

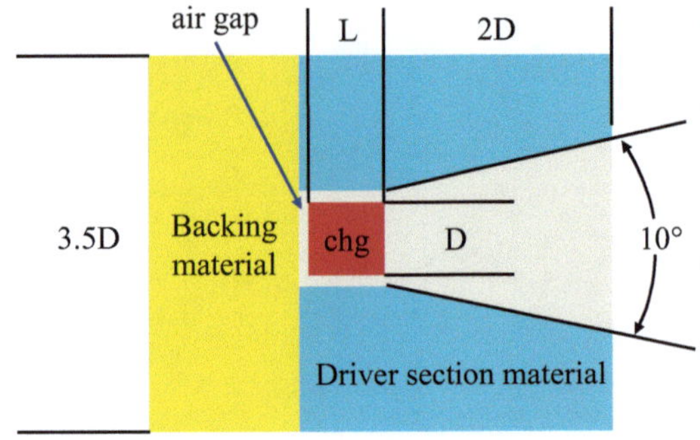

Fig. 22.2 Schematic of shock tube driver section

Table 22.1 Computational test matrix for 10° shock tube study

Configuration	Driver section material	Backing material	Air gap (mm)	Charge geometry	Charge L/D	Charge diameter (mm)	Charge mass (g)
Baseline	4340 steel	Air	None	Cylinder	1.0	16.0	5.0
50% mass (L/D = 1)						12.7	2.5
50% mass (L/D = 0.5)					0.5	16.0	
150% mass (L/D = 1)					1.0	18.3	7.5
150% mass (L/D = 1.5)					1.5	16.0	
200% mass (L/D = 1)					1.0	20.1	10.0
200% mass (L/D = 2)					2.0	16.0	
L/D = 2						12.7	5.0
Frustum				10° conical frustum	1.25[a]	16.0	
Air gap			2.0		1.0		
Steel back		32 mm 4340 steel		Cylinder			
Al back		32 mm 7075 Al	None				
Al driver	7075 Al	Air					

[a]measured with respect to the smallest diameter (reported as "charge diameter")

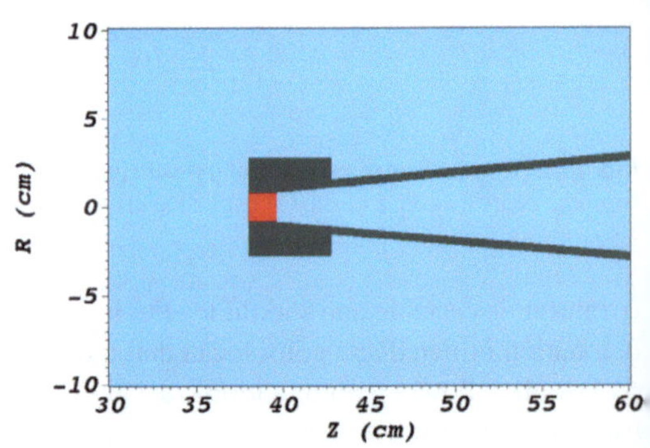

Fig. 22.3 Zoomed-in view of the ALE3D shock tube entry section geometry for the baseline shock tube design (the *red* material represents pentolite, the *grey* represents the steel shock tube, and the *blue* represents air)

22.3 Computational Results

The influence of various driver section (e.g., geometry and material) and explosive charge (e.g., geometry and size) modifications on the downstream loading profile is investigated in the following paragraphs for an explosively driven 10° conical shock tube. Figure 22.2 depicts a schematic of a general shock tube driver section and Table 22.1 lists the computational test matrix used for this investigation. As an example, Fig. 22.3 shows a zoomed-in view of the ALE3D geometry used for the baseline design, with the design parameters defined in Table 22.1 for the various iterations.

Figure 22.4 presents the ALE3D pressure profile for the baseline driver section design with a 5 g driver charge (see Table 22.1) obtained 2.0 m downstream of the nearest charge surface and halfway between the axis and shock tube wall (approximately 93 mm off-axis for this baseline design). Also included in the figure is the modified Friedlander waveform for a 544 g (1.2 lb) pentolite charge obtained from ConWep (as mentioned in Sect. 22.1, the modified Friedlander parameters were obtained by inputting ALE3D's peak pressure and distance from the charge to the gauge into ConWep). The shock

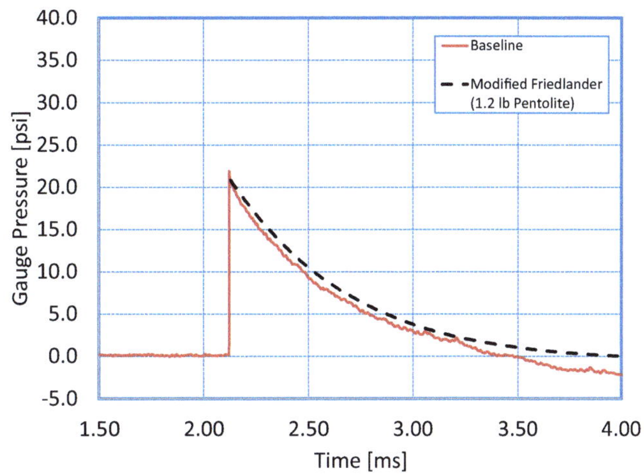

Fig. 22.4 Pressure loading measured 2.0 m downstream of the driver charge surface for the baseline design of the 10° shock tube driver section and 5 g driver charge compared with a modified Friedlander waveform obtained from CONWEP for a 544 g (1.2 lb) pentolite charge

tube waveform appears to diverge from the modified Friedlander around 3.2 ms; this divergence is likely due to the entry of the tube being open rather than due to release waves from the exit of the tube since the shock has only just arrived at the exit, another 0.5 m downstream from the point at which the data is gathered, around 3.2 ms. Even with this divergence, the loading profile of this baseline case does, in general, provide a good approximation of a modified Friedlander waveform.

Figure 22.5 shows the resultant air blast as a function of shock tube driver explosive mass. For each mass (other than the baseline), two configurations are shown: Fig. 22.5a shows the aspect ratio being held constant (L/D = 1) while the diameter and length are modified, whereas Fig. 22.5b shows the diameter being held constant (D = 16.0 mm) while the charge length is modified. Both configurations result in similar peak pressures relative to the baseline case; however, a comparison of Fig. 22.5 indicates that the longer charges (L/D > 1) do a worse job of approximating the desired modified Friedlander waveform. For the controlled aspect ratio pressure profiles of Fig. 22.5a, the peak pressures appear to be increasing at approximately half the rate of the mass (i.e., doubling the mass yields a roughly 50% increase in peak pressure). In general, Fig. 22.5 indicates that a shock tube with a fixed conical angle is sufficiently flexible to produce a range of loading profiles approximating a modified Friedlander waveform simply by increasing the explosive mass; furhtermore, it demonstrates that controlling the aspect ratio to have an L/D = 1 or less results in a better approximation of the desired modified Friedlander waveform.

The influence of placing a backing material behind the driver charge is investigated in Fig. 22.6. In this case, a 32-mm-thick disc of either 4340 steel or 7075 Al is placed directly behind the explosive charge (see Fig. 22.2) to minimize energy losses due to product gases escaping out the back of the shock tube; the diameters of these backing discs are the same as the outer diameter of the driver section (i.e., 3.5 times the explosive charge diameter). Figure 22.6 shows that the pressure peak calculated at the gauge location 2.0 m downstream from the charge surface is increased substantially (nearly 70% for the steel backing) relative to the baseline case where the product gases freely escape. The aluminum backing material results in a slightly lower peak pressure relative to using a steel backing material. The shock structure does appear to be significantly noisier when using either aluminum or steel as a backing material and the loading profile deviates somewhat from a smooth modified Friedlander waveform. In fact, the shock profiles when using a backing material other than air (Fig. 22.6) look qualitatively similar to the shock profiles from having long charges (e.g., the L/D > 1 cases in Fig. 22.5b). One possibility for this similarity is that both designs mitigate the influence of release waves from behind—in the first case due to the relatively high shock impedance backing material and, in the second case, due to the detonation through a relatively longer, narrow channel—and, for similar reasons, generate more secondary shocks in the driver section relative to the baseline case that are then propagated downstream through the shock tube channel section.

Finally, a number of other modifications, all using a 5 g explosive charge, were investigated and are presented in Fig. 22.7. Using a frustum geometry for the driver charge (and coincident driver section) instead of the more easily manufactured conical geometry increased the peak pressures by 6%. Introducing an air gap between the explosive and driver section walls in the cylindrical case also resulted in higher pressures (5.0% higher peak) relative to the baseline. Increasing the cylindrical charge's L/D from 1 to 2 resulted in a more than 15% decrease in peak pressure relative to the baseline design, while changing the driver section's material from steel to aluminum decreased the peak pressure by 23.5%. Fabricating the entire shock tube out of aluminum results in practically the same downstream pressure profile (not shown here) as simply making the driver

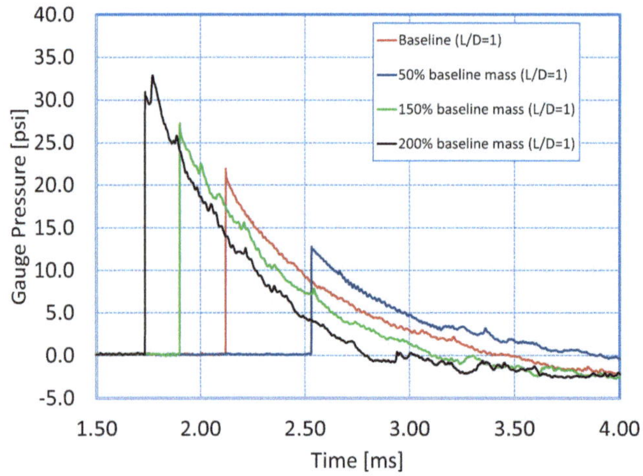

(a)

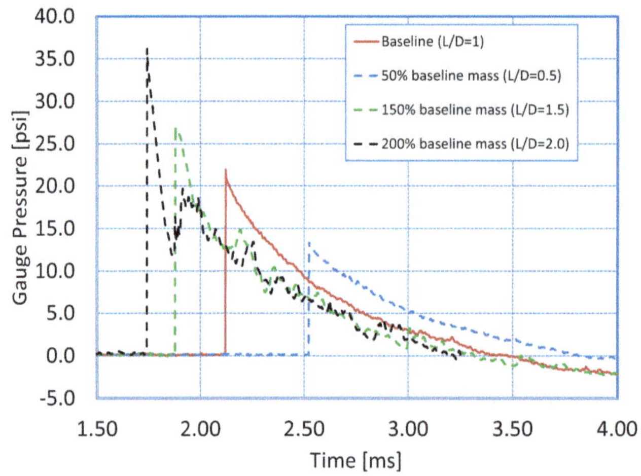

(b)

Fig. 22.5 Influence of driver charge mass on the pressure profile of the 10° shock tube measured 2.0 m downstream of the driver charge surface (**a**) fixed aspect ratio (L/D = 1) and (**b**) fixed charge diameter (D = 16.0 mm)

section aluminum; this result implies, as might be expected, that the channel section primarily serves to propagate the shock waves downstream and does not contribute notably to the downstream pressure profile.

One seemingly counter-intuitive observation from Fig. 22.7 is that the introduction of an air gap between the explosive charge and the radial wall of the driver section yields an increase in the downstream pressure relative to the baseline design, which has the explosive charge in contact with the wall. The reasoning for this phenomenon is as follows. The primary blast wave generated when the driver section is in contact with the detonating charge will indeed be slightly stronger than the primary blast wave generated when there is an air gap between the detonating charge and the radial driver section walls; this difference in the primary shock wave strength is to be expected due to release waves at the explosive-air interface when an air gap is present. However, there are also reflected shock waves that are generated behind the detonation wave, some of which combine downstream with the primary blast wave generated from the detonation front, and it is this combined shock wave that is measured downstream.

When the driver section wall is in contact with the detonating charge, reflected shock waves are generated at the wall-explosive interface (and transmitted back through the product gases) from the velocity component of the curved detonation wave that is normal to the radial wall; much of these reflections get lost out the open end of the shock tube into the surrounding air. In contrast, when an air gap is introduced between the explosive charge and the wall of the driver section, release waves decrease the pressures at the explosive's radial surface, as would be expected, but the impact of the expanding product gases with the wall generates a reflected shock wave back through the explosive products (see the top half of Fig. 22.8), which

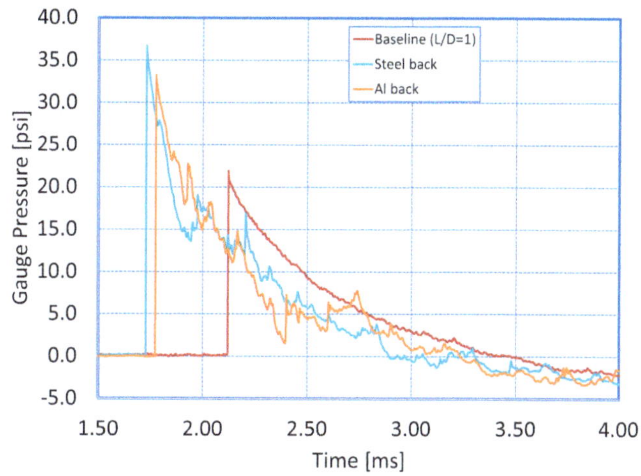

Fig. 22.6 Influence of a backing plate behind the 5 g driver charge on the pressure profile of the 10° shock tube measured 2.0 m downstream of the driver charge surface

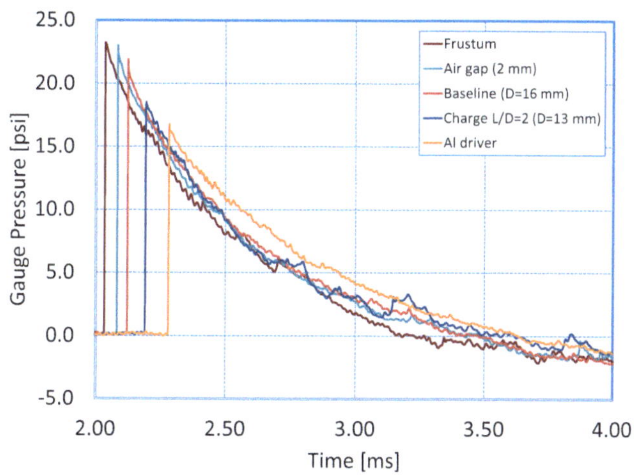

Fig. 22.7 Influence of various modifications to the driver section and 5 g driver charge on the pressure profile of the 10° shock tube measured 2.0 m downstream of the driver charge surface

then gets transmitted downstream and eventually combines with the primary air blast. The shock waves reflected back into the product gases when an air gap is present are moving toward the axis and down the shock tube due to the direction of the expanding gases when they impact and, therefore, fewer of these shock waves are lost out of the open end of the shock tube. In either case, the secondary shock reflection (and subsequent reflections off the axis and shock tube wall) gets transmitted into the air downstream and eventually combines with the primary air blast due to the detonation wave.

Table 22.2 lists the peak overpressures for all the various designs (see also Table 22.1) and the percent difference of these peak pressures relative to the baseline. Using a steel backing disc in the driver section resulted in the highest peak pressure relative to the baseline (higher even than doubling the explosive mass) whereas using an Al driver or halving the explosive mass resulted in the lowest peak pressures.

22.4 Conclusions

This paper has detailed a computational study focused on the design choices related to explosively driven conical shock tubes. Earlier experimental work done by the author using a 17° shock tube was compared with ALE3D computational data to demonstrate the ability of the computational approach to capture the main features of interest (e.g., the pressure loading profiles and impulse per unit area). This computational approach was then used to conduct a parametric study on the influence

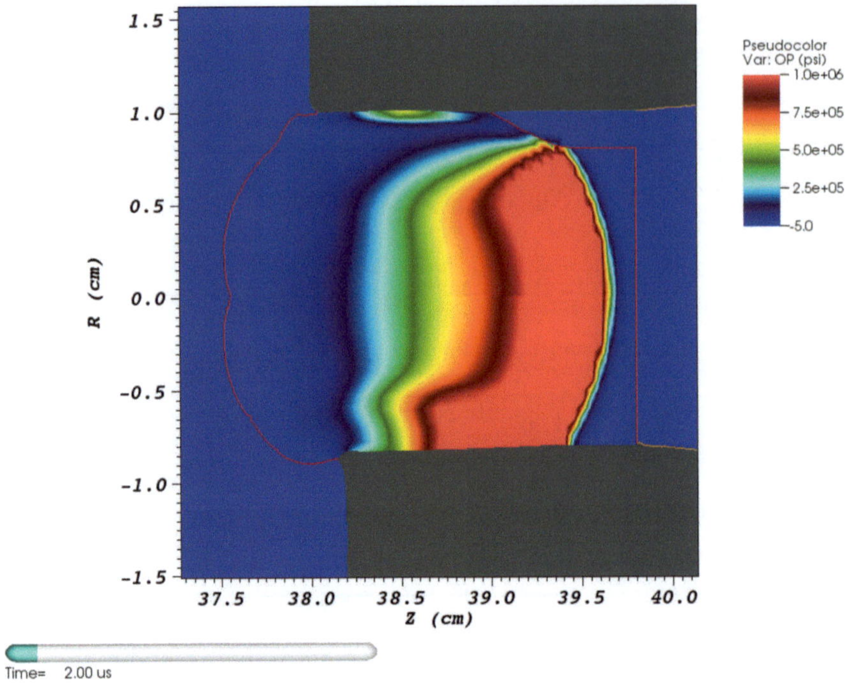

Fig. 22.8 Pressure contours at 2 microseconds after detonation showing the detonation wave and shock reflections through the explosive driver charge (*red boundary line*) for the baseline case with no air gap between the driver charge and the steel walls (*bottom half*) and for the case with a 2 mm air gap (*top half*)

Table 22.2 Peak Pressures (gauge) in 10° shock tube study

Configuration	Charge mass (g)	Peak overpressure @ 2 m (psi)	% diff from baseline
Baseline	5.0	21.92	0.0
50% mass (L/D = 1)	2.5	12.77	−41.7
50% mass (L/D = 0.5)		13.33	−39.2
150% mass (L/D = 1)	7.5	27.29	24.5
150% mass (L/D = 1.5)		27.02	23.3
200% mass (L/D = 1)	10.0	32.85	49.9
200% mass (L/D = 2.0)		36.14	64.9
L/D = 2	5.0	18.54	−15.4
Frustum		23.23	6.0
Air gap		23.02	5.0
Steel back		36.72	67.6
Al back		33.27	51.8
Al driver		16.76	−23.5

of various driver section (e.g., geometry and material) and explosive charge (e.g., geometry and size) modifications on the downstream loading profile for an explosively driven 10° conical shock tube. The study was conducted to investigate the influence of these parameters on the air blast 2.0 m downstream of the explosive charge and to determine shock tube driver modification options that might be used to tailor the air blast arriving downstream at a test location. This work will be used to inform future design choices required when tailoring the blast wave generated by the explosively driven conical shock tube.

Bibliography

1. H. Davy, Philosophical Transactions of the Royal Society of London. **106,** 1 (1816)
2. Courtney, A.C., Andrusiv, L.P., Courtney, M.W.: Oxy-acetylene driven laboratory scale shock tubes for studying blast wave effects. Rev. Sci. Instrum. **83**(4), 045111 (2012)
3. Stewart, J.B.: Considerations for Explosively Driven Conical Shock Tube Design: Computations and Experiments (ARL-TR-7953). Army Research Laboratory, Aberdeen Proving Ground (2017)
4. Stewart, J.B., Pecora, C.: Explosively driven air blast in a conical shock tube. Rev. Sci. Instrum. **86**(3), 035108 (2015)
5. Nichols III, A.L.: An Arbitrary Lagrange/Eulerian 2D and 3D Code System, Version 4.24.x, Volumes 1 and 2; LLNL-SM-662355. Lawrence Livermore National Laboratories, Livermore (2014)
6. Friedlander, F.G.: The diffraction of sound pulses. I. Diffraction by a semi-infinite plate. Proc. Roy. Soc. Lond. A. **186**, 322–344 (1946)
7. Dewey, J.M.: The Shape of the Blast Wave: Studies of the Friedlander Equation. In Proceedings of the 21st International Symposium on Military Aspects of Blast and Shock (2010)
8. D.W. Hyde, Microcomputer programs CONWEP and FUNPRO, applications of TM 5-855-1, fundamentals of protective design for conventional weapons (user's guide). Tech. Rep. WES/IR/SL-88-1, Army Corps of Engineer Waterways Experiment Station Structures Lab, Vicks-burg, MS (1988)

Chapter 23
Unidirectional Carbon-Epoxy Composite Plates Subjected to Extreme Marine Environment

C. Javier, J. LeBlanc, and A. Shukla

Abstract An experimental investigation was conducted to understand the dynamic behavior of Carbon-Epoxy (CE) composite plates after prolonged exposure to accelerated aging conditions. Specimens were submerged in an elevated temperature deionized water bath of 3.5% salt content for 0, 35 and 70 days, simulating saline marine environments. Two cases of unidirectional carbon fiber orientation were studied; composite plates of orientation [0°, 90°] symmetric, and [±45°] symmetric. Blast experiments were performed on bi-axially clamped specimens using a shock tube apparatus to investigate the dynamic behavior of the material. All experiments were conducted for each exposure time and fiber orientation scheme.

Keywords Weathered composites • Accelerated aging • Material degradation • Shock loading • Digital image correlation

23.1 Introduction

The marine community has an interest in using composite materials for the construction of underwater vehicles and several components for off-shore structures due to their high strength to weight ratio and increased corrosion resistance. During operation, these structures are subjected to aggressive marine environments that can degrade their performance over time. The diffusion of water into an epoxy matrix has been studied extensively [1], as well as the effect of salt water on composite materials [2]. The effect of water diffusion into an epoxy matrix is of particular concern when composite structures are subjected to dynamic loading. Therefore, there is a need to investigate how composite materials respond to shock events once the material has been exposed to marine environments.

23.2 Experimental Details

A total of four plies of [0°, 90°] carbon fiber, and [±45°] carbon fiber were used to manufacture the CE plates. The plates were submerged in a 3.5% salt water solution at 65 °C for 0, 35 and 70 days, in order to simulate marine environments. The temperature was maintained at 65 °C to accelerate water diffusion into the matrix. After exposure to the salt water bath, specimens were bi-axially clamped and subjected to a controlled and concentrated air blast using a shock tube. Figure 23.1 shows a schematic of the experimental setup.

The shock tube is 8 m in length and is composed of four separate sections: driver section, driven section, converging conical section, and reduced diameter muzzle. A Mylar diaphragm separates the driver and driven sections. The driver section is pressurized using helium gas. Under critical pressure, the diaphragm bursts, releasing a pressure wave. The pressure wave develops into a shock wave as it travels down the length of the shock tube. The shock wave impacts the specimen and the pressure from the impact is reflected back into the muzzle. The reflected pressure is the loading that the specimen experiences. Two high speed cameras were used to employ 3D DIC and track the out of plane displacement of the composite panels during shock loading. Images were taken at 20,000 frames per second.

C. Javier (✉) • A. Shukla
Dynamic Photo-Mechanics Laboratory, Department of Mechanical, Industrial and Systems Engineering University of Rhode Island, Kingston, RI, 02881, USA
e-mail: carlos_javier@my.uri.edu

J. LeBlanc
Naval Undersea Warfare Center, Newport, RI, USA

J. Kimberley et al. (eds.), *Dynamic Behavior of Materials, Volume 1*, Conference Proceedings of the Society for Experimental Mechanics Series, DOI 10.1007/978-3-319-62956-8_23

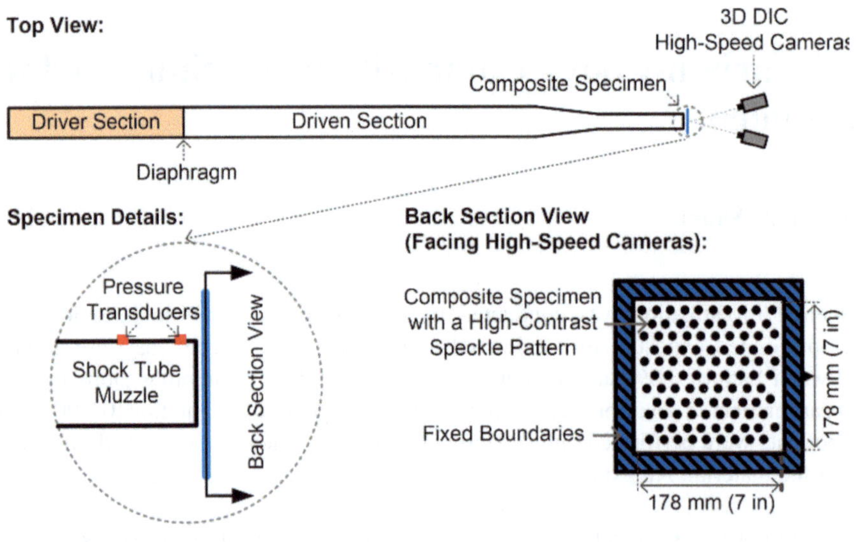

Fig. 23.1 Top view of shock tube experimental setup

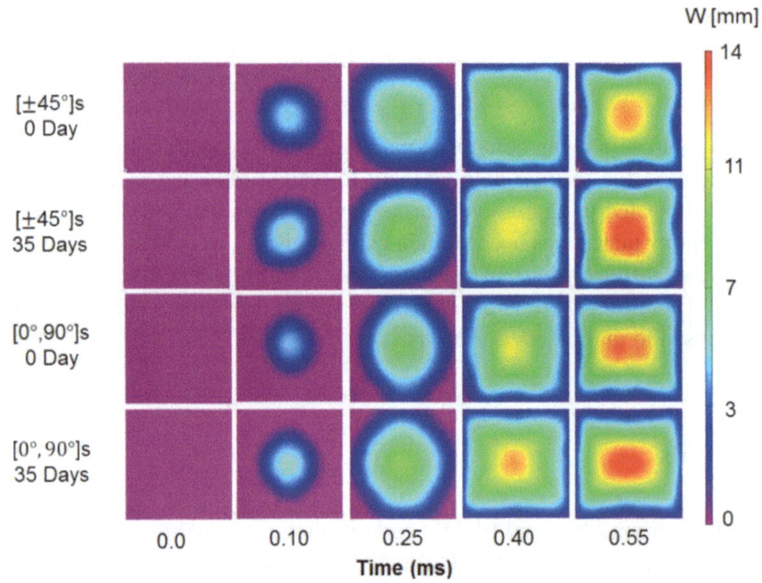

Fig. 23.2 Full field out of plane displacement evolution for the 0 and 35 day weathering cases, for both fiber orientations

23.3 Results

Due to similarities between the 35 and 70 day exposure time, only the 35 day case is presented. The 35 day weathered specimens displayed dramatically different behavior after being subjected to a dynamic shock loading, as seen in Fig. 23.2. The shock wave arrives at 0.10 ms, at which point the specimens begin to observe out of plane displacements. At 0.25 ms, the area of deflection has increased for all cases. It can be seen that the area of deflection extends in the direction of the fiber orientation. At 0.40 ms, the area of deflection has propagated to most of the CE plate. However, the 35 day weathering case for the [0°, 90°] plates developed a region of high deflection at the center of the specimen. Maximum out of plane displacement occurs at 0.55 ms. The maximum out of plane displacement occurs at the center of each specimen, where the shock tube is positioned. The 35 day weathering case for the [0°, 90°] plates exhibited similar out of plane displacement than the [±45°] fiber orientation case. Due to similarities between the 0 and 35 cases for both fiber orientations, it is concluded that fiber orientation did not play a major role in the diffusion of water into the epoxy matrix.

Acknowledgements The authors acknowledge the financial support provided by NEEC (NUWC and NSWC) under contract number N00174-16-C-0012 and the Naval Undersea Warfare Center, Newport RI.

References

1. Powers, D.A.: Interaction of water with epoxy. Sandia Laboratories, Albuquerque (2009)
2. Kootsookos, A., Mouritz, A.P.: Seawater durability of glass- and carbon-polymer composites. Compos. Sci. Technol. **64**, 1503–1511 (2004)

C. Javier is a PhD student at the University of Rhode Island working on the dynamic response of composite materials after material degradation due to extreme marine environments.

Chapter 24
Hybrid Computational and Experimental Approach to Identify the Dynamic Initiation Fracture Toughness at High Loading Rate

Ali Fahad Fahem and Addis Kidane

Abstract In this work, a new hybrid computational and experimental method is proposed which can be used to extract the dynamic fracture parameters in engineering materials at high loading rate. Torsional Hopkinson Bar is used to load an aluminum spiral notched specimen in mode I fracture configuration at different loading rate. A tubular specimen with spiral crack at 45° made of Aluminum 6061-T6, was used for the experiment. Using the experimental measured torque value as input computation analysis is performed to identify the dynamic initiation fracture toughness of the material. For the computational analysis, finite element is implemented in Abaqus. The result show that the dynamic stress intensity factor is loading rate sensitive, giving $K_{Ic}^{d} \cong (1.2 \mp 0.2)\, K_{Ic}$. The crack initiation time is also influenced by the rate of loading.

Keywords TSHB • Dynamic initiation fracture toughness • High loading rate • 3D–digital image correlation • FEM

24.1 Introduction

Intermediate and high strain rate properties of materials are a subject of interest in the mechanical researcher community for long time. These types of loading commonly exists in most materials used for day-to-day activities. The behavior of most engineering materials changes with strain rate [1]. So far, the mechanics community have developed computational and experimental methods to identify the behavior and properties of materials with a wide range of strain rates. Fracture mechanics is one branch of solid mechanics which started with the work of Griffith in 1920 [2]. Today, there are few successful methods to estimate the fracture toughness of materials, K_{Ic}, at quasi-static condition. On the other hand, determining the dynamic fracture initiation toughness of materials under dynamic loading conditions is still a great challenge. Currently, there is no a single standard method to identify the dynamic fracture toughness of materials, though there are many mathematical models and analysis for dynamic fracture of materials [3–6].

Recently, a new approach has proposed to evaluate the quasi-static and the dynamic fracture toughness of materials by loading under pure torsion load [7–10]. One of the main challenge with the technique is, there is no a closed form equation that can be used to extract fracture parameters directly from experimental results. Instead, a combined experimental and numerical method is adapted to measure and extract the fracture initiation toughness of materials.

In this work, a recent result on the dynamic fracture toughness $KI_{C,}^{d}$ of aluminum obtained from the proposed method is presented. An aluminum specimen with 45° spiral crack was loaded under pure torsion using a torsional split Hopkinson apparatus. The torsional load creates a mode I fracture loading on the crack line that is pre-notched at 45°. The torque measured in the experiment was used as input for FE model performed in ABAQUS and the fracture parameters were extracted.

A.F. Fahem (✉)
Department of Mechanical Engineering, University of South Carolina, Columbia, SC, 29208, USA

Department of Mechanical Engineering, University of Al-Qadisiyah, Al-Diwaniyah, Qadisiyah Province, Iraq
e-mail: afahem@email.sc.edu; fahem.usc@gmail.com

A. Kidane
Department of Mechanical Engineering, University of South Carolina, 300 Main Street, Columbia, SC, 29208, USA
e-mail: kidanea@cec.sc.edu

J. Kimberley et al. (eds.), *Dynamic Behavior of Materials, Volume 1*, Conference Proceedings of the Society for Experimental Mechanics Series, DOI 10.1007/978-3-319-62956-8_24

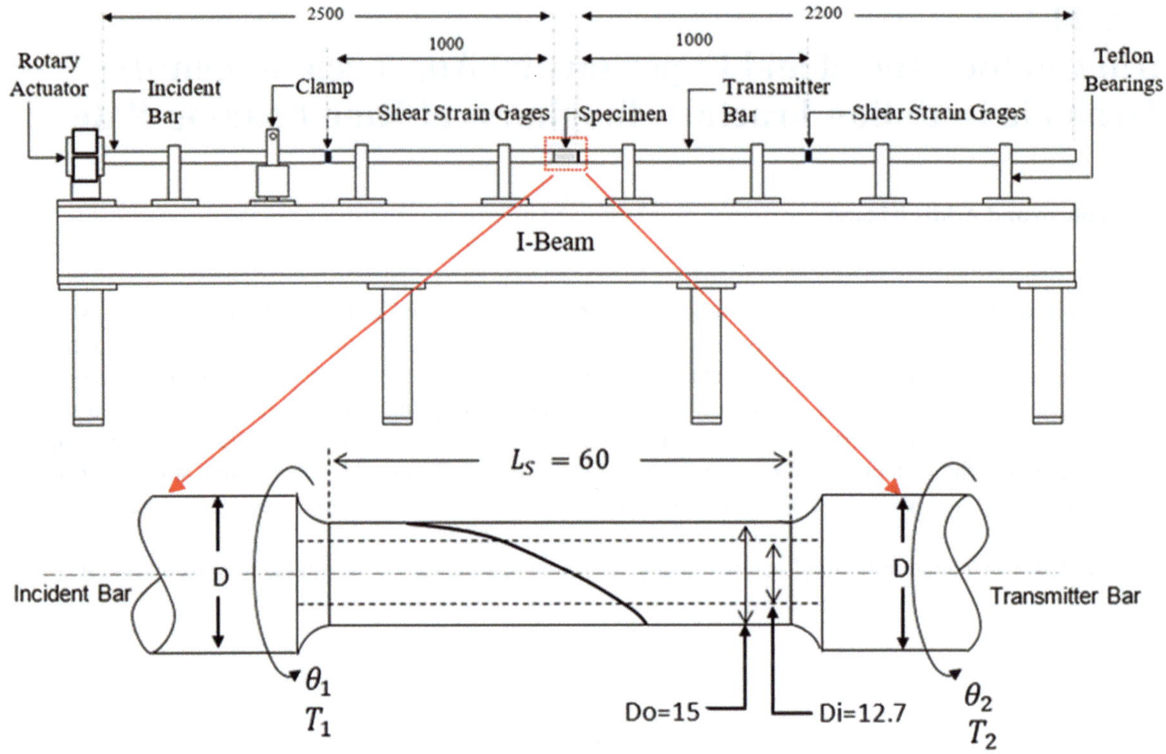

Fig. 24.1 Schematic of torsional split Hopkinson bar (TSHP) and spiral notch specimen

24.2 Experimental Setup

A Torsional Split Hopkinson Bar (TSHB) shown in Fig. 24.1, is used to generate dynamic torsional load. The details of the device and its working principle are available in literature [11, 12], however for completeness, the overview of the details will be presented.

The TSHB apparatus consists of incident and transmitter bars made of Aluminum 7075-T651, clamping mechanism and a torque pulley. The entire system is supported on an I-beam frame. The clamp system is used to hold the input bar at a specified position and store torsional elastic energy between the clamp and the rotary actuator. The rotary actuator determines how much elastic wave can be generated and hence determines the strain rate and applied stress on the specimen. In order to generate the dynamic torsional load, the clamp system will be released suddenly, allowing the stored torsional load to propagate along the bar through the specimen. Strain gages positioned at a mid-length of the incident and transmitter bars will record the incident, transmitted and reflected strain waves. In addition, 3D digital image correlation is used to measure the full field deformation of the specimen directly.

A tubular Aluminum 6061-T6 specimen with a 45° spiral crack shown in Fig. 24.1 was used for the experiment. The specimen has outside diameter of 18.5 mm, inside diameter of 12.7 mm and gage length of 60 mm. A spiral crack was made on the outer surface of the tube by using 4-axis CNC machine with a crack depth of 1.9 mm, crack ligament of 1 mm, and 45° with respect to the specimen axis

24.3 3D Digital Image Correlation

3D–Digital image correlation was used to measure full field deformation near the crack on the curved surface. Two high-speed cameras, SAX2 by Photon Inc., were used at a frame rate of 100,000 frames/sec. The region of interest around the crack face was speckled with white and black flat paint, which gave minimally reflective finish once dried as shown Fig. 24.2. This method will enable us to measure the deformation and strain on the specimen directly. Measuring the crack initiation time accurately has been difficult in dynamic fracture experiment [13]. The displacement and strain fields obtained for the DIC are used to measure the crack initiation time accurately.

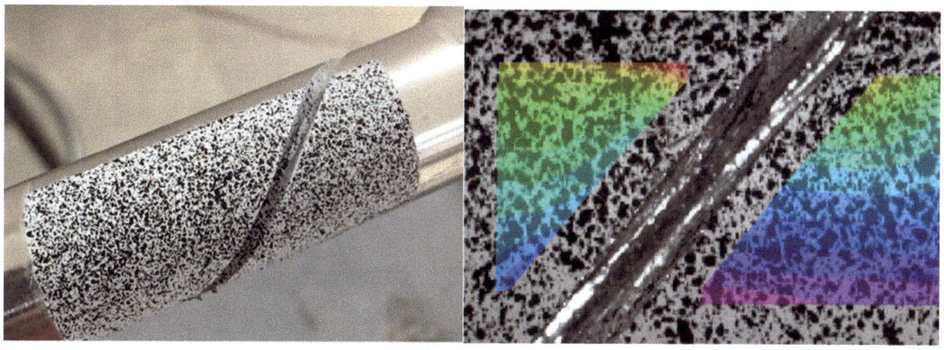

Fig. 24.2 Typical speckle pattern and strain field around the spiral notch of the specimen

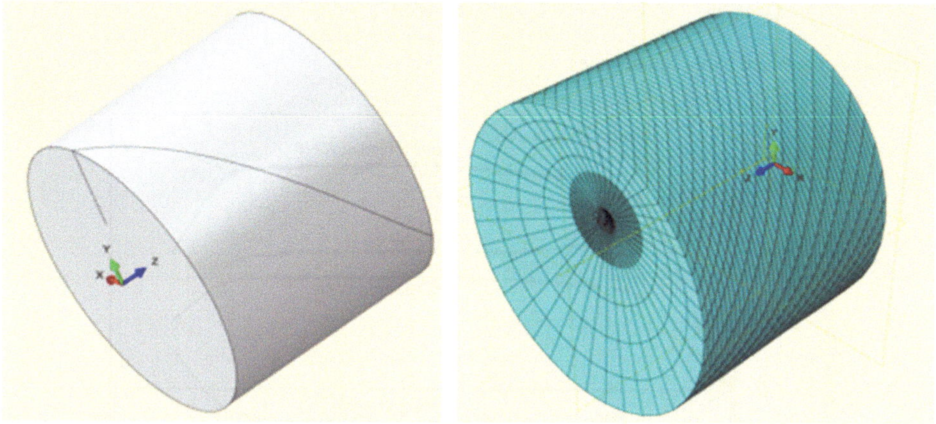

Fig. 24.3 Finite element model

24.4 Finite Element Setup

All the stress and strain waves measured experimentally were measured on the outer surface of the specimen and it is difficult to measure the local stress close to the crack tip in this type of specimen. So, the Finite element analysis was used to get the stress and strain near the crack tip and extract the fracture parameters. A three-dimensional element, C3D20H, was used as shown in Fig. 24.3. Since the torsional load is constant along the gage length, only the quarter of the gage length was modeled with 90° spiral crack to reduce the number of elements. The modeling parameters and information such as, boundary conditions, number of elements, element type, etc. are available for an interested reader with extensive details in Fahem 2016 [8].

24.5 Results and Discussion

Typical strain signals measured by the strain gages attached on the bars are shown in Fig. 24.4. Incident, reflected and transmitted strain waves are extracted and shown in Fig. 24.5.

The strain signals shown in Fig. 24.5. are used to calculate the torque applied on the specimen by using one-dimensional wave equations shown in Eq. (24.1). The effective impulse torque used in the analysis is the average of the torque on a both side of the specimen given by Eq. (24.1).

$$T_1(t) = \frac{2GJ}{D}\left(\gamma_I(t) + \gamma_R(t)\right), \quad T_2(t) = \frac{2GJ}{D}\left(\gamma_T(t)\right) \tag{24.1}$$

The torque calculated from the incident signals using Eq. (24.1) are shown in Fig. 24.6.

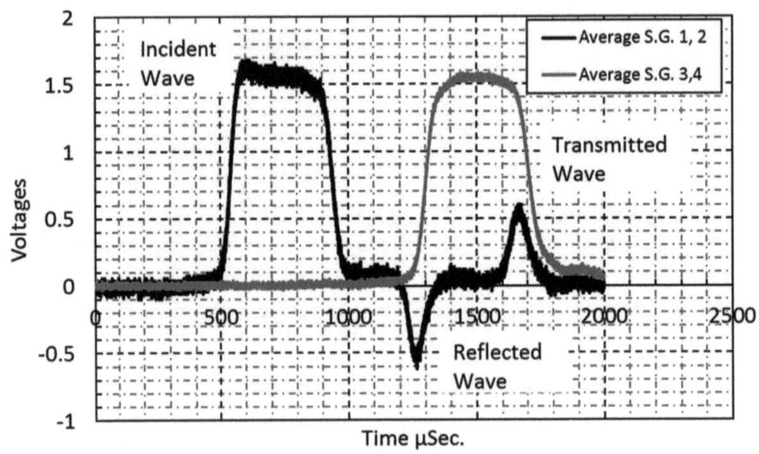

Fig. 24.4 Typical waveform as measured by the strain gages

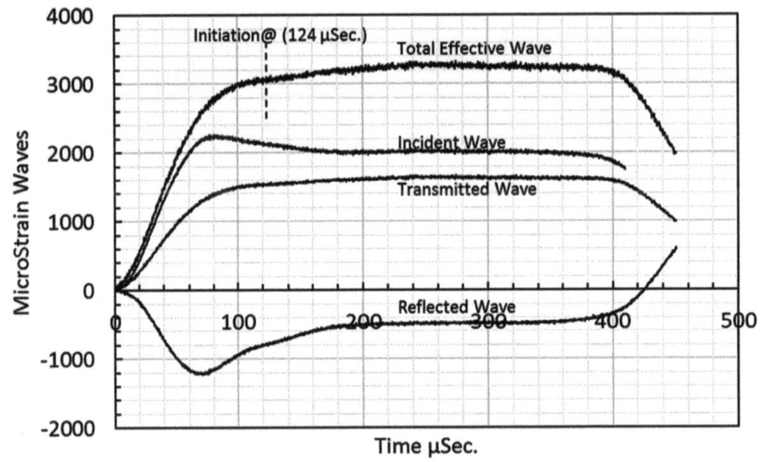

Fig. 24.5 Merge of a typical incident, reflected and transmitted strain

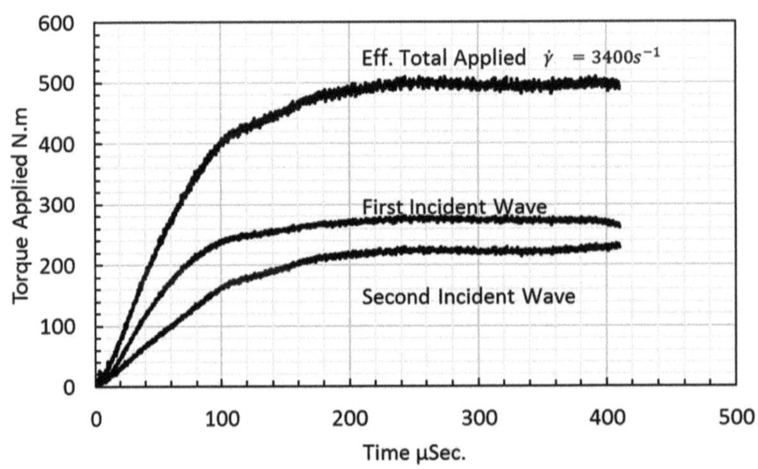

Fig. 24.6 Torsional impulse load measured

Using the torque value presented in Fig. 24.6, as input data, the fracture parameters are extracted using FM model discussed above. Overall, as shown in Fig. 24.7, the shear stress distribution in the mid layer of a cross section area is non-liner. Also, the stress is symmetric about a line connecting the bar center with the crack tip.

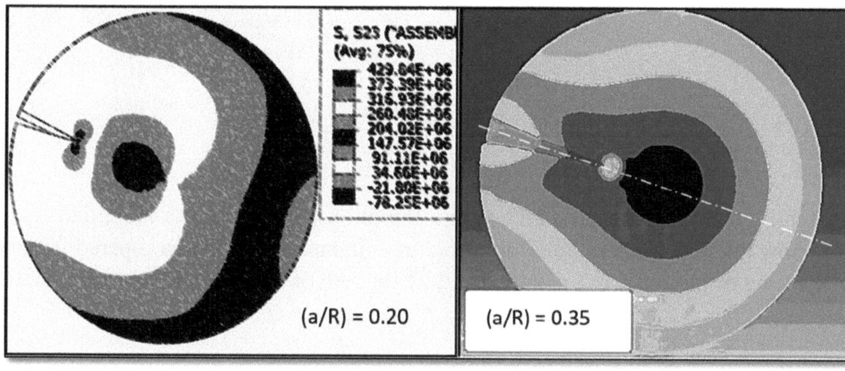

Fig. 24.7 Stress distribution near the crack tip for two different crack size configurations of Aluminum 6061

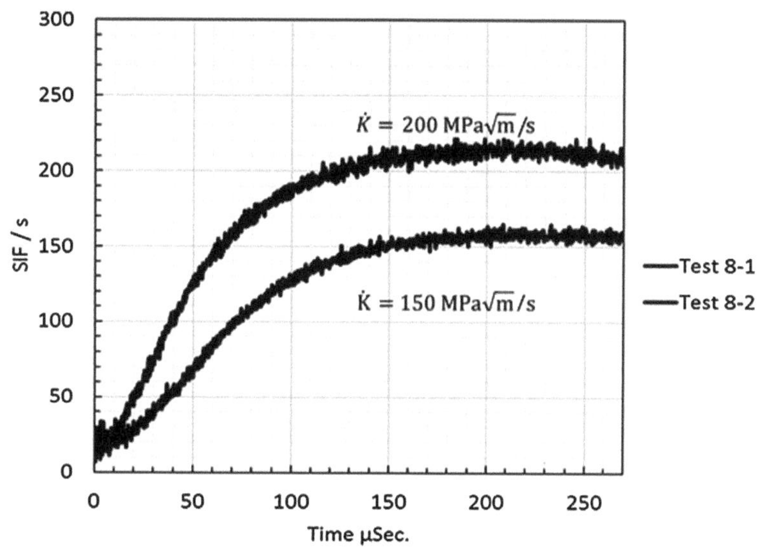

Fig. 24.8 Stress intensity factor- time for aluminum 6061

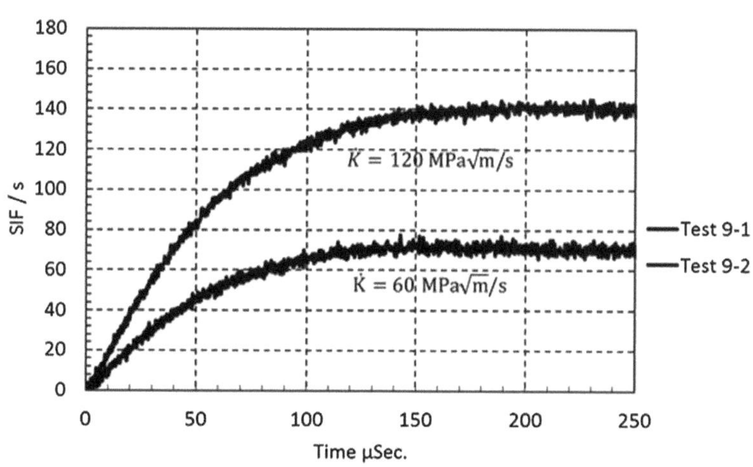

Fig. 24.9 Stress intensity factor- time for aluminum 6061

Figures 24.8 and 24.9 shows the stress intensity factor for different experiment conducted at different rate of loading. As clearly shown, the stress intensity factor is strain rate sensitive, increasing with loading rate increase.

Table 24.1 Loading rate and dynamic initiation fracture toughness Alu. 6061_T6

$\dot{K}$ $(MPa\sqrt{m})/s$	t_f μSec.	K_{Ic}^d $(MPa\sqrt{m})$	(K_{Ic}^d/K_{Ic})
120	110	32	1.1
150	170	36.1	1.245
200	160	43.492	1.5

From the DIC results and the DSIF's trend, the Dynamic Initiation toughness was identified for each loading rate. In general, the experiments show that the dynamic initiation fracture toughness is higher compared with the quasi-static value, $K_{Ic}^d \cong (1.2 \mp 0.2)\, K_{Ic}$. The time of initiation (t_f) is influenced by the rate of loading too, as shown in Table 24.1.

24.6 Conclusion

A dynamic mode I fracture of Aluminum 6061-T6 is investigated using a Torsional Hopkinson Bar and spiral notch specimen. The torque applied on the specimen is measured by using the well-known wave equation in a Hopkinson Bar. The torque is then used as input for the FM model. The fracture parameters are extracted directly from the FE model. A digital image correlation method was used to measure a full-field displacement and identify the fracture time It was demonstrated that the proposed hybrid numerical-experimental approach is visible to measure the mode I fracture initiation toughness of materials under dynamic loading conditions. It was noticed that the aluminum material used in this work was rate sensitive, showing high stress intensity factor at a higher loading rate.

References

1. Meyers, M.A.: Dynamic Behavior of Materials. Wiley, New York (1994)
2. Griffith, A.A.: The phenomena of rupture and flow in solids. Phil. Trans. Roy. Soc. London. Ser. A. **221**, 163–198 (1920)
3. Knauss, W.G.: CI of T, Williams ML (Universit. of P (1985) Dynamic Fracture. doi: 10.1016/B978-008044352-2/50011-9
4. Freund L.B.: (Brown U). Dynamic Fracture Mechanics. Cambridge University Press (1990). doi: 10.1007/s13398-014-0173-7.2
5. Morozov, N., Petrov, Y.: Dynamics of Fractture. Springer (2000). doi: 10.1007/978-3-7091-1777-4
6. Ravi-Chandar, K.: Dynamic Fracture. University of Texas, Austin (2004). doi:10.1016/B978-008044352-2/50011-9
7. Wang J-A., Oak R., Lin K.C Oak R.: Fracture toughness determination using spiral-grooved cylindrical specimen and pure torsional loading. United States Pat Appl Publ. doi: US2010/0311130Al
8. Fahem, A.F., Kidane, A.: A general approach to evaluate the dynamic fracture toughness of materials. Strain. **1**, 229–236 (2016). doi:10.1007/978-1-4614-0216-9
9. Kidane, A., Wang J-A: New method for dynamic fracture toughness determination using torsion hopkinson pressure bar. Conf Proc Soc Exp Mech Ser. (2013). doi:10.1007/978-1-4614-4238-7
10. Wang, J.-A., Kidane, A.: A new approach to determine the quasi-static and dynamic fracture toughness of engineering materials, dynamic behavior of materials. Proc Soc Exp Mech Ser. **1**, 545–551 (2013)
11. Weerasooriya, T.: THE MTL Torsional Split-Hopkinson Bar, p. 32. MTL TR 90-27, U.S. Army Materials Technology Laboratory, Watertown (1990)
12. Naik, N.K., Asmelash, A., Kavala, V.R., Ch, V.: Interlaminar shear properties of polymer matrix composites : strain rate effect. Mech Mater. **39**, 1043–1052 (2007). doi:10.1016/j.mechmat.2007.05.0
13. Kidane, Shukla, A.: Quasi-static and dynamic fracture initiation toughness of Ti/TiB layered functionally graded material under thermo-mechanical loading. Eng. Frac. Mec. **77**, 479–491 (2010)

Chapter 25
High-Rate Mechanical Response of Aluminum Using Miniature Kolsky Bar Techniques

Daniel T. Casem, Jonathan P. Ligda, Brian E. Schuster, and Shane Mims

Abstract The compressive stress-strain response of 99.999% pure aluminum at strain-rates ranging from 5 k/s to 1 M/s is presented. Experiments were conducted using optically instrumented miniature Kolsky bars. Samples were machined from 25 um thick rolled annealed foils using a laser mill and a focused ion beam. They were oriented such that the compression axis of each sample is along the direction of the foil thickness so that the final dimensions of each cylindrical sample was nominally 25 um long and 50 um in diameter. The grain size of the material, as well as its texture, was determined by EBSD. The results are compared to published pressure shear plate impact data on the same material at similar strain-rates. It is found that the mini-Kolsky bar data contains a large amount of scatter, which is attributed to the relatively large grain size compared to the small sample volume. In spite of this scatter, it is observed that the flow stresses measured using the miniature Kolsky bar methods are substantially lower than those measured with pressure shear plate impact even at similar strain-rates. Potential reasons for this discrepancy include texture effects in the samples and pressure hardening that is present in the pressure shear experiment but absent in the Kolsky bar.

Keywords Kolsky Bar • Split Hopkinson Pressure Bar • High strain rate • Micro-compression • Aluminum

25.1 Introduction

High strain-rate mechanical properties are typically obtained using the Split Hopkinson Pressure Bar, or Kolsky Bar, the operation of which is well documented (e.g., [1]). Common arrangements in compression use bars made of steel or aluminum with diameters between 12 and 25 mm, and can be used to obtain strain-rates in the range of 1000/s to 10,000/s. To achieve higher strain-rates, smaller bars must be used. The analysis of a sample deformed by a Kolsky bar depends upon the existence of one-dimensional (uniaxial stress) linear elastic wave propagation within the bars. In practical terms this means that the wavelengths of the stress waves in the bars need to be substantially greater than the bar diameter. For very high strain-rate experiments, these wavelengths are very short, and therefore require bars of suitably small diameters. In addition, requirements for a state of quasi-static equilibrium within the sample requires small sample sizes. This is because small samples equilibrate faster when loaded at a given strain-rate. Therefore small bar sizes are required as they are more compatible with small samples.

At the US Army Research Laboratory we have developed Kolsky bars with diameters as small as 127 um and have used them to test metallic samples with dimensions as small as 25 um at strain-rates exceeding 1 M/s. Interferometers are used for instrumentation to obtain the data normally obtained with strain gages which are impractical at these size scales. Because of the small size of the samples, we refer to these as micro Kolsky bars (uKB) but it is emphasized that the fundamentals of their operation are otherwise the same as conventional bars. These and related systems have been discussed previously [2–7].

D.T. Casem (✉)
US Army Research Laboratory, RDRL-WMP-C, Aberdeen Proving Ground, MA, 21005, USA
e-mail: daniel.t.casem.civ@mail.mil

J.P. Ligda
US Army Research Laboratory, RDRL-WMM-F, Aberdeen Proving Ground, MA, 21005, USA

B.E. Schuster
US Army Research Laboratory, RDRL-WML-H, Aberdeen Proving Ground, MA, 21005, USA

S. Mims
Army Educational Outreach Program, RDRL-WMP-C, Aberdeen Proving Ground, MA, 21005, USA

J. Kimberley et al. (eds.), *Dynamic Behavior of Materials, Volume 1*, Conference Proceedings of the Society for Experimental Mechanics Series, DOI 10.1007/978-3-319-62956-8_25

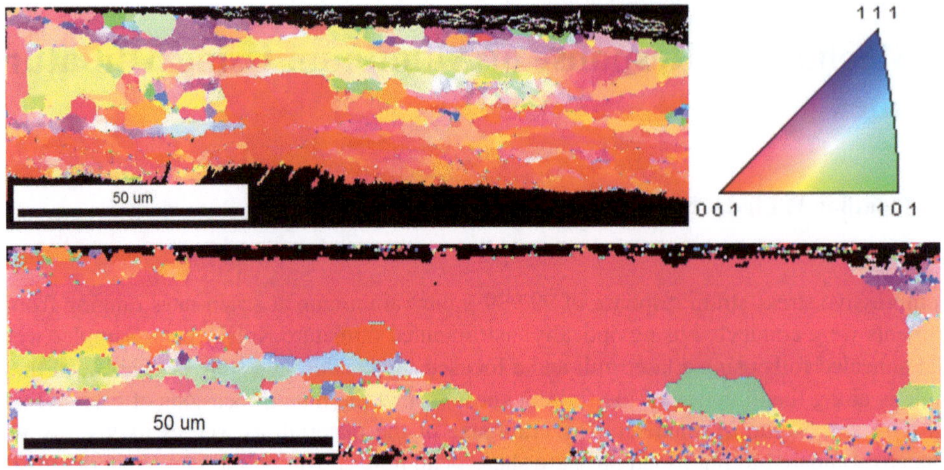

Fig. 25.1 EBSD scans of the edge of the foil (the thickness direction is vertical)

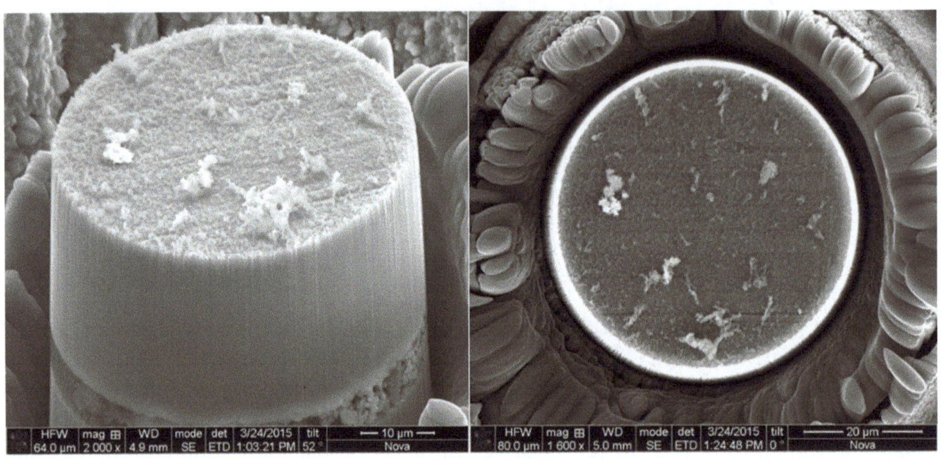

Fig. 25.2 SEM image of a typical sample. In this case, the height is 28.18 um and the diameter is 50.47 um

This paper presents a study on 99.999% pure aluminum. Strain-rates over a range of 5 k/s to 1 M/s were obtained using two different miniature bars; details of both are given below. This data is compared to data obtained on the same material using the Pressure Shear Plate Impact technique (PSPI) at overlapping strain-rates [8].

25.2 Material

The material studied here was in the form of nominally 25 um thick 99.999% pure annealed aluminum foil. Actual thickness was closer to 29 um. It was provided to us by the authors of [8], and was the remaining stock of material that they used for their pressure shear experiments. EBSD scans of the edge of the foil are shown in Fig. 25.1. The rolling texture and the large, elongated grains are apparent, with the lower image showing an exceptional case. Cylindrical samples were machined from the foil 50 um in diameter such that the axis of the cylinder (the compression axis) is along the thickness direction. Only the cylindrical surface was machined, i.e., the ends of the samples are the original surfaces of the foil faces, leading to a length to diameter ratio of about 1:2. Initial machining was completed using a femtosecond pulsed laser mill. The final stage of the machining was done with a Focused Ion Beam (FIB). A typical sample is shown in Fig. 25.2. Of particular note is the large grain sizes in relation to the sample size, which would be present in the samples in a random fashion (no characterization of the grain size or orientation within each individual sample was made).

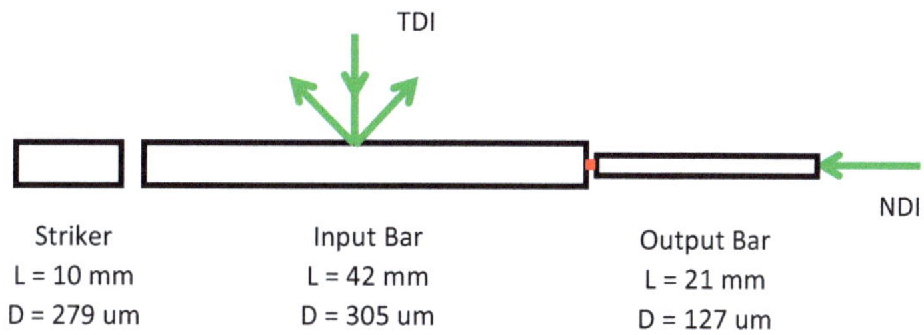

Fig. 25.3 The bar used for the high rate experiments (100 k/s to 1 M/s). The bars and striker are made from high strength tool steel and the aluminum sample is shown in red. Because of the small sizes, the bars are instrumented with interferometers. Not to scale

25.3 High-Rate Bar

Two uKBs were used for these experiments. The first was used for strain-rates between 100 k/s and 1 M/s, and is shown in Fig. 25.3 (not to scale). The bars and striker are made from high strength tool steel. Because of the small bar diameters, interferometers are used to make measurements of the stress waves in the bars rather than strain gages. A transverse displacement interferometer (TDI) is used at the mid-point of the input bar and a normal displacement interferometer (NDI) is used at the free-end of the output bar. This is explained in more detail in [4, 5]. The input bar is 42 mm long and 305 um in diameter. The output bar is 21 mm long and 127 um in diameter. The TDI requires a diffraction grating which is machined into the side of the input bar with a FIB. This requires a small flat on the side of the bar, which is polished along its entire length to avoid any impedance (area) changes along the bar. For this reason, the cross-sectional area of the input bar is slightly reduced, and is the reason for using a slightly undersized projectile (279 um diameter).

Conventionally the input and output bars have the same diameter (and impedance). This is not the case here, nor in the low rate bar discussed next, and requires some explanation. The input bar diameter of 305 um is considerably larger than the 127 um output bar. A small diameter (127 um) was preferred because the associated low impedance matched the samples in terms of their mechanical response; the bars possessed both the temporal resolution for the high strain-rates and the force sensitivity to measure the very low forces generated when compressing the sample. However, we encountered difficulties with projectile impacts at these small scales (see [7]). Much better alignment is achieved with projectiles/strikers around 300 um in diameter, and were still able to carry the high frequency signal components (rise times as small as 0.1 us were consistently measured) needed for the high rate experiments. The only downside of using the larger input bar is that we lose the ability to accurately measure the force between the input bar and the sample and experimentally check specimen equilibrium. However based on simulation results (not included here) we do not believe this to be an issue. We therefore rely only on the output bar for the measurement of specimen load (and hence specimen stress). A similar approach was taken in [9].

Positioning of samples and bars are achieved with various mechanical and piezoelectric stages and grippers. All samples are lubricated with vacuum grease.

25.4 Low-Rate Bar

Strain-rates in the range from 5 k/s and 100 k/s were achieved using the uKB shown in Fig. 25.4. Because of the lower strain-rates, longer duration loading pulses are needed, which requires longer bars and strikers. Because of the difficulty in machining long, small diameter bars precisely, a much larger 3.18 mm diameter, 305 mm long input bar was used. As above, we do not use this bar for a measurement of the force applied to the sample. To retain an adequately sensitive measure of force, a 283 um diameter, 102 mm long output bar is used. A smaller diameter output bar such as the 127 um bar used previously would have been preferred but was not available with the longer length needed for these experiments.

Because of the large diameter of the input bar, strain gages are used instead of the TDI on the input bar (Kyowa KSP-2-1 K-E4). Because of the small sample gage length, very low magnitude input pulses are needed to achieve strain-rates in the desired range, on the order of 40–60 u. For this reason, semi-conductor gages are used rather than standard foil gages; the former are approximately 70 times more sensitive than the latter.

The small diameter aluminum striker is used simply because it proved difficult to fire an input bar-matched striker (steel, 3.18 mm diameter) at the low-speeds necessary to generate the desired input pulses. In other words, it was easier to fire a

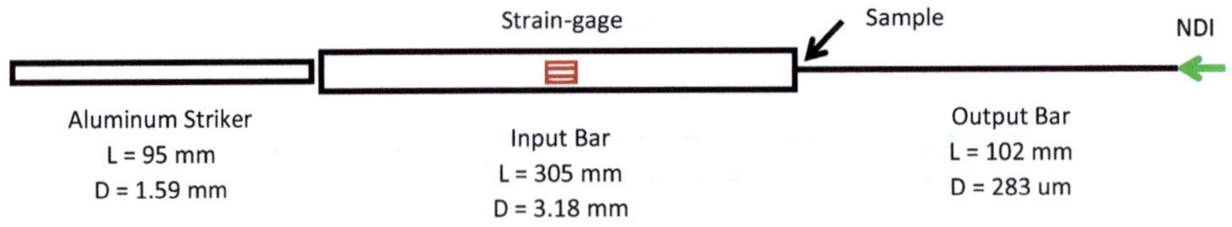

Fig. 25.4 The bar used for the "low" rate experiments (5 k/s to 100 k/s). The bars are made from high strength tool steel and the striker from 7075-T6 aluminum. Because of the larger input bar diameter, strain gages can be used. The sample is not shown. Not to scale

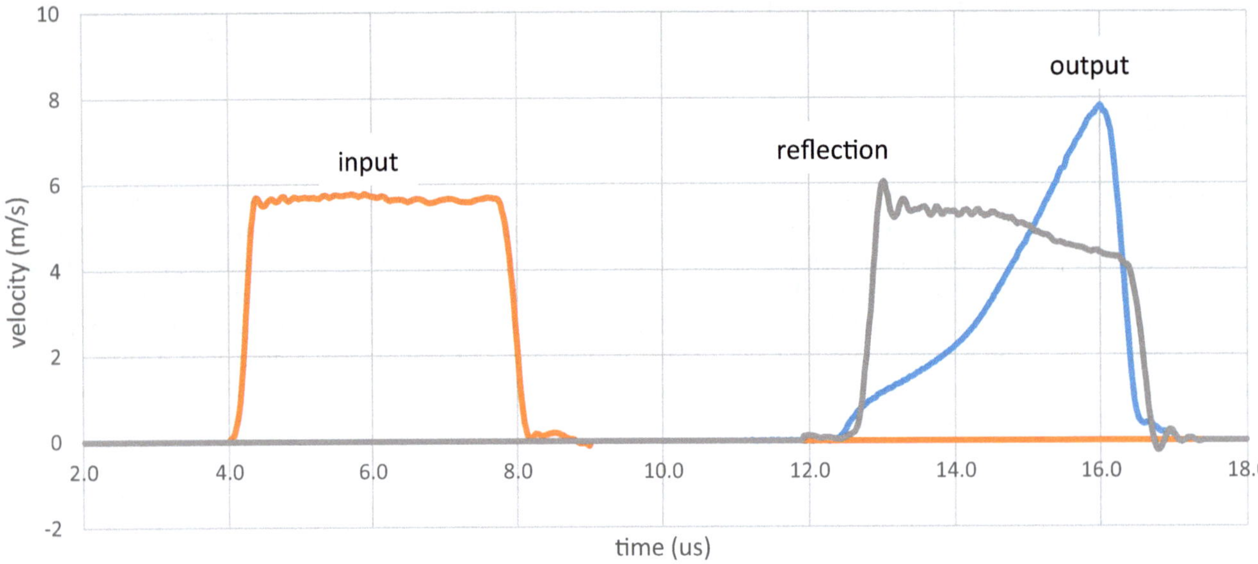

Fig. 25.5 Pulses measured for a test with the high rate bar. Note they are in terms of velocity, as measured by the interferometers (this is why the reflected pulse, which is tensile, is positive, i.e., particle motion in the same direction as that of the input pulse)

lower impedance striker at a higher speed than a higher impedance striker at a low speed. The fact that the low impedance striker rebounds from the input bar back into the gun barrel at a substantial portion of the firing velocity did not cause problems in our gun system. However, it should be considered by readers who may wish to try a similar approach.

25.5 Results

Results from a high-rate experiment, about 500 k/s, are shown in Figs. 25.5 and 25.6. Figure 25.5 shows the particle velocity measured by the interferometers. Positive velocities are downrange, i.e., in the direction the projectile is fired. These measurements are similar to typical strain measurements except that a tensile reflection is positive in the same sense as the other two pulses. This is not the case for strain measurements, and may appear unusual to readers accustomed to that data. The stress-strain curve, along with the strain-rate, for this sample is shown in Fig. 25.6.

Results from a low-rate experiment are shown in Figs. 25.7 and 25.8. Figure 25.7 shows the stress waves, in this case measured by strain gages on the input bar and an interferometer on the output bar. The velocity data of the interferometer has been converted to strain for easy comparison. (Note because of the diameter mismatch of the bars, we do not expect $\varepsilon_i + \varepsilon_r = \varepsilon_0$, which assumes bars of equal impedance; again this is mentioned for readers used to more conventional systems). The stress-strain data from this experiment, along with the rate, is shown in Fig. 25.8.

Figure 25.9 shows a summary of all of the uKB experiments performed on this material, using both bar arrangements, in terms of true stress at 0.2 true strain, plotted against rate of true strain at 0.2 true strain. Also plotted is data for 1100-O aluminum, using conventional servo-hydraulic and Kolsky bar methods on macro-scale (L = D = 5 mm) samples. This is included for comparison, as data in this range of strain-rates was unavailable with the 99.999% pure aluminum which is the subject of this study. A shear strength measured from [8], from a PSPI experiment at about 5 GPa, is also shown.

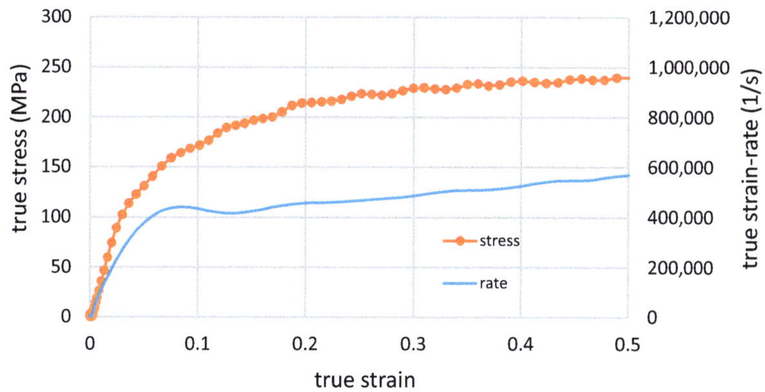

Fig. 25.6 Stress-strain curve for a 99.999% pure aluminum micro-compression sample using the high-rate bar

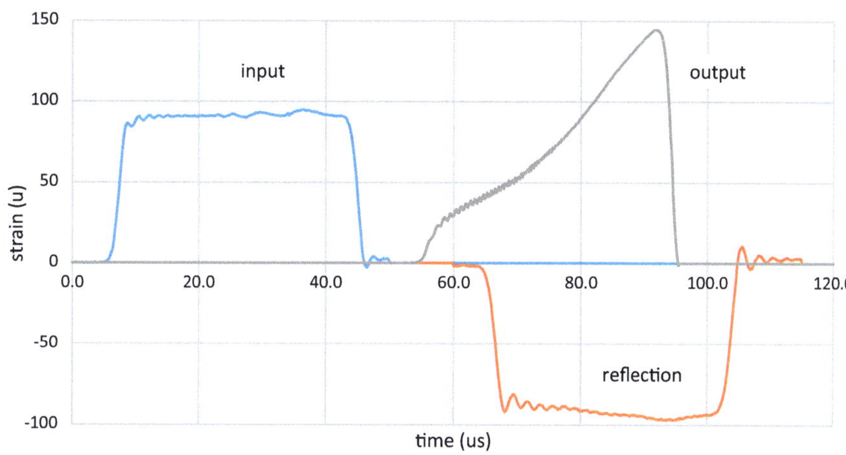

Fig. 25.7 Pulses from the low-rate bar. The input and reflected pulses are from the strain gages at the midpoint of the input bar, and the output pulse is from the NDI. The NDI more directly measures particle velocity of the free-end; here it is shown in terms of strain for comparison to the other two pulses

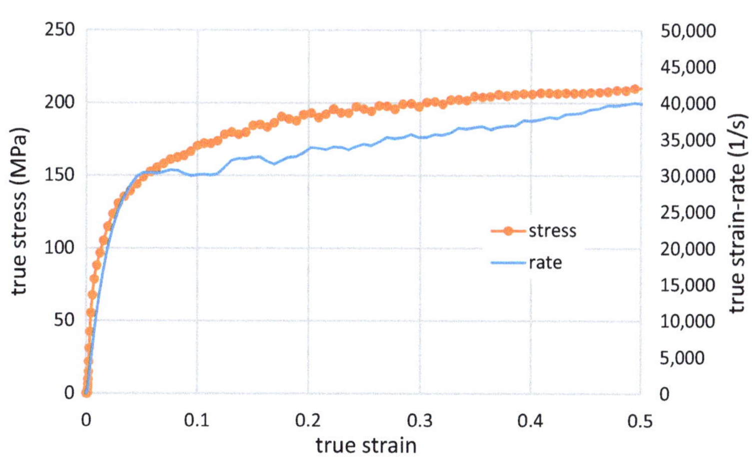

Fig. 25.8 Stress-strain curve for a 99.999% pure aluminum micro-compression sample using the low-rate bar

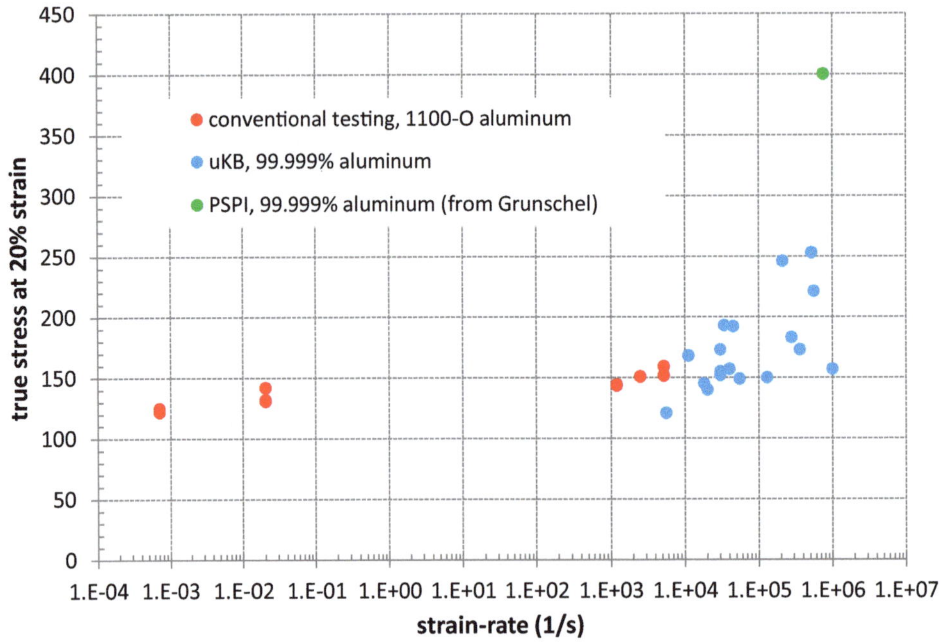

Fig. 25.9 Stress vs. strain-rate for 99.999% aluminum using uKB experiments and PSPI. Data for 1100-O aluminum, using conventional testing methods, is shown for comparison, along with a measurement from Grunschel et al. [8]

25.6 Discussion and Conclusion

A major goal of this research is to compare the results from uniaxial stress uKB experiments to those from the PSPI experiments, which has a large superimposed pressure, since similar strain-rates are achieved. This was the main reason for using the 25 um thick foil supplied by the authors of [8], since this was the same material used in the PSPI. The results clearly show (1) large scatter in the uKB data and (2) a large discrepancy in the shear strength between the two methods. The main problem with the uKB data is that overall the samples do not contain enough grains to be representative of continuum properties. The PSPI does not suffer from this problem because, although the foil may not contain many grains through its thickness, the large lateral extents of the PSPI specimen lead to an averaging of the response over many grains at the measurement location. This likely leads to the large amount of scatter in the uKB data, and makes any conclusion regarding the strain-rate hardening of this material from that data inconclusive. That being said, the large difference between the strength measured by PSPI, at 400 MPa, and the range measured by the uKB, between 150 and 250 MPa, at similar rates, cannot be ignored. There are reasons to expect this discrepancy, for example the orientation of the material as loaded in the different experiments, or a pressure hardening effect present in the PSPI (e.g., [10, 11]) but absent in the uKB. We are currently planning future experiments in this area that will use larger uKB samples, on the order of 150 um, possibly machined from bulk material instead of rolled foils. This should greatly reduce the scatter in the uKB experiments and lead to more reliable results.

Acknowledgments We are grateful to Professor Rodney Clifton, Tonia Jiao, and Steve Grunschel for their advice and useful comments, and for providing the material for these experiments.

References

1. Chen, W., Song, B.: Split Hopkinson (Kolsky) Bar. Springer, New York (2011)
2. Casem DT.: "A small diameter kolsky bar for high-rate compression," Proc. of the 2009 SEM Annual Conference and Exposition on Experimental and Applied Mechanics, Albuquerque, NM, 1–4 June 2009
3. Casem, D.T., Grunschel, S.E., Schuster, B.E.: Normal and transverse displacement interferometers applied to small diameter Kolsky bars. Exp. Mech. **52**(2), 173–184 (2012)

4. Huskins, E.L., Casem, D.T.: Compensation of bending waves in an optically instrumented miniature kolsky bar. J. Dyn. Behav. Mater, **1**(1), 65–69 (2012)
5. Casem, D.T., Zellner, M.: Kolsky bar wave separation using a photon doppler velocimeter. Exp. Mech. **53**, 1467–1473 (2013)
6. Avinadav, C., Ashuach, Y., Kreif, R.: Interferometry-based Kolsky bar apparatus. Rev. Sci. Instrum. **82**, 073908 (2011)
7. Casem, D.T., Huskins, E.L., Ligda, J.., Schuster, B.E.: A kolsky bar for high-rate micro-compression – preliminary results. Proc. 2015 SEM Annual Conference and Exposition, Costa Mesa, Ca, June (2015)
8. Grunschel, SE, Clifton, RJ, Jiao, T.: Shearing resistance of aluminum at high strain rates and at temperatures approaching melt, Shock Compression of Condensed Matter, 2011, AIP Conference Proceedings, Volume: 1426, Elert, ML. Buttler, WT. Borg, JP. et al., (2012)
9. Guo, Z., Casem, D., et al.: Transverse compression of two high-performance ballistic fibers. Text. Res. J. **85**(5), 502–511 (2016)
10. Huang, H., Asay, J.R.: Compressive strength measurements in aluminum for shock compression over the stress range of 4-22 GPa. J. Appl. Phys. **98**(3), (2005). doi:10.1063/1.2001729
11. Vogler, T.J.: On measuring the strength of metals at ultrahigh strain rates. J. Appl. Phys. **106**(5), (2009). doi:10.1063/1.3204777

Chapter 26
Application of 3D Digital Image Correlation In Ballistic Testing

Phillip Jannotti and Brian E. Schuster

Abstract Digital image correlation has been identified as a promising technique for use in terminal ballistic research; however, the limited resolution of ultra-high-speed cameras requires coarse speckle patterns not consistently achievable by conventional methods like spray painting. Thus, a robust and automated speckling technique that can apply a speckle pattern with appropriate speckle size and density is desired. The current study outlines an efficient, repeatable means of applying a high-quality black-on-white speckle pattern to test specimens. The speckle patterns were designed digitally and reproduced on the samples using a 3-axis mill coupled to a permanent marker with a custom-made, spring-actuated marker holder. By automating the process, the resulting speckle pattern is significantly more consistent compared to traditional specking techniques like spray painting or using a permanent marker manually. It also allows for precise control of the speckle size, density, and speckle pattern randomness. The method outlined in this work is capable of easily achieving 0.2–4 mm speckles and can generate speckles at a rate of 1–2 speckles per second. Example patterns ranging in size from 10 × 10 mm to 200 × 200 mm are shown which required as little as 5–10 min to complete (each containing 600 speckles). High-velocity impact experiments were then performed on 6.35 mm thick aluminum plates that were speckled using the described technique to demonstrate the high-rate, localized deformation that takes place during an impact event. It emphasizes the extreme spatial and temporal difficulties of using digital image correlation for ballistic research and the need for a tailored speckle pattern.

Keywords Digital image correlation • DIC • Speckle pattern • Terminal ballistics • Deformation

26.1 Introduction

A critical need exists to understand the mechanical response of materials during impact events. The typical data collected during terminal ballistic research is the ballistic limit, or V_{50}, which represents the velocity at which a particular armor/material is defeated by a particular projectile 50% of the time. Little emphasis is commonly placed on extracting quantitative data which describes the material behavior or state during the impact. On the other hand, fundamental studies which thoroughly characterize the mechanical response of materials often involves quasi-static testing of simple sample geometries under ideal loading conditions. Such studies must draw correlations between observed failure mechanics and the actual ballistic response. Even dynamic mechanical testing, which examines materials at more relevant strain rates, is unable to probe the material deformation at stress states which reflect the inhomogeneous, multi-axial character of a ballistic event. This emphasizes the need to develop time-resolved impact experiments which identify the relevant physics and quantify the deformation behavior, e.g., displacement and velocity histories, strain evolution, etc.

Significant progress has been made in coupling in situ diagnostics with traditional ballistic testing. High-speed photography and flash X-ray radiography have become a standard technique employed to help visualize the real-time failure mechanics of armor materials and systems. Unfortunately, there is limited ability for quantification. More recently, photonic Doppler velocimetry (PDV) has become a popular point measurement tool for examining displacement- and velocity-time histories based on Doppler shifted laser light. Interpreting the data can sometimes be complicated due to the complex nature of the loading. Finally, digital image correlation (DIC), especially stereo or 3D DIC, has seen widespread usage in a variety of quasi-static and dynamic studies for more than a decade. It offers the ability to make full-field measurements of displacement, velocity, and strain. However, the application of 3D DIC to terminal ballistics has largely been limited by camera technology

P. Jannotti (✉) • B.E. Schuster
U.S. Army Research Laboratory, APG, MD, 21005, USA
e-mail: phillip.a.jannotti.civ@mail.mil

J. Kimberley et al. (eds.), *Dynamic Behavior of Materials, Volume 1*, Conference Proceedings of the Society for Experimental Mechanics Series, DOI 10.1007/978-3-319-62956-8_26

due to spatial and temporal concerns (resolution, frame rate, and record time) [1–3]. Currently, a number of ultra-high-speed CMOS camera systems, like the Shimadzu HPV-X2 and Specialized Imaging Kirana, have become popular options for DIC research boasting an adequate combination of resolution (400 × 250 pixels or 924 × 768 pixels), frame rate (5 Mfps), and record time (128 or 180 frames).

The basic concept of 3D DIC is to precisely track a speckle pattern with multiple cameras [4]. With proper knowledge of the relative positioning of the cameras in space, achieved through calibration, the deformation history of the pattern can be used to extract displacement, velocity, and strain evolution. In addition to advanced high-speed cameras, a high-quality, high-contrast speckle pattern is a necessity for DIC analysis. Typical speckle patterns consists of black dots on an even white background. The method used to apply the speckle pattern ranges widely and is largely considered an "art". The most popular method of speckling a target is by spray painting which can be quick and easy, however, the results are largely inconsistent and thus should be avoided if possible. This is due to variability in speckle size, poor contrast, unpredictable speckle sizes and speckle densities. Many advocate for the use of permanent markers to manually apply speckles to a material surface, but this is time-consuming and produces patterns that are not optimal or easily reproduced. Because different speckle patterns invariably affect measurement resolution and accuracy, methods of speckling which produce consistent and reliable results are preferred. Wherever possible, efforts should be made to remove the art (variability) aspect of speckling that makes pattern quality high dependent on the skill of the individual, and replace it with automation and design. Some improvements can be made with the use of control consistency with masks (or stencils) and 'stamp-like' rollers, but these methods can suffer from paint bleed and difficult adjustment of speckle parameters (size and density), respectively.

The focus of this work is to discuss a methodology. A 3-axis mill with a custom-made, spring-actuated permanent marker holder is used to produce speckle patterns which enables automated, high-quality, and consistent speckling of samples for producing high-quality speckle patterns that can be tailored for the specific requirements of the testing. A 3-axis mill with a custom-made, spring-actuated permanent marker holder is used to produce speckle patterns in an automated and repeatable fashion. The process of designing and applying the patterns is outlined, and example patterns are shown. Lastly, high-velocity impact experiments were performed on 6.35 mm thick aluminum plates which were patterned using the described method. The results demonstrate the spatial and temporal considerations when applying DIC to terminal ballistics, but also shows the utility of DIC.

26.2 Methods

A commercially-available 3-axis mill (Inventables X-Carve) was used to control a permanent marker, which was held in a custom-made fixture (see Fig. 26.1). The mill has a 1000 × 1000 mm X-Y work area and 65 mm Z- range. Also, there is one motor per axis controlled via a motor shield. The marker holder consisted of a polycarbonate tube which mounted

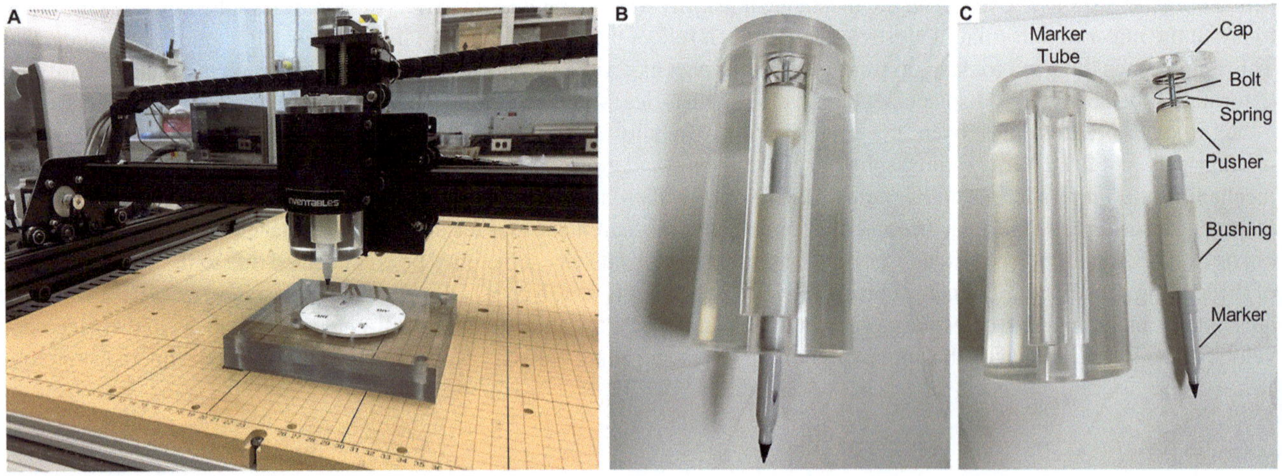

Fig. 26.1 (**a**) Image of automated speckle generator which is composed of a 3-axis mill with spring-actuated marker holder mounted to the Z-axis motor, (**b**) marker holder assembly and (**c**) an exploded view of the marker holder

to the Z-motor of the mill. The marker was press-fit into a polyethylene bushing, which allowed the marker to easily slide vertically in the tube, but with minimal free play to improve speckling accuracy. A second bushing, called a pusher, sat on top of the marker. A spring placed between the top cap and pusher allowed the marker dot sizes to be adjusted based on spring weight and pre-load, which varied the applied load on the marker and increased or decreased the speckle size. A bolt from the cap to the pusher was used to vary the applied preload on the spring by changing the gap between the two components. The plunge depth and rate as well as dwell time of the marker can also be manipulated to achieve the desired speckle size. Finally, permanent markers with varying tip sizes ranging from 0.2 to 4.0 mm were used to provide additional flexibility. Depending on the imaging system being used for the DIC experiment, it is possible to optimize the setup to produce ideally sized speckles [5, 6]. For example, if an appropriate speckle size is chosen to be approximately 3 pixels across, then for a camera sensor with 400 × 250 pixels and FOV of 40 × 25 mm, the speckle can be made 0.3 mm across.

Once the speckle size and parameters to produce a given size speckle were determined, a digital speckle pattern was made. A speckle generating software (Speckle Generator, Correlated Solutions) was used to produce a pattern of 50:50 black/white coverage. The digital pattern was improved through image processing and saved as a high-quality bitmap. The bitmap was converted into a toolpath (Aspire Software, Vectric Ltd.). Prior to applying the speckle pattern, the aluminum (6061-T6) test samples were scuffed using 600-grit SiC paper and cleaned thoroughly. The samples were then painted with matte white primer (Rust-Oleum(R) Professional Aluminum Primer). The samples were then mounted in the mill and the toolpath was sent to the motors via a G-code parser (GRBL, Open-Source program).

26.3 Results

Image analysis was used to assess the spatial resolution of the patterning technique and the resulting speckle size. An example of the image analysis method is shown for a speckle pattern with regular spacing of 2 mm which spanned approximately 50 × 50 mm (see Fig. 26.2). The spatial resolution was determined by autocorrelation. Figure 26.2b shows normalized 2D autocorrelation data which denotes the correlation of the reference image with a shifted copy of itself. When the speckles in the shifted copy overlap speckles in the reference image a sharp signal peak is observed. By determining the distance between the peaks, the speckle spacing was computed and compared with the expected spacing. The patterning spatial resolution was found to be around 3%, or as low as 50 microns. Lastly, when varying parameters to achieve a certain speckle size, the speckle size was assessed using a binary distance transform of the speckle pattern image. By converting the speckle image to binary (white background and black speckles), the distance transform gave the radial distance from edge to center. For example, the average speckle size in Fig. 26.2a was found to be 0.67 mm in diameter, as shown in 26.2c.

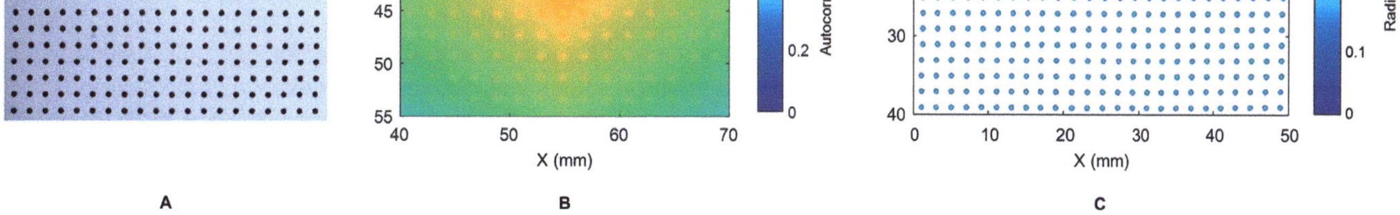

Fig. 26.2 Data analysis of a speckle pattern. (**a**) Black-on-white speckle pattern with 2 mm dot spacing, (**b**) autocorrelation plot, and (**c**) distance transform plot with inset showing a magnified view of the data

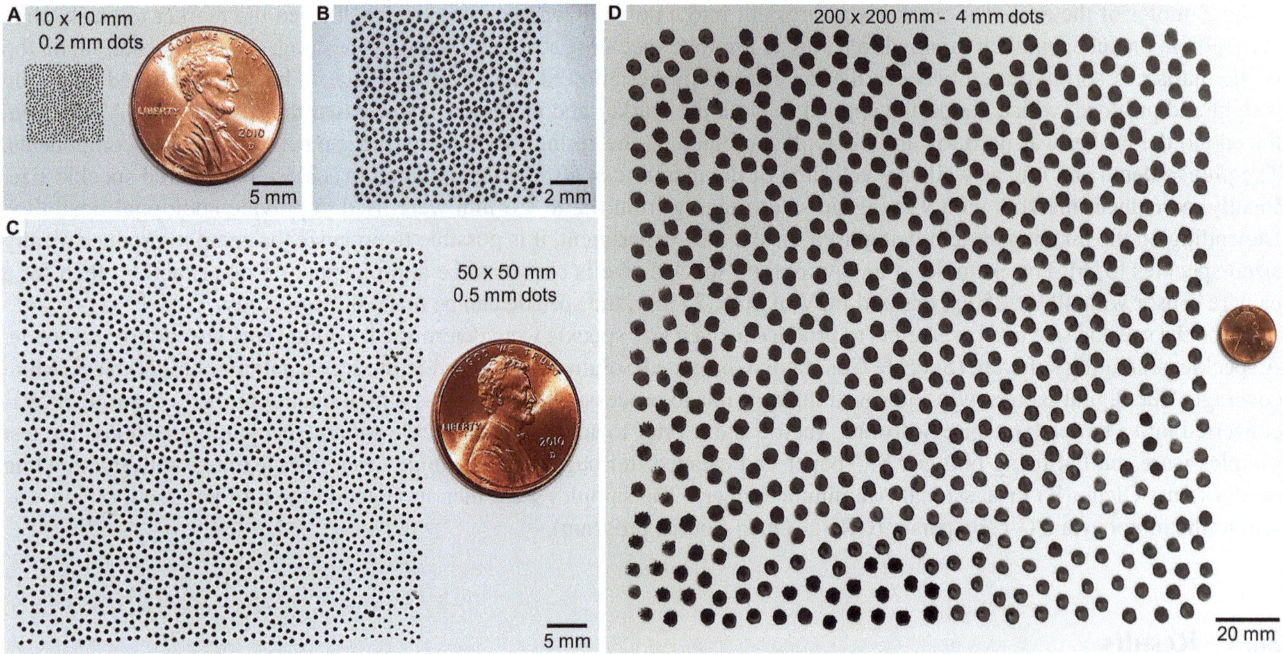

Fig. 26.3 Images of speckle patterns. (**a**) 10 × 10 mm pattern, (**b**) magnified view of 10 × 10 mm pattern, (**c**) 50 × 50 mm pattern, and (**d**) 200 × 200 mm pattern. In (**a**), (**c**), and (**d**) a U.S. penny is placed next to the patterns for size comparison

Once the patterning resolution and speckle size generation were understood, randomized speckle patterns were generated. Speckle patterns ranging from 10 × 10 mm to 200 × 200 mm are shown in Fig. 26.3. The patterns shown in Fig. 26.3a, d represent the smallest and largest patterns that can be reliably produced using a permanent marker. Figure 26.3c represents a typical pattern used when impacting targets with a 0.25–0.50-caliber projectile. A U.S. penny is shown for size comparison. Note that the patterns in Fig. 26.3a, c, d utilized a 0.4, 1.0, and 4.0 mm-tip marker, respectively. It was found that speckle sizes of 0.2–4.0 mm were achievable. In addition to enabling automated and repeatable speckle patterns, the technique described in this work allows for complex patterning. For example, in Fig. 26.4, five speckle patterns were applied to one sample, all centered on a planned test location. This allowed greater flexibility when designing experiments than previously possible. Another advantage is the ability to add supplemental markings in the speckle pattern, such as fiducials and lettering (see the inset in Fig. 26.4). The complexity of these patterns is not possible using manual patterning techniques.

In order to demonstrate the utility of the speckle patterning technique, terminal ballistic experiments were performed. The tests consisted of 6.35 mm thick aluminum plates impacted by 12.7 mm diameter copper spheres at 1200 m/s. For reference, the cameras used in this work were Specialized Imaging Kiranas recording at 2 Mfps equipped with Nikon 200 mm macro lenses. The cameras were placed at approximately 25° apart. Other details of the experiment are given in Table 26.1. Figure 26.5 shows selected high-speed images which illustrates the highly localized deformation that takes place in tens of microseconds. The bulge was 18 mm across prior to failure, and displaced nearly 9 mm in less than 15 microseconds. Beyond this point, several issues complicated the DIC analysis including limited depth-of-field (DOF) and specular reflection. Unfortunately, if the FOV and stereo angle cannot change, little can be done regarding the DOF. Perhaps, the penetrator/target dimensions can be scaled down slightly to reduce the maximum displacement. Specular reflection can be mitigated by implementing cross-polarization of the illumination source and the cameras. Specifically, a linear polarizer is placed on the light source, and another linear polarizer is placed on the camera lens rotated 90° relative to the light source polarizer. Because specular reflections maintain polarization, crossed linear polarizers eliminates glare.

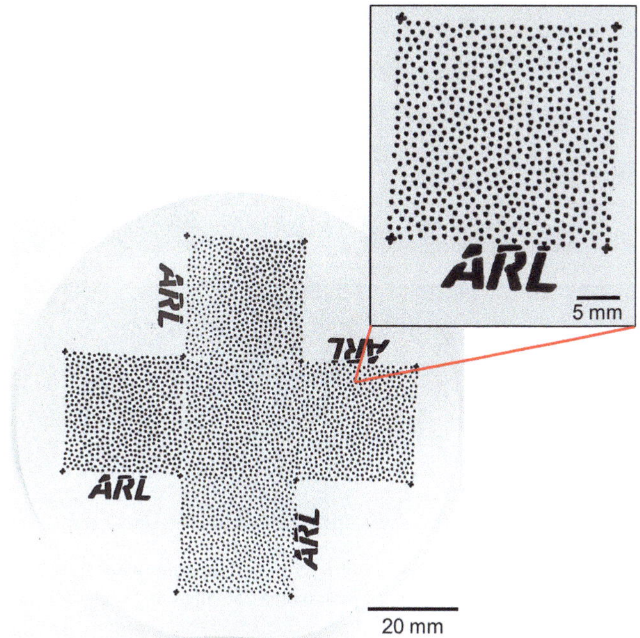

Fig. 26.4 Image of a metal disk which contains five individual patterns with fiducials and logos. The inset shows a magnified view of a typical 25 × 25 mm speckle pattern which includes fiducials placed at the corners of the pattern as well as a company logo all drawn by automated-patterning

Table 26.1 DIC report detailing specifics about the data analysis

Camera	Specialized Imaging Kirana
Sensor/digitalization	924 × 768/8-bit
Camera noise	0.9%
Lens/imaging distance (mm)	200 mm F-mount macro/1150 mm
Sampling rate	2 MHz
Pixel/mm	7 pixels = 1 mm
Speckle size (pixels)	5–7
ROI (mm)	132 × 110 mm
Speckles across bulge section	14
Subset/step	29/4
Interpolation/correlation criterion	Optimized 8-tap/NSSD
Pre-smoothing	None
Displacement noise	0.04 pixels, 5 μm

Figure 26.6 demonstrates the strain and strain rate at failure for the deforming 6.35 mm Al plate. Clearly the maximum von Mises strain occurred at the base of the rear surface bulge. This represents the point of greatest strain localization. The circumferential and radial tensile strains were maximum at the tip of the bulge, while the radial compressive strain was maximum at the base of the bulge. Also note the substantial strain rates experienced by the deforming metal plate during ballistic impact, reaching 10^4–10^5/s. This is several orders of magnitude above strain rates achieved by typical high rate mechanical testing techniques like Kolsky bar testing which is around 10^2–10^3/s.

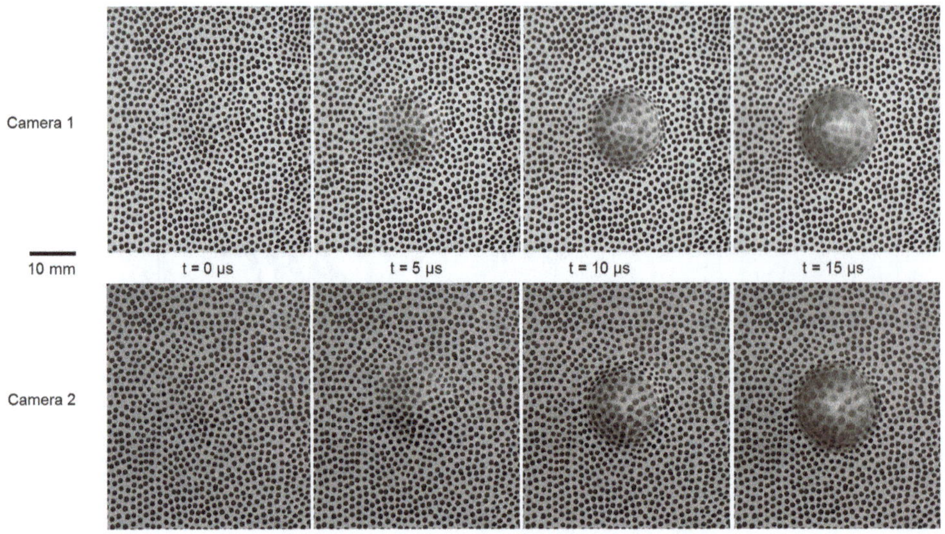

Fig. 26.5 Selected high-speed images acquired during penetration of 6.35 mm aluminum plates by 12.7 mm copper spheres fired at 1200 m/s. The images were acquired using Specialized Imaging Kiranas at 2 Mfps, with a 25° angle between each camera

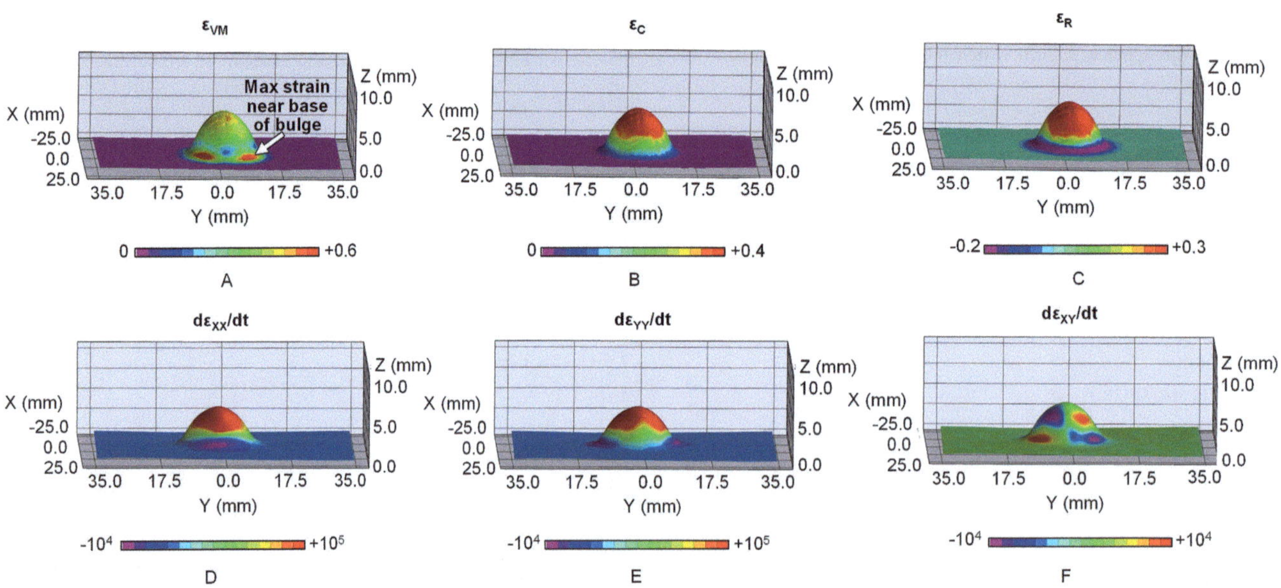

Fig. 26.6 Selected strain data collected by post-processing the speckle images acquired during the ballistic event at t = 10 us. (**a**) von Mises strain, (**b**) circumferential strain, (**c**) radial strain, strain rates in (**d**) X and (**e**) Y, and (**f**) shear strain rates

26.4 Conclusions

A methodology is presented for applying speckle patterns to samples intended for DIC experiments. A desired speckle size and pattern density can be generated in an automated fashion using a permanent marker coupled to a 3-axis mill by a spring-actuated holder. The speckling technique was found to be capable of rapidly applying speckles ranging in size from 0.2–10.0 mm with a lateral spatial resolution of approximately 50 microns. With those considerations in mind, the marker tip size, plunge depth and rate, spring preload, spring weight, and dwell time can all be manipulated to achieve a the designed speckle pattern. The resolution can be improved with a higher precision mill, but the observed resolution was found to be appropriate for the size and complexity of patterns used for most high-speed cameras and ballistic impact experiments.

References

1. Jannotti, P., Schuster, B., Doney, R., Walter, T., Andrews, D.: Instrumented penetration of metal alloys during high-velocity impacts. In: Dynamic Behavior of Materials, Volume 1, pp. 139–145. Springer International Publishing (2017)
2. Reu, P., Miller, T.: The application of high-speed digital image correlation. J. Strain Anal. Eng. Des. **43**(8), 673–688 (2008)
3. Kumar, P., Shukla, A.: Dynamic response of glass panels subjected to shock loading. J. Non-Cryst. Solids. **357**(24), 3917–3923 (2011)
4. Reu, P.: Introduction to digital image correlation: best practices and applications. Exp. Tech. **36**(1), 3–4 (2012)
5. Lecompte, D., Smits, A., Bossuyt, S., Sol, S., Vantomme, J., Van Hemelrijck, D., Habraken, A.M.: Quality assessment of speckle patterns for digital image correlation. Opt. Lasers Eng. **44**(11), 1132–1145 (2006)
6. Reu, P., Sweatt, W., Miller, T., Fleming, D.: Camera system resolution and its influence on digital image correlation. Exp. Mech. **55**(1), 9–25 (2015)

Phillip Jannotti received his BS and PhD in Mechanical Engineering from the University of Florida.
He works at US Army Research Laboratory.
His research focuses on examining the high-rate failure response of ceramics and metals through the use of in situ diagnostic techniques.

Chapter 27
Compaction Wave Propagation Characteristics in Polymer Bonded Explosives at Macro-Meso Scale

Suraj Ravindran, Peter Malchow, Addis Tessema, and Addis Kidane

Abstract Gas-gun direct impact experiments are performed to characterize the compaction wave propagation in polymer bonded explosives (PBX). The digital image correlation based experiment setup is developed to capture the full-field deformation field at a temporal resolution of 200 ns with a spatial resolution that can vary from 8 μm/pixel to 100 μm/pixel. The experiment is conducted at an impact velocity of 56 m/s for both macro and meso-scale. The compaction velocity is calculated from the macro-scale full-field measurement. From the displacement field, using inertia stress calculation, the stress across the specimen is calculated at the macro scale. The meso-scale deformation mechanisms associated with the dissipation of the compaction wave was identified to be the plastic deformation of the binder and fracture of the crystals.

Keywords Compaction • PBX • Direct impact • Shock

27.1 Introduction

Polymer bonded explosives (PBX) are the class of granular composites with high explosive solid crystals dispersed in a lean binder matrix. The impact-induced explosion of this material is a complex process that involves multiple length scale features [1–3]. It is believed that the failure initiation of this materials that lead to complete failure is started in the order of few microns. Therefore, full-field deformation measurements at high spatial and temporal resolution are essential in understanding the multiscale failure mechanisms that leads to such events.

Recently, high resolution-high speed deformation measurement techniques were developed with the help of digital image correlation (DIC) to obtain the displacement field at multiple length scales [4–8]. This experimental method is promising in measuring strain field at small length scales at high temporal resolution. With the help of this technique, recently it was shown that the strain localization plays an important role in the deformation behavior of polymer bonded explosives [9, 10].

In this study, results from direct impact experiments at macro and meso-scale is presented. The main objective of the work is to understand compaction wave propagation in polymer bonded sugar at multiple length scales. The compaction wave stresses are estimated using information from the macro-scale experiment and a recently developed non-parametric method [11–13]. In addition, the compaction wave velocity is calculated by locating the compaction front in the spatial stress profile.

27.2 Materials and Methods

In this study, polymer bonded sugar (PBS), a mechanical simulant of the polymer bonded explosives was used. The specimen constituents were sugar crystals and plasticized hydroxyl terminated polybutadiene (HPTB). Constituents proportion used in PBS is shown in Table 27.1. The specimen was prepared by using the following steps: first, the HTPB was mixed with TDI and DOS in the proportion indicated in Table 27.1, followed by adding sugar crystals. The mixture was kept in an oven for 6–7 h in a vacuum assisted oven at 80 °C. The partially cured mixture was then cold pressed into billets of diameter 25.4 mm in a steel mold at 30 MPa. These billets were kept in an oven at 80 °C for 120 h for the complete curing of the samples. Specimens of dimensions 20 × 18 × 18 mm were extracted from the billets using a milling machine. Finally, these specimens were polished in order to remove the deformed layer from the surface of the specimen.

S. Ravindran (✉) • P. Malchow • A. Tessema • A. Kidane
Department of Mechanical Engineering, University of South Carolina, 300 Main Street, Columbia, SC, 29208, USA
e-mail: muthiram@email.sc.edu

J. Kimberley et al. (eds.), *Dynamic Behavior of Materials, Volume 1*, Conference Proceedings of the Society for Experimental Mechanics Series, DOI 10.1007/978-3-319-62956-8_27

Table 27.1 Material constituents and composition

Material constituents	Composition (% of weight)
Sugar crystal	87.5
Hydroxyl-terminated polybutadiene (HTPB)	9.00
Di-octyl sebacate (DOS)	2.40
Toluene diisocyanate (TDI)	1.10

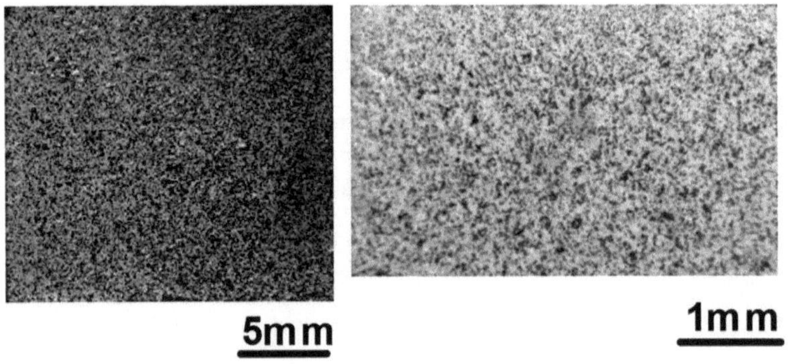

Fig. 27.1 The speckles obtained for macro and meso-scale experiments

Table 27.2 Imaging and post-processing parameters

Macro-scale imaging parameters	Macro-scale postprocessing parameters	Meso-scale imaging parameters	Meso-scale postprocessing parameters
Magnification −75 μm/pixel	Subset size – 13 × 13 Pixel	Magnification −10.66 μm/pixel	Subset size – 9 × 9 Pixel
Field of view – 30 × 18.75 mm	Step size – 3	Field of view – 4.2 × 2.6 mm	Step size – 1
Inerframe time — 0.5 μs	Filter size – 9	Inerframe time — 0.5 μs	Filter size – 9

Specimen density was measured to be 1.34 gm/cm^3. Speckles need to be applied in order to facilitate the specimen for DIC measurements on the surface. For the macro-scale and meso-scale measurements, a speckle size of 225–375 and 30–50 μm were required. The macro scale speckles were obtained using a spray paint. For macro scale speckles, a thin layer of the white base coat was applied on the surface of the sample, followed by a black paint spray on the surface. Whereas, the meso-scale speckles were prepared by using toner powder. In order to make meso-scale speckles, the following procedure was developed: first, a thin layer of white paint was applied on the surface, before the drying of the paint, toner powder was applied by using an airbrush. After the complete drying of the paint, the particles appeared to be strongly adhered to the surface. The macro-scale and meso-scale speckles obtained using this method is shown in Fig. 27.1.

A complete schematic of the experimental setup that was developed for this study is shown in Fig. 27.2. The experimental setup is comprised of two components, (1) loading device, (2) high speed camera with the optical assembly. The loading of the sample was performed using a modified configuration of the split Hopkinson bar setup (SHPB) as shown in Fig. 27.2. Experiments were conducted by shooting a projectile made of polycarbonate at the specimen. Speckles have been applied at the tip of the projectile to measure the velocity of the projectile *in-situ*. A high-speed imaging system for macro-scale measurement consists of a camera from Shimadzu (HPV-X2) with 100 mm Tikona lens. Whereas, for meso-scale measurement, a high magnification long distance extension tube from Navitar was used. Illumination for the macro-scale was provided using a Metal Arc Lamp (Lumen-200). The high magnification imaging requires a higher amount of light, therefore a flash light from Photogenic was used in the meso-scale experiment. The images for both the experiments were captured at two million frames/second. The complete details of post-processing and imaging parameters for macro and meso-scale measurements are given in Table 27.2.

27.2.1 *Full-Field Stress Calculation*

The compaction wave forms upon impact and propagates along the specimen. During the compaction wave propagation, the stress in the material is purely due to the material acceleration (inertia stress). In order to calculate the inertia stress, we

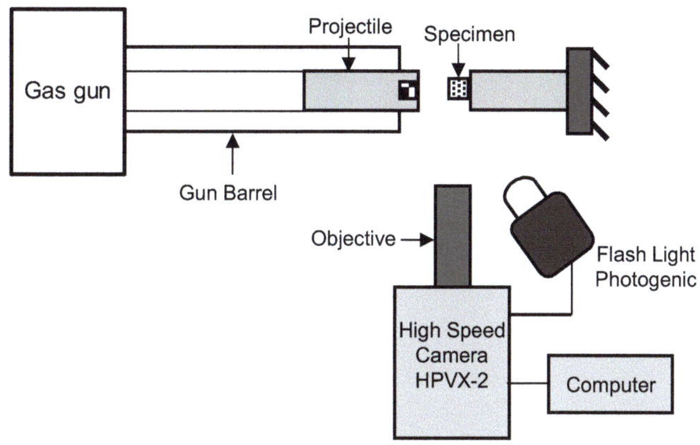

Fig. 27.2 Schematic and complete experimental setup

Fig. 27.3 Depiction of the direct impact of PBS. Loading direction is X_1

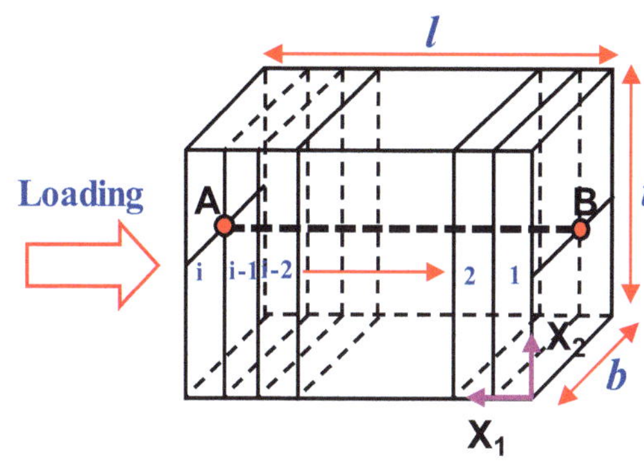

consider the impact scenario as shown in Fig. 27.3. Assuming, dynamic stress equilibrium (i.e. momentum conservation), the nominal inertia stress can be calculated. Interestingly, along line AB at the center as shown in Fig. 27.3, uniaxial strain assumption is valid until the lateral relief wave reaches the center. Assuming axial surface displacement is equal to the spatial displacement along AB, the axial stress along line AB can be calculated using Eq. 27.1. First Piola Kirchoff's stress is equal to true stress (Cauchy's stress) for uniaxial strain assumption.

First Piola-Kirchoff stress,

$$P_{11}(X_1, t) = \int_0^{X_1} \rho_o \frac{\partial^2 u}{\partial t^2} dX_1 \tag{27.1}$$

Where ρ_0 is the initial density, $\frac{\partial^2 u}{\partial t^2}$ is the acceleration that can be obtained by numerically differentiating the displacement field obtained from DIC.

27.3 Results and Discussion

27.3.1 Full-Field Axial Strain

The full-field contour plots of axial strain evolution with time is shown in Fig. 27.4. It is clearly seen that a compaction wave is formed and propagates along the loading direction. Axial strain behind the compaction front is close to 7% whereas the region in front of the compaction front is undisturbed, yielding an axial strain close to 0%. The compaction wave is planar at t = 5 μs, however, it transforms to a non-planar front as it traverses across the specimen. This is due to the lateral relief wave generation at the top and bottom edges of the specimen on the impact face.

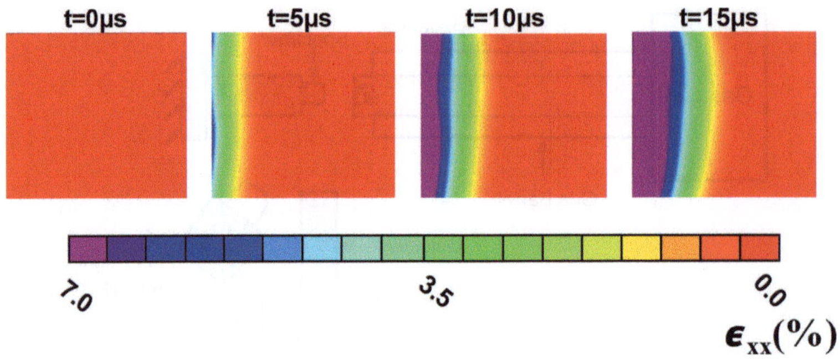

Fig. 27.4 Full-field axial strain with time

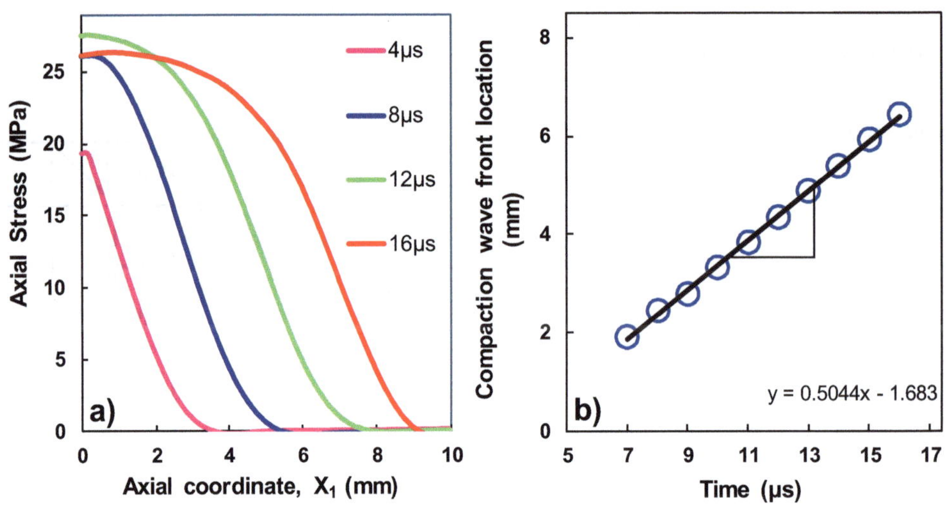

Fig. 27.5 (**a**) Axial stress along the undeformed axial coordinates (X_1), (**b**) Compaction wave front location with time

27.3.2 Axial Stress

The axial stress along line AB (see Fig. 27.3) is calculated using Eq. 27.1. Axial stress along the axial coordinate (X_1) is shown in Fig. 27.5. Interestingly the stress behind the compaction front is nearly constant and close to 25 MPa, and stress ahead of the front is nearly zero. Compaction wave front was located by finding the maximum stress gradient location of the stress and plotted in Fig. 27.5b. The slope of the linear fit gives the compaction velocity of 505 m/s.

The axial stress at a different axial location along line AB (see Fig. 27.3) are plotted with time in Fig. 27.6. at $X_1 = 0$ mm, the rise time is close to 9 μs, which increases to 12–13 μs at $X_1 = 6$ mm. In addition, at $X_1 = 0,1,$ and 2 mm the peak stress is dropping after t = 16 μs which indicates the arrival of the lateral relief at the center of the specimen. Interestingly, the peak stress is dropping with distance, which shows the dissipation of the energy during the propagation of the compaction wave. We discuss the energy dissipation mechanisms in the upcoming section.

27.3.3 Mesoscale Deformation Mechanisms

The contour plot of local axial strain field is shown in Fig. 27.7. A highly heterogenous strain field was seen with high strain localization within the compaction band. Ahead of the of the finite thickness wave front, the strain is nearly zero. In order to compare the strain field with the corresponding microstructure, the microstructure is marked with region A and B. Region A shows the location of a crystal and location B is a polymer binder area between the crystals. Interestingly, the axial strain on crystal at region A shows a significantly high value, close to 15%. Conversely, the sugar crystal is brittle in nature, as result,

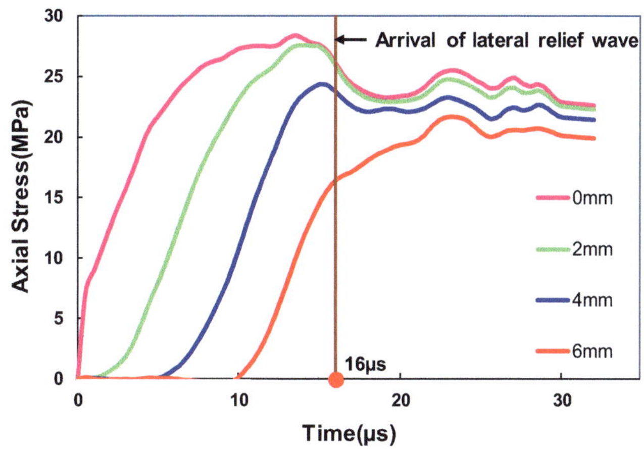

Fig. 27.6 Axial stress at different axial locations (X_1) with time

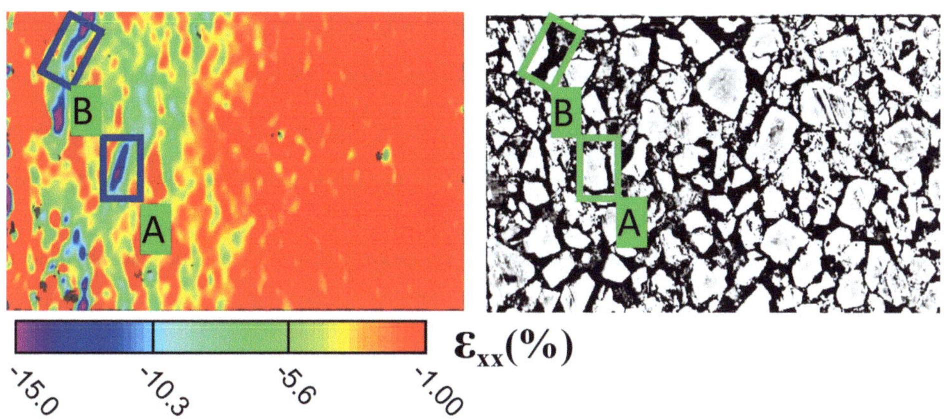

Fig. 27.7 The meso-scale local strain field and corresponding microstructure

a low strain is expected. Therefore, the high strain in crystals is due to the microcracks in the crystals as a result of high compressive force transfer between the crystal. The observed large compressive strain in the region B is due to the plastic deformation of the soft polymer binder. The energy dissipation mechanisms during the compaction wave propagation are, therefore, crystal fracture and plastic deformation of the polymer binder. This could be the main reason for the peak stress drop while the compaction wave travels across the specimen.

27.4 Summary

An experimental setup was developed to perform the macro and meso-scale deformation measurement under dynamic loading conditions. The macro-scale measurement shows the propagation of compaction wave front at 505 m/s. Compaction wave stress was estimated using the displacement field obtained from DIC. It was seen that the compaction wave strength decreases as it propagates across the specimen due to dissipation of energy. The energy dissipation mechanisms such as crystal fracture and plastic deformation of the binder were identified from the meso-scale deformation measurement.

Acknowledgement The financial support of Air Force Office of Scientific Research (AFOSR) under Grant No. FA9550-14-1-0209 is gratefully acknowledged.

References

1. Barua, A., Horie, Y., Zhou, M.: Microstructural level response of HMX–Estane polymer-bonded explosive under effects of transient stress waves. In: Proc. R. Soc. London A Math. Phys. Eng. Sci., pp. 3725–3744. The Royal Society (2012)
2. Bouma RHB, van der Heijden AEDM (1999) Simulations of Deformation Processes in Energetic Materials, INTECH, Janeza Trdine 9 51000 Rijeka, Croatia - EUROPEAN UNION
3. Baer, M.R.: Modeling heterogeneous energetic materials at the mesoscale. Thermochim. Acta. **384**, 351–367 (2002)
4. Sutton, M.A., Orteu, J.J., Schreier, H.: Image Correlation for Shape, Motion and Deformation Measurements: Basic Concepts, Theory and Applications. Springer Science & Business Media, New York, NY (2009)
5. Bodelot, L., Escobedo-Diaz, J.P., Trujillo, C.P., et al.: Microstructural changes and in-situ observation of localization in OFHC copper under dynamic loading. Int. J. Plast. **74**, 58–74 (2015)
6. Ravindran, S., Tessema, A., Kidane, A., Sutton, M.A.: Meso-scale deformation mechanisms of polymer bonded energetic materials under dynamic loading. In: Mechanics of Composite and Multi-Functional Materials, vol. **7**, pp. 451–456. Springer, Cham, (2016)
7. Ravindran, S., Tessema, A., Kidane, A.: Note: Dynamic meso-scale full field surface deformation measurement of heterogeneous materials. Rev. Sci. Instrum. **87**, 36108 (2016)
8. Ravindran, S., Tessema, A., Kidane, A.: Local deformation and failure mechanisms of polymer bonded energetic materials subjected to high strain rate loading. J. Dyn. Behav. Mater. **2**, 146–156 (2016)
9. Ravindran, S., Tessema, A., Kidane, A.: Effect of Crystal Density on Dynamic Deformation Behavior of PBX. In: Challenges Mech. Time Depend. Mater. Vol. **2**, pp. 87–92. Springer, Cham, (2017)
10. Ravindran, S., Tessema, A., Kidane, A.: Multiscale damage evolution in polymer bonded sugar under dynamic loading. Mech Mater **114**, 97–106 (2017)
11. Othman, R., Aloui, S., Poitou, A.: Identification of non-homogeneous stress fields in dynamic experiments with a non-parametric method. Polym. Test. **29**, 616–623 (2010)
12. Koohbor, B., Kidane, A., Lu, W.-Y., Sutton, M.A.: Investigation of the dynamic stress–strain response of compressible polymeric foam using a non-parametric analysis. Int. J. Impact Eng. **91**, 170–182 (2016)
13. Pierron, F., Zhu, H., Siviour, C.: Beyond Hopkinson's bar. Philos. Trans. R. Soc. Lond. A Math. Phys. Eng. Sci. **372**, 20130195 (2014)

Suraj Ravindran is a PhD student at University of Carolina working on Multiscale Behavior of Energetic Materials under Dr. Addis Kidane.

Chapter 28
Direct Compression Loading Using the Pre-stretched Bar Technique: Application to High Strains Under Moderate Strain Rates

G. Haugou, H. Morvan, and N. Leconte

Abstract The mechanical properties of ductile materials are of high interest in the domain of energy absorption, high speed forming processes, machining processes and ballistics. The range of the intermediate strain rates has always been required for the main testing devices, particularly for classical Split Hopkinson Pressure Bars. But due the transient aspect of the loading conditions, high ductility levels and moderate rates of strain are not easily to attain. As the length of the striker is mostly limited to 1 m due to practical reasons, the duration time of metallic strikers is theoretically close to 0.4 ms and then limited to 0.04 in total strain at 100/s. However, it is admitted that a long duration time is required to determine mechanical responses when high plastic strains are observed under real-life conditions. The technique of the pre-stretched bar is a good candidate to determine high strain responses of metallic materials up to fracture for the range of moderate rates of strain. The main challenge consists in the initiation of the elastic waves' system to load samples under direct compression. The authors propose here a new configuration of the pre-stretched technique where the pre-stretched part is preserved but the compression loading generated through the sample is radically modified with comparison to the tensile testing version developed by Albertini in the 70′. The strategy is based on FEM approach to generate a representative model of the classical pre-stretched bars combined with analytical calculations. The experimental tests are performed on samples extracted from Aluminum and brass alloys so as to validate the modified pre-stretched bars designed for compression loadings at moderate strain rates.

Keywords Pre-stretched bar • Compression • Moderate strain rate • High ductility • FEM modeling

28.1 Introduction

Kolsky bars apparatus – composed of two cylindrical bars equipped of strain bridges – generate a transient elastic waves' system once the striker hits the incident bar. The elastic waves' system is used to determine the typical response of the sample sandwiched between the incident and transmitted bars [1–3]. The duration time τ of the incident wave depends on the wave speed C and the length L of the striker, whatever the speed of the striker. Most of the time, the length L of the striker is limited to 1 m at most due to practical restrictions (the mass of the striker, friction with the launcher). Some authors have proposed disposals to launch long projectiles (over 2.5 m) by using a U-shaped projectile coupled with a hydraulic system [4]. For a conventional 1 m long metallic striker, the duration of the pulse generated by the impact (assuming that the wave speed is the same in the striker and in the incident bar) is generally limited to 0.4 ms. Based on this limitation in time to load the sample, the total strain of the sample is directly limited by the duration time that depends on the strain rate level. For a strain rate of 100/s, the total strain is limited to 0.04. Therefore, it is generally admitted that Kolsky bars cannot be employed or adapted in order to characterize high ductility metals up to fracture at moderate strain rates. Fortunately, the pre-stretched bar technique is an alternative method that allows to load specimens up to fracture thanks to higher duration times than those generated by strikers' impact. Until now, this technique is useful for tensile loadings of rounded or flat metallic alloys samples [5–9]. Here, the authors propose to generate directly compression testing conditions although the challenge has recently been achieved. However, a difficulty concerns the fracture of the sacrificial disk of the locking system depending on the preload value [10]. The present design gives the possibility to avoid any inversed set up subjected to discussions (loss of stiffness on the connectors, impedance mismatches) and also ensures samples' loading by simply varying the preload value. It is not possible to generate a compression elastic wave due to risks of bending of the incident bar.

G. Haugou (✉) • H. Morvan • N. Leconte
Laboratory of Industrial and Human Automation Control, Mechanical Engineering and Computer Science, UMR CNRS 8201, University of Valenciennes, Le Mont-Houy, 59313, Valenciennes, France
e-mail: gregory.haugou@univ-valenciennes.fr

J. Kimberley et al. (eds.), *Dynamic Behavior of Materials, Volume 1*, Conference Proceedings of the Society for Experimental Mechanics Series, DOI 10.1007/978-3-319-62956-8_28

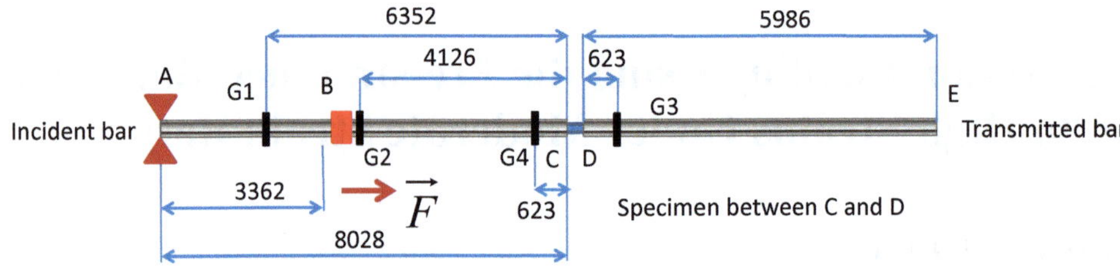

Fig. 28.1 Scheme of the pre-stretched split Hopkinson compression bar (pre-stretched SHPB)

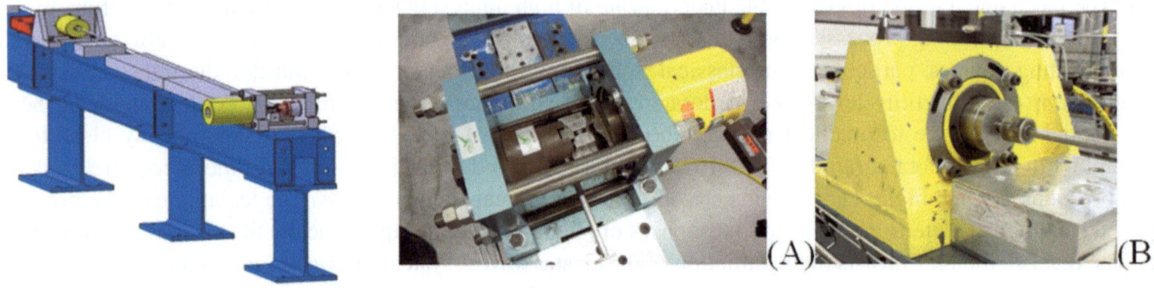

Fig. 28.2 (**a**) Pre-stretched part of the split Hopkinson compression bar (**b**) Front view of the points A and B

28.2 Design of the Pre-stretched Compression Testing

The technique of the pre-stretched bar is a not a common set up in comparison with classical SHPB developed for different loadings modes [11–15]. In fact, the pre-stretched technique allows testing materials with high ductility. The principle of compression loading generated using an incident wave developed in tension is detailed hereafter. The base technique is kept, but the location of both locking and tensioning mechanisms [7] is modified as illustrated in Fig. 28.1. The locking mechanism is positioned at point A, and it consists in locking any displacement of the incident bar. Once the fixed condition is applied to the incident bar, a second mechanism placed at point B put the incident bar in tension up to a nominal value F. Therefore, a hydraulic jack is in contact with a cylindrical connector welded on the incident bar. As the free end of the bar is locked, the incident bar stores elastic energy between points A and B. The cylindrical specimen is sandwiched between point C and D before being tested. The condition to load the sample under compression consists in the release of the elastic energy stored in the pre-stretched zone by the break of a fuse at the position A [7]. The force F is defined as a function of the mechanical and geometrical properties of the specimen and the wished strain rate. The amplitude of the reflected and transmitted pulses depends also on these statements.

Practically, the cylindrical bars are aligned using guides spaced with a gap of 400 mm threaded on a rigid framework composed of I-beams (Figs. 28.2a, b). The bars are equipped with full strain bridges so as to capture incident, reflected and transmitted pulses generated during the test. Gauge G1 is used to measure the force of the pre-stretched zone whereas gauges G2 and G3 are useful for the detection of incident, reflected and transmitted, respectively. The position of gauge G4 is suitable for the check of forces' equilibrium during the loading of the sample. The bars are calibrated as velocity and force sensors so as to calculate precisely the behavior laws by using the three typical elastic waves. The typical pulses are recorded using a high speed numerical oscilloscope. The amplitude of the incident pulse depend on the force F, the section S and the Young's modulus E of the bar according to Hooke's and Hugoniot's laws (relation 28.1).

$$\varepsilon_{INC}(t) = \frac{4 \cdot F}{\pi \cdot D^2 \cdot E} \tag{28.1}$$

Its duration time τ is a function of the length L of the pre-stretched part of the incident bar and the wave speed propagating along the incident bar and noticed C (see relation 28.2).

$$\tau = \frac{2 \cdot L}{C} \tag{28.2}$$

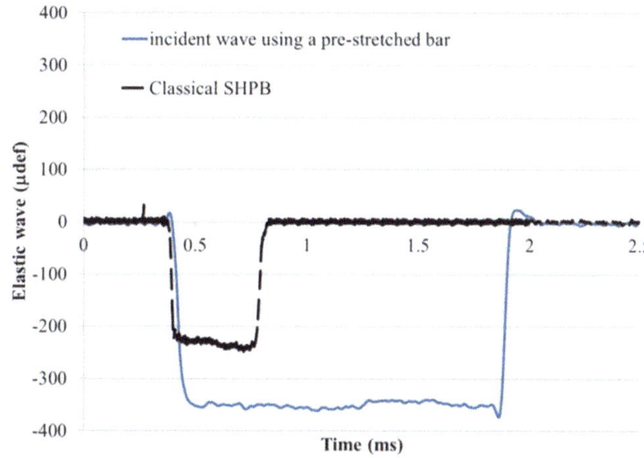

Fig. 28.3 Incident pulse of pre-stretched bar and classical SHPB 400/s for an Al. sample

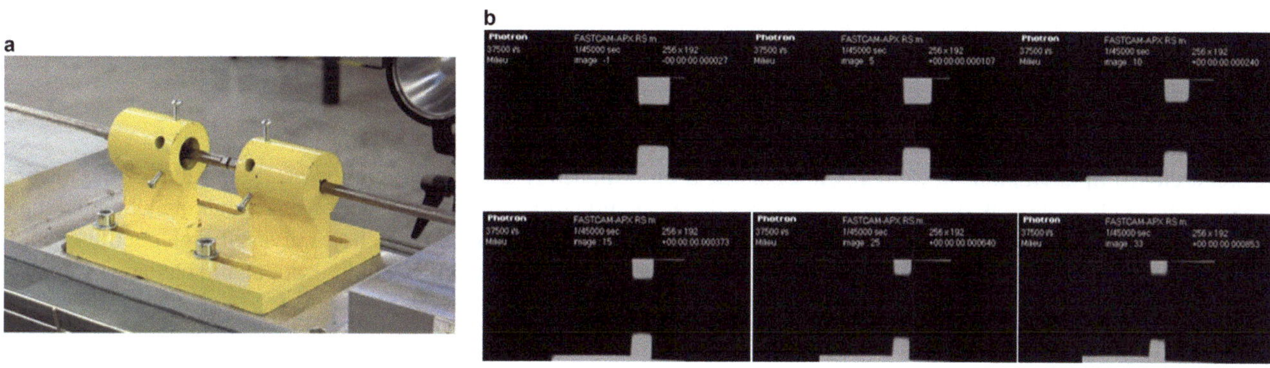

Fig. 28.4 (**a**) View of the cylindrical connectors (**b**) High speed imaging of the compression tests

Here, the length of this part is equivalent to a striker with a length of 3.4 m, thus a duration time of 1.5 ms. As a consequence, it is now permitted to reach a total strain ε of 0.14 at 100/s according to relation (28.3).

$$\varepsilon = \tau \cdot \dot{\varepsilon} \tag{28.3}$$

28.3 Preliminary Tests and Results

A comparison of incident pulses generated by conventional SHPB and pre-stretched SHPB is done and presented in Fig. 28.3 for compression tests at 400/s. The duration time is multiplied here by 3 in comparison with Classical SHPB testing and gives the possibility to reach high ductility level at the moderate strain rate theoretically announced in relation 28.3.

A preliminary test consists in the check of the influence of the connectors threaded at the end of each bar as depicted in Fig. 28.4a. To this end, the bars are placed in contact without sample, and the aim is to ensure that the incident wave is completely transmitted to the output bar without reflected disturbances. Here, the error is estimated lower than 0.5%. High speed imaging is a useful tool to extend the behavior law. As shown in Fig. 28.4b, the retro lighting provides a shadow of the sample so as to catch its evolution in function of time up to 110, 000 frames/s (Photron APX RS3000 – 128 × 32pi).

A set of compression tests have been performed on Aluminum [140–2800]/s and Brass [500–2200]/s using the pre-stretched Split Hopkinson Compression Bars and a classical SHBP (Fig. 28.5a). The samples have a diameter equal to 4.05 mm and 2.6 mm in thickness (average values). The analysis of the elastic waves' system is done using David© software [16] where the governing equations are classically implemented for the calculation of the mechanical properties of the tested alloys with associated strain rates (Fig. 28.5a). In Fig. 28.5b, the strain/stress flow curves obtained for both Hopkinson devices are compared, in the particular case of Aluminum alloy.

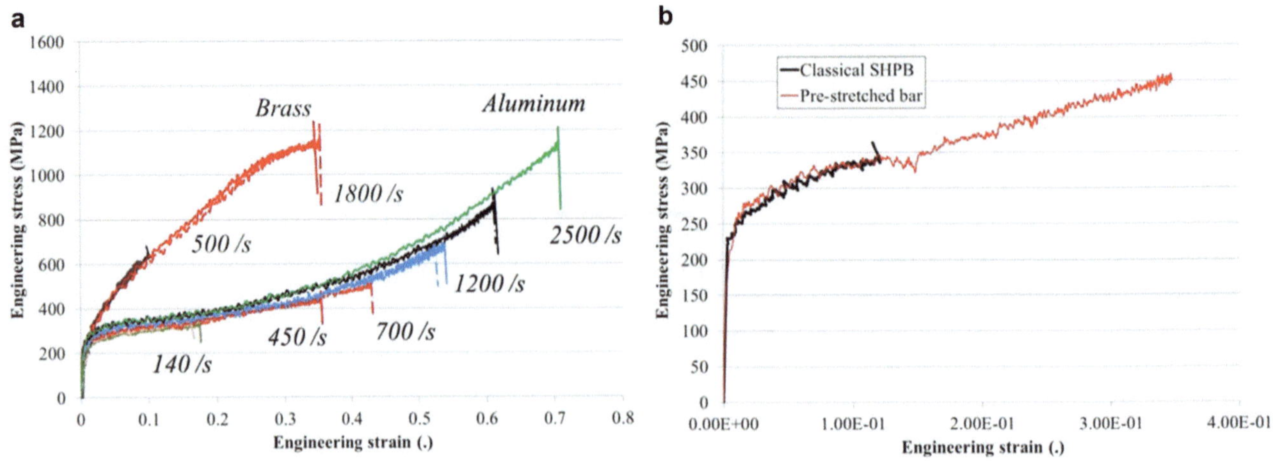

Fig. 28.5 (**a**) Flow curves of Aluminum and Brass coupons at moderate and high strain rates (**b**) Flow curves of Aluminum at 450/s (Pre-stretched bar and classical SHPB)

28.4 Concluding Remarks

A pre-stretched Split Hopkinson Compression bar has been designed. It allows varying the pre load value of the loading system. The controlled break of a brittle fuse ensures the generation of an incident tensile wave propagating along the input bar. The duration time of this pulse gives the opportunity of determine the compressive flow curves of high ductility metals at the early 140/s from a total strain of 0.2 and an energy storage of few 50 MJ/m^3; which is not attainable using Classical SHPB.

28.5 Acknowledgments

The present research work has been supported financially by the International Campus on Safety and Intermodality in Transportation (http://www.cisit.org), the Nord-Pas-de-Calais Region, the European Community, the Regional Delegation for Research and Technology, the Ministry of Higher Education and Research, and the National Centre for Scientific Research. The authors gratefully acknowledge also D. Lesueur for the preparation of the Hopkinson devices' platform developed at the L.A.M.I.H. laboratory (University of Valenciennes – France).

References

1. Kolsky, H.: Stress Waves in Solids. Dover Publications Inc., New York (1963)
2. Chen, W., Song, B.: Split Hopkinson (Kolsky) Bar, Design, Testing and Applications. Springer, New York, NY (2011)
3. XXXX
4. Gerlach, R., Kettenbeil, C., Petrinic, N.: A new split Hopkinson tensile bar design. Int. J. Impact Eng. **50**, 63–67 (2012)
5. Albertini, C., Montagnani, M.: Mechanical properties at high strain rates. Inst. Phys. Conf. Ser. **21**, 22–32 (1974)
6. Børvik, T., Clausen, A.H., Hopperstad, O.S., Langseth, M.: Perforation of AA5083-H116 aluminium plates with conical-nose steel projectiles – experimental study. Int. J. Impact Eng. **30**(4), 367–384 (2004)
7. Haugou, G., Morvan, H., Leconte, N.: Design of a pre-stretched tension Hopkinson bar device: configuration, tail corrections, and numerical validation. Int. J. Impact Eng. **97**(4), 89–101 (2016)
8. Saletti, D.: Measurement of phase transformation properties under moderate impact tensile loading in a NiTi alloy. Mech. Mater. **65**, 1–11 (2013)
9. Cadoni, E., Dotta, M., Forni, D., Tesio, N., Albertini, C.: Mechanical behaviour of quenched and self-tempered reinforcing steel in tension under high strain rate. Mater. Des. **49**, 657–666 (2013)
10. Mancini, E., Sasso, M., Rossi, M., Chiappini, G., Newaz, G., Amodio, D.: Design of an innovative system for wave generation in direct tension–compression split Hopkinson bar. J. Dyn. Behav. Mater. **1**, 201–213 (2015)
11. Lindholm, U.S., Yeakley, L.M.: Some experiments with the split Hopkinson pressure bar. Exp. Mech. **12**, 317–355 (1964)

12. Davies, E.D.H., Hunter, S.C.: The dynamic compression testing of solids by the method of the split Hopkinson pressure bar test. J. Mech. Phys. Solids. **11**, 155–179 (1963)
13. Nicholas, T., Lawson, J.E.: On the determination of the mechanical properties of materials at high shear-strain rates. J. Mech. Phys. Solids. **20**(2), 57–**64** (1972)
14. Harding, J.: Review of dynamic testing techniques and materials behaviour. Mechanical Properties at High Rates of Strain, Institute of Physics Conference Series, vol. 21 (1974)
15. Klepaczko, J.R.: An experimental technique for shear testing at high and very high strain rates – the case of the mild steel. Int. J. Impact Eng. **15**, 25–39 (1994)
16. Gary, G., Degreef, V.: DAVID, Users' Manual Version, Labview Version. LMS Polytechnique, Palaiseau. version 12341 (2008)

G. Haugou is an assistant professor since 10 years at the university of Valenciennes (France). He works in the field of the experimental mechanics. He focusses his research in the development of Hopkinson bar techniques based on FE modeling. He is the (co) author of 37 papers.

Chapter 29
Compaction Wave Characteristics of Polymeric Foams Under Dynamic Loading

Suraj Ravindran, Behrad Koohbor, Peter Malchow, and Addis Kidane

Abstract The continuum scale shock wave theory has been extensively used to characterize the compaction wave propagation in polymeric foams. However, experimental full-field measurements are scarce to validate such shock wave models. In this study, dynamic experiments with the help of ultra-high speed imaging in conjunction with digital image correlation (DIC) to characterize the compaction wave propagation in polymeric foams. A direct impact is applied on a polymer foam sample using a Hopkinson bar apparatus. The experiment is conducted by shooting a projectile directly at the specimen at an impact velocity of 90 m/s. Typical characteristics of the compaction wave such as compaction wave velocity, compaction wave profile, full-field particle velocities are obtained from the DIC measurements. Furthermore, using a non-parametric method, inertia stress and hence the full-field axial stress is calculated across the compaction wave.

Keywords Compaction • Foams • Direct impact • Shock

29.1 Introduction

High energy absorption characteristics makes foam materials attractive to packaging and impact damage protection applications [1]. Designing strong structures for impact protection requires a detailed characterization of foams under different impact loading conditions. Numerous theoretical, numerical and experimental studies have been performed in the past decade to predict the material behavior of the foams under impact loading. A well-known rate independent rigid perfectly plastic locking model (RPPL) was developed and used to predict the strength enhancement in wood under impact loading [2]. This model has been extensively used to predict the behavior of foams under dynamic loading conditions. Most of the experiments to validate such models are devised with the help of boundary measurements [3] and could not be used to diagnose and observe the deformation mechanisms. In addition, in the case of a compaction wave propagation, the spatial variation of the stresses makes it difficult to estimate the stress and strain induced in the specimen with the boundary measurement techniques.

With the advent of digital image correlation, (DIC) [4] it enables us to measure the full-field deformation fields at a variable spatial and temporal resolution [5, 6]. Recently, the displacement field from DIC has been used to estimate the inertia stress in order to get the constitutive response of solid foams, metal and rubber [7–9]. This technique is promising in estimating stress across the specimen and has been extensively used in rigid foams [10–12]. However, the work was focused on obtaining the stress-strain relation and never dealt with the compaction profile. In this study, the compaction wave stress is estimated across the specimen length by adopting the method presented in [10]. Furthermore, other compaction wave parameters such as particle velocities, elastic wave velocity and compaction wave speeds are estimated.

29.2 Materials and Methods

Specimens of dimension 21 × 14.5 × 14.5 mm^3 was extracted from a block of foam received from General Plastics. The nominal density of the specimen is about 154 kg/m^3. Prior to the experiment, specimens were polished using silicon carbide papers of grit size from 300–1200 μm to have a good surface finish for speckling. The final dimensions of the samples are shown in Fig. 29.1. Considering the resolution of the imaging system, a speckle size of 200–400 μm was required for DIC

S. Ravindran (✉) • B. Koohbor • P. Malchow • A. Kidane
Department of Mechanical Engineering, University of South Carolina, 300 Main Street, Columbia, SC, 29208, USA
e-mail: muthiram@email.sc.edu

J. Kimberley et al. (eds.), *Dynamic Behavior of Materials, Volume 1*, Conference Proceedings of the Society for Experimental Mechanics Series, DOI 10.1007/978-3-319-62956-8_29

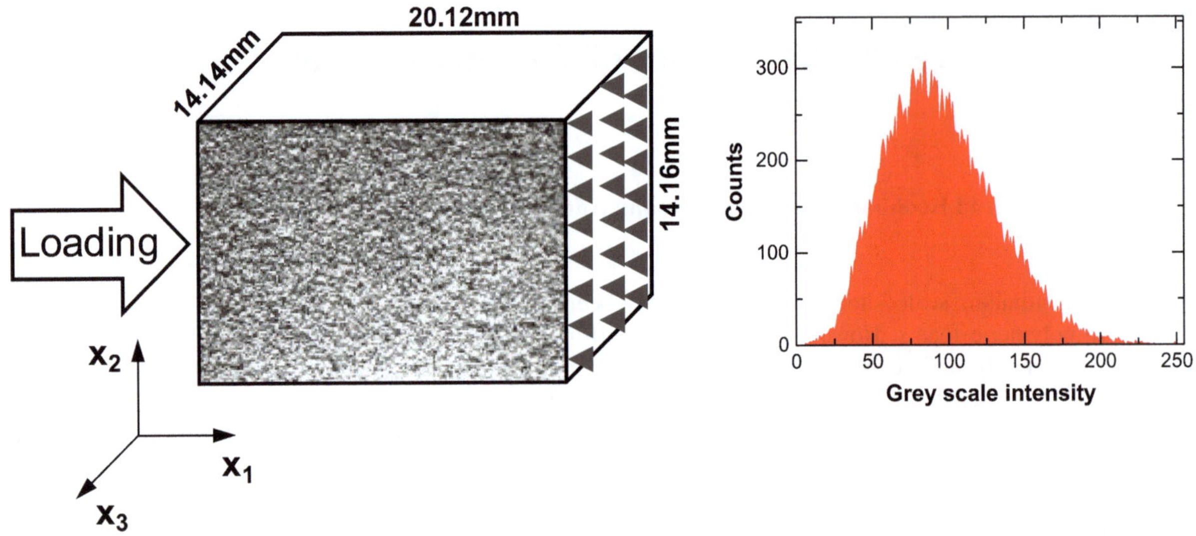

Fig. 29.1 Specimen dimension, speckle and gray scale intensity data are shown

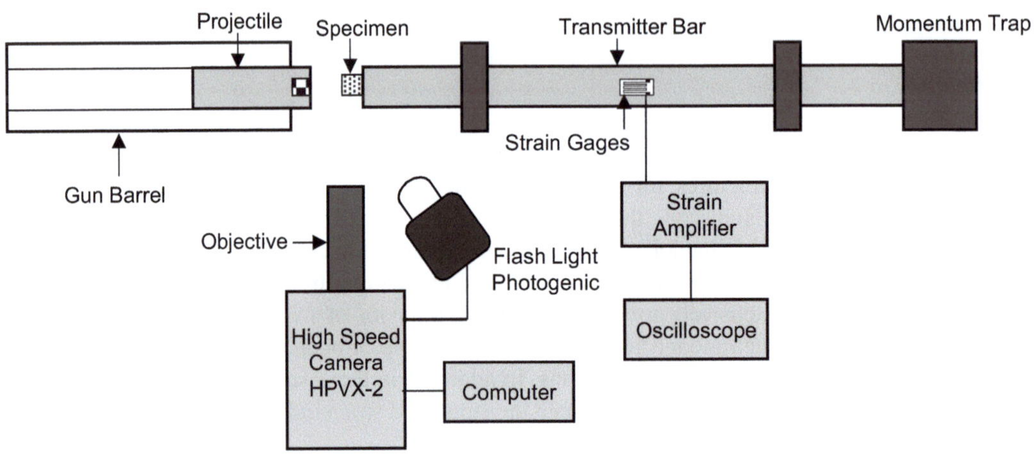

Fig. 29.2 Schematic and complete experimental setup

Table 29.1 Imaging and post-processing parameters

Imaging parameters	Post-processing parameters
Field of view – 30 × 18.75 mm^2	Subset size – 13 × 13 $Pixel^2$
Pixel to length ratio – 75 μm/pixel	Step size – 3
Inerframe time – 1.5 μs	Matching algorithm – zero normalized

measurements. The speckling of the specimen was done by using an airbrush. First, a thin layer of white paint was applied on the surface of the specimen followed by spraying black paint on the surface. The speckling pattern obtained using this method and the grey scale intensity curve are shown in Fig. 29.1.

A schematic of the experimental setup used in this study is shown in Fig. 29.2. The experimental setup is comprised of a loading device and a high-speed imaging system. The loading of the sample was performed in a direct impact configuration of the split Hopkinson bar setup (SHPB). In this type of experimental setup, the specimen is loaded by propelling a projectile at the specimen using a gas-gun. An 88 mm long projectile made of polycarbonate was used in this study. The projectile velocity was close to 90 m/s. It should be noted here, no measurements from the strain gage was used in this study, since we were interested in estimating only the inertia stress during the compaction wave propagation. The high-speed imaging system consisted of a high-speed camera from Shimadzu (Model HPVX-2) attached with a 100 mm Tikona optical lens. Illumination was provided using a Metal Arc Lamp (Lumen-200). The images were captured at 666666 frames/second. The details of post-processing and imaging parameters are given in Table 29.1.

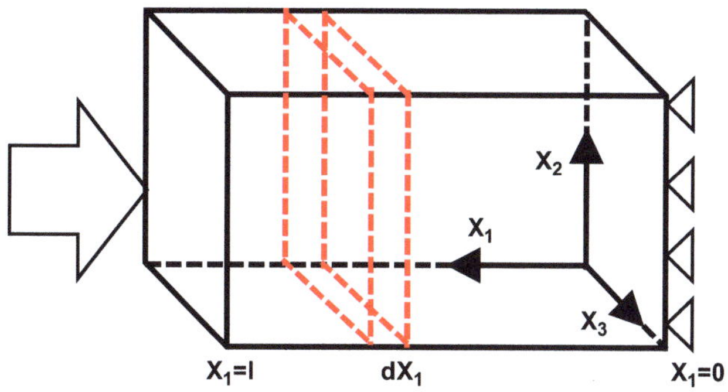

Fig. 29.3 Depication of the direct impact of foam

29.2.1 Data Analysis

Figure 29.3 depicts the direct impact scenario and the coordinates adopted in the inertia stress calculation. When the projectile impacts the specimen, a compaction wave is generated and propagates across the material. During the propagation of the compaction wave the stresses in the material is predominantly due to inertia. Using dynamic stress equilibrium equations (i.e. momentum conservation), the nominal inertia stress can be calculated as given by Eq. 29.1.

First Piola-Kirchoff stress,

$$P_{11}\left(X_1, t\right) = \int_0^{X_1} \rho_o \frac{\partial^2 u}{\partial t^2} dX_1 \tag{29.1}$$

Where, ρ_0 is the initial density, $\frac{\partial^2 u}{\partial t^2}$ is the Lagrangian acceleration. Since the foam used in this study is low density, a uniaxial strain assumption will be valid across the cross section. In that case, First Piola Kirchoff's stress is equal to true stress (Cauchy's stress). The Lagrangian acceleration can be obtained numerically by differentiating the displacement field obtained from DIC with respect to time.

29.3 Results and Discussion

29.3.1 Particle Velocity

The time evolution of the full-field particle velocity in the X_1-direction is shown in Fig. 29.4. It illustrates the propagation of the compaction wave along the loading direction. The compaction wave front is planar for the entire duration of the propagation of the wave in the material. The planar front allows us to use a 1-D compaction wave propagation analysis to study the characteristics of the compaction wave. Behind the wave front, the particle velocity is close to 90 m/s and it is nearly constant after t = 24 μs. Ahead of the compaction front, the particle velocity is not zero indicating the propagation of an elastic precursor before the generation of the compaction front. Elastic wave front propagation and its velocity are discussed in the next section.

29.3.2 Particle Velocity Along Axial Location and Wave Speeds

In order to see the profile of the particle velocity, as well as to calculate the velocity of the elastic precursor and compaction wave, we have plotted the particle velocity along the undeformed axial locations as seen in the Fig. 29.5. Interestingly, the particle velocity has two wave profiles in the beginning (t = 12, 15, and 18 μs), and this is due to an elastic precursor propagation ahead of the compaction front. After the precursor reaches the distal end of the sample, it reflects to the material.

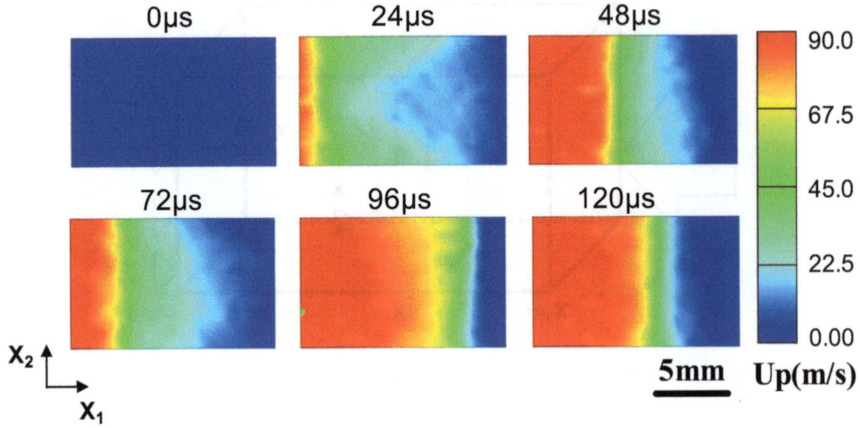

Fig. 29.4 Full-field particle velocity with time

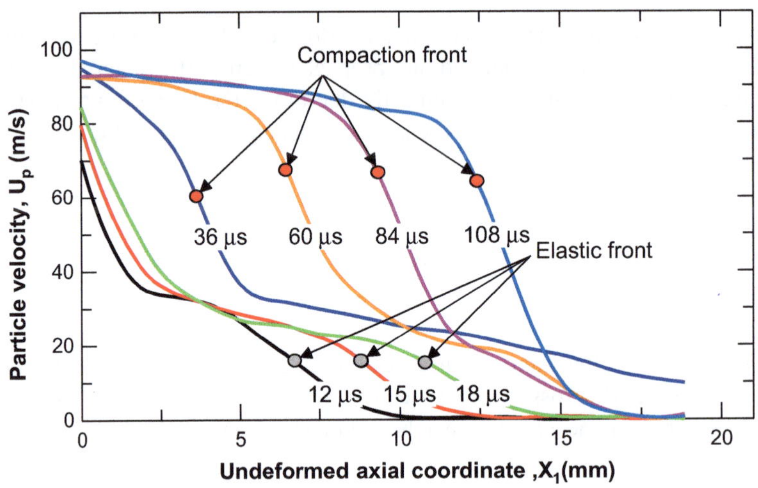

Fig. 29.5 Particle velocity along the axial direction at time t = 12, 15, 18, 36, 60, 84, and 108 μs

The compaction wave profile formed after the elastic precursor is shown at t = 36, 60, 84, and 108 μs. The particle velocity is constant behind the front indicating a complete compaction of the material.

In order to calculate the velocity of the elastic precursor and the compaction wave, the elastic wave front and compaction wave fronts are obtained. The location of the elastic front is obtained by finding the maximum of the absolute value of the particle velocity gradient ($\max(|dU_p/dX_1|)$). The elastic front marked in Fig. 29.5 shows a particle velocity in the range of 15–20 m/s. The location of the elastic front (X_e) with time is shown in the Fig. 29.6a and the elastic precursor velocity was calculated as 740 m/s. Particle velocity behind the compaction front is nearly constant indicating the attainment of a quasi-steady condition. The compaction front location with time is plotted in Fig. 29.6b and the slope of the linear fit gives the compaction wave velocity of 140 m/s.

29.3.3 Inertia Stress

Inertia stress is calculated using Eq. 29.1. The inertia stress along the axial coordinates are shown in Fig. 29.7. At t = 24 μs, a two-wave profile is observed due to the propagation of the elastic precursor. After, t = 24 μs, the wave reflects from the distal end as a tensile wave, see t = 48 μs. This reflected elastic wave releases the stress from 2.4 MPa to 0.5 MPa at the impact face. The elastic precursor decays its amplitude due to multiple reflections. At the same time, it continues movement of the projectile to further compresses the material which increases the inertia stresses to 1.5 MPa from 0.5 MPa at the impact end, see t = 72, 96 and 108 μs.

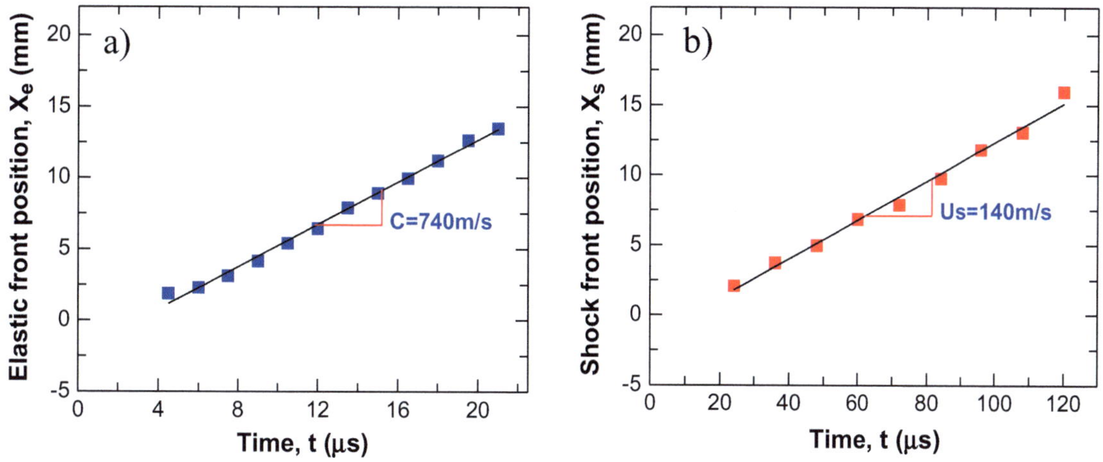

Fig. 29.6 (**a**) The elastic wave front location at different times. The slope of the curve gives the longitudinal wave velocity of 740 m/s, (**b**) Compaction front location with time. Slope of the linear fit give a compaction wave velocity of 140 m/s

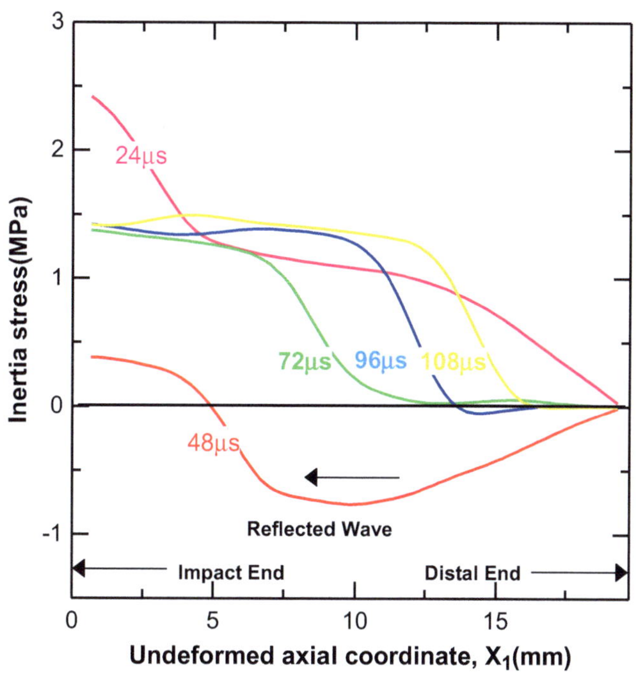

Fig. 29.7 Inertia stress along the undeformed axial coordinates

29.4 Summary

An experimental setup was designed to characterize foams under impact loading. Direct impact experiments were conducted in a low-density foam to describe the compaction wave propagation in the material. The particle velocity profile shows a two-wave structure during the propagation of the elastic precursor. The elastic precursor and shock velocity was calculated to be 740 m/s and 140 m/s respectively. Also, the inertia stress was calculated from the full-field displacement obtained from DIC.

Acknowledgement The College of Engineering and Computing and the Department of Mechanical Engineering at the University of South Carolina is gratefully acknowledged.

References

1. Gibson, L.J., Ashby, M.F.: Cellular Solids: Structure and Properties. Cambridge university press (1999), The Pitt Building, Trumpington street, Cambridge, United Kingdom
2. Reid, S.R., Peng, C.: Dynamic uniaxial crushing of wood. Int. J. Impact Eng. **19**, 531–570 (1997)
3. Elnasri, I., Pattofatto, S., Zhao, H., et al.: Shock enhancement of cellular structures under impact loading: part I experiments. J. Mech. Phys. Solids. **55**, 2652–2671 (2007)
4. Sutton, M.A., Orteu, J.J., Schreier, H.: Image Correlation for Shape, Motion and Deformation Measurements: Basic Concepts,Theory and Applications. Springer Science & Business Media (2009), Springer Science+Business Media, LLC, 233 Spring Street, New York, NY 10013, USA
5. Ravindran, S., Tessema, A., Kidane, A.: Local deformation and failure mechanisms of polymer bonded energetic materials subjected to high strain rate loading. J. Dyn. Behav. Mater. **2**, 146–156 (2016)
6. Ravindran, S., Tessema, A., Kidane, A.: Note: dynamic meso-scale full field surface deformation measurement of heterogeneous materials. Rev. Sci. Instrum. **87**, 36108 (2016)
7. Othman, R., Aloui, S., Poitou, A.: Identification of non-homogeneous stress fields in dynamic experiments with a non-parametric method. Polym. Test. **29**, 616–623 (2010)
8. Koohbor, B., Kidane, A., Lu, W.-Y., Sutton, M.A.: Investigation of the dynamic stress–strain response of compressible polymeric foam using a non-parametric analysis. Int. J. Impact Eng. **91**, 170–182 (2016)
9. Pierron, F., Zhu, H., Siviour, C.: Beyond Hopkinson's bar. Philos. Trans. R. Soc Lond. A Math. Phys. Eng. Sci. **372**, 20130195 (2014)
10. Koohbor, B., Kidane, A., Lu, W.-Y.: Effect of specimen size, compressibility and inertia on the response of rigid polymer foams subjected to high velocity direct impact loading. Int. J. Impact Eng. **98**, 62–74 (2016)
11. Koohbor, B., Kidane, A., Lu, W.-Y.: Characterizing the constitutive response and energy absorption of rigid polymeric foams subjected to intermediate-velocity impact. Polym. Test. **54**, 48–58 (2016)
12. Koohbor, B., Mallon, S., Kidane, A., Lu, W.-Y.: The deformation and failure response of closed-cell PMDI foams subjected to dynamic impact loading. Polym. Test. **44**, 112–124 (2015)

Suraj Ravindran is a PhD student at University of Carolina working on Multiscale Behavior of Energetic Materials under Dr. Addis Kidane.

Chapter 30
Comparison of Numerical Simulations with Experiments of Blast-Induced Pressure Wave Impact on a Surrogate Head Model

Rohan Banton, Thuvan Piehler, Nicole Zander, Richard Benjamin, and Josh Duckworth

Abstract There is an urgent need to protect the warfighter from blast-induced injuries such as traumatic brain injury and its subsequent cascading effects. To that end the current research has adopted a systematic numerical approach to investigate the intracranial pressure loading on a surrogate brain model impacted by pressure waves from an RDX explosive. A tetrahedron meshed model of a human head form filled with representative brain simulant material was placed in an Eulerian air domain and subjected to simulated blast waves. The numerical results were compared against experiments using 1.7 grams of RDX explosive placed 180 mm from a surrogate head form filled with brain simulant. Both the experimental and simulated results revealed dominant compressive pressure loading and volumetric tensions in the anterior regions of the brain simulant. Further, calculated and experimental results also revealed focal regions of elevated pressure loading in the posterior region resulting from reflected waves from the back of the skull. This study was performed as a basis for understanding the internal wave mechanics and their impacts on the brain simulant subjected to real blast effects and to serve as a genesis to improve protective head equipment for the warfighter.

Keywords Blast traumatic brain injury • Intracranial pressure waves • RDX explosive • skull • surrogate brain

30.1 Introduction

Blast induced injuries leading to traumatic brain injuries (TBI) in military personnel have gained heightened awareness in the recent conflicts in Iraq and Afghanistan. This is primarily due to the asymmetrical nature of the conflicts and the extensive use of improvised explosive devices (IEDs) [1]. Other factors contributing to the rise in blast traumatic brain injuries (bTBI) include improvements in personnel armor, as well as trauma evacuation and emergency care, allowing survival of previously fatal injuries [2]. Injuries related to blast have been categorized as (1) Primary blast injuries – injuries associated with direct blast impact to the head and body, (2) Secondary blast injuries – injuries resulting from blast associated fragment or debris, (3) Tertiary blast injuries – injuries resulting from the blast victim being thrown against stationary objects, and (4) Quaternary blast injuries – injuries develop from a variety of physical processes associated with explosive detonation such as thermal, toxic detonation products [1–3]. From the four stated categories of related blast injuries, the mechanisms leading to injuries in categories (2) and (3) are comparable to mechanical trauma that have been studied through clinical [4], experimental and computational research such as injuries occurring from a fall [5], automotive collisions [6], and contact sports [7]. The mechanism of traumatic brain injuries from categories (1) and (2) are rarely understood. In particular, the role of primary blast impact to the head is still unclear.

To unravel the mechanism leading to bTBI from primary blast waves, researchers have reconstituted (within the laboratory) the free field conditions resulting in blast wave evolution from explosive detonation. Among the experimental techniques employed were compress-driven air shock tube [8–10], explosively-driven shock tube [11], and laboratory scale real explosive blast experiments [12]. These techniques were used to study animal models, surrogate head models, and post mortem human specimens, subject to blast impact [9–15]. While the results produced from these experiments have provided insightful information of the pressure loading and strains experienced in the brain, only a few selected regions of interest were investigated, leaving open the possibility that key findings and recorded information were missed from unexplored spatial regions of the brain.

R. Banton (✉) • T. Piehler • N. Zander • R. Benjamin
U.S. Army Research Laboratory, Aberdeen Proving Ground, Adelphi, MD, USA
e-mail: rohan.j.banton.civ@mail.mil

J. Duckworth
The Uniformed Services University of the Health Sciences, Bethesda, MD, USA

J. Kimberley et al. (eds.), *Dynamic Behavior of Materials, Volume 1*, Conference Proceedings of the Society for Experimental Mechanics Series, DOI 10.1007/978-3-319-62956-8_30

To aid in the understanding of the intracranial wave mechanics in the spatial regions of the brain, computational efforts were also performed using a range of head forms. Several teams have utilized simplified surrogate head models composed of water/gel used as brain simulant enclosed with polycarbonate or Poly(methyl methacrylate) (PMMA) skull [14, 16] to study egg-shaped and spherical-shaped head forms. These numerical models with surrogate head forms were reasonably validated against shock tube generated blast wave experiments. However, they lack the appropriate human head geometry which governs the flow dynamics around the head, which in turns affects the net mechanical load on the head, thus further influencing the intracranial wave dynamics in the brain. More biofidelic head forms generated from segmentation of magnetic resonance images (MRI) consisting of scalp, skin, bone, CSF, brain showing gray and white matter, and nasal cavity have emerged [15, 17–19]. While these detailed models are more representative of the human head forms and have become more anatomically correct, they are rarely validated against blast experimental data. Limited validation of these detailed model are starting to emerge to include the work of Ganpule et al. [15] where a finite element (FE) model of a human head generated from MRI data consisting of skin, skull, subarachnoidal space, and brain was developed and validated against surface pressure impact to the skull of a dummy head placed in a compress-driven shock tube. Wang et al. [20] also developed an anatomically correct FE head model from MRI data that was validated against blunt frontal impact on cadaveric head [21]. Subsequent simulations with the FE head model were performed showing frontal, back, and lateral blast loading impact to head using simulated TNT explosive of mass ranging from 250–450 g placed at a standoff distance of 1 m.

To address the need for more direct comparison of computationally developed head forms subject to blast impact with experimental data we adopted the following approach. We developed a tetrahedron meshed model of a human head form filled with representative brain simulant material placed in an Eulerian air domain and subjected it to simulated blast waves. The numerical results were compared against blast experiments using spherical cyclotrimethylene trinitramine Class 5 (C3H6N6O6; RDX Class V, 1.7 g weight) charges. The following sections provide a brief descriptions of the experimental and computational approach followed by comparison of the numerical model results with the experimental data.

30.1.1 Experimental Procedure

Figure 30.1, shows the experimental setup for detonation of 1.7 gram of RDX explosive charge placed 180 mm from a PMMA surrogate head model. The head model was rotated downward 20° to create a horizontal alignment between the RDX charge and the impact point at the nasion on the skull. Two free-field air-blast pencil probes (ICB model 137A23, PCB Piezotronics Inc., Depew, NY) were placed at the same distances from the charge to measure the pressure loading imparted to the head. The evolution of the experimental blast waves from the charge to the surrogate head was captured using high-speed shadowgraph imaging for visualization using the techniques outlined and developed by G. Settles [22]. The blast conditions were selected to be within the pressure loading threshold for survivability according to the Bowen curves [23] These curves correlate the probability of fatality to specific blast levels and durations of exposure for a human standing in the path of an explosive blast [23]. In this experiment, the pressure loading produced by the charge on the surrogate head at the selected standoff distance falls within the survivability level on the Bowen curve. That is the probability for fatality was less than 1%. However, while the blast loading at this distance produce low fatality chances, the impact loading to the head still produce intracranial pressures which could lead to mild traumatic brain injuries. We investigated the intracranial pressure in the surrogate head by placing piezoelectric high-frequency dynamic pressure sensors (ICP model 102A, PCB Piezotronics Inc., Depew, NY) at protruding depths of 5 cm into the surrogate head model at location in the front left, top and back of the skull (see Fig. 30.1).

Figure 30.2, shows a hemispherical cutaway of the surrogate head filled with a biogel used as brain simulant. The biogel was composed of Poly(styrene-*b*-ethylene-*co*-butylene-*b*-styrene) triblock copolymer (SEBS) mixed with mineral oil [24]. SEBS powder and mineral oil were mixed at a polymer loading of 5 vol %, heated to 120 °C and intermittently mixed for 3 h to obtain a homogeneous solution without any visible undissolved SEBS powder. The sample was cooled to room temperature to verify gel formation. The gel was then heated to 60 °C to melt the gel and enable pouring into the pre-formed skull. After filling, the skull was held at 60 °C for 2 h to remove air bubbles entrapped during the filling process. The skull was then cooled to room temperature and any remaining openings were filled with a silicone sealant to prevent the gel from creeping out of the skull and reducing the final fill level. All the blast experiments performed on the surrogate head model were conducted in a 6- × 10- × 6-m blast room certified for testing up to 25 g of explosive.

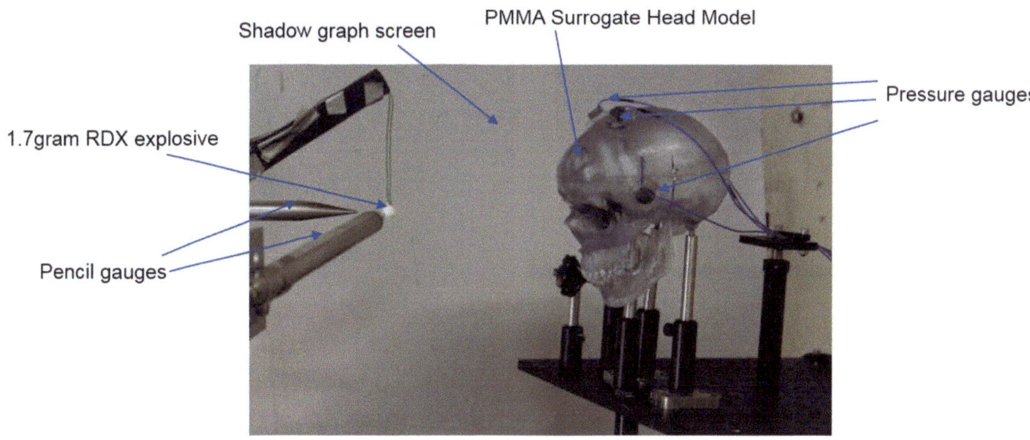

Fig. 30.1 Experimental setup for RDX blast propagation onto PMMA head model with instrumented pressure gauges. RDX explosive (1.7 gram) was positioned to hang in air while maintaining a level standoff distance of 180 mm from the nasion point of the PMMA head model. Pencil gauges were positioned 180 mm from the RDX charge to capture the pressure loading delivered to the head model

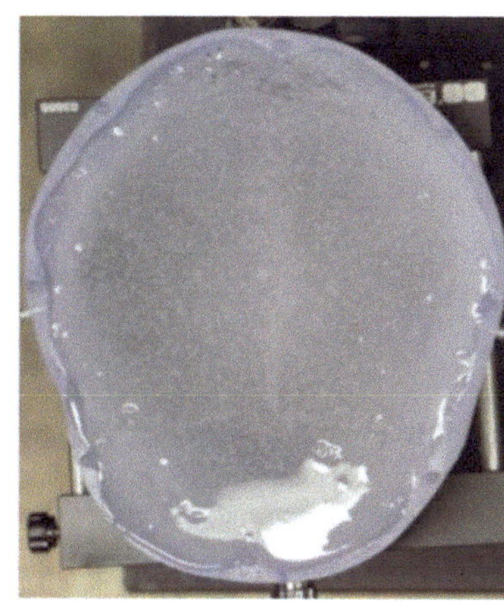

Fig. 30.2 Hemi-spherical cutaway showing surrogate head model filled with biogel

30.1.2 Computational Procedure

Figure 30.3a, shows the midsagittal view of the surrogate head model used in the blast simulations for direct comparison to the blast experiments performed (Fig. 30.3b). To obtain the computational geometry, 3D laser scan was performed on the experimental PMMA skull model (Fig. 30.3b). The 3D laser scan data was processed and converted to Solid Works CAD geometry where the internal contours of the skull was used to generate the surrogate brain part. A tetrahedron mesh overlay on the skull and brain surrogate head model was later performed using Sandia National Laboratory Cubit software package.

To carry out the blast simulations of the head impact, the shock physics code CTH was employed [25]. This code utilizes a finite volume solution scheme that solves the conservation equations of mass, momentum and energy for each material inserted in a fixed spatial mesh. CTH uses a two-step, second order accurate Eulerian solution algorithm. The governing equations are solved first using a Lagrangian step followed by a second remap step. In the Lagrangian step, a Lagrangian representation of the governing equations is solved explicitly across each timestep. The initial mesh is distorted to allow for material motion while not permitting mass flux across the cell boundaries. Subsequently, a remap step is performed where the distorted cells from the prior step are remapped back to the initial fixed mesh.

The complete material insertion into the CTH code included the head model (which consists of the surrogate skull and brain), the explosive charge and the surrounding air. To capture the material response of the head model to the blast impact appropriate equation-of-state (EOS) and strength models were utilized. The EOS representations for the skull and brain surrogate utilized an elastic compressible equation of-state model for the volumetric response [25] and a linear elastic perfectly plastic constitutive strength model for the skull and a Maxwell viscoelastic model for the deviatoric (shear) response for the brain [2, 7].

At the start of the simulation (Fig. 30.3a) the head model was placed at a standoff distance of 180 mm from an RDX charge in surrounding air. A programmed burn initiation model was used to start the detonation of the RDX explosive while

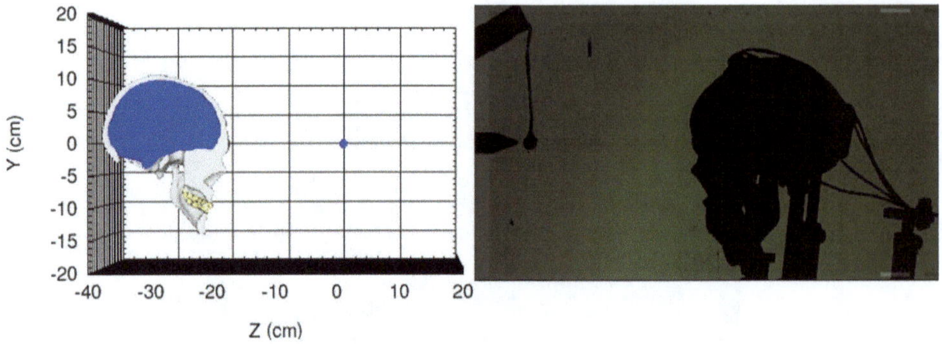

(a) (b)

Fig. 30.3 (**a**) Blast simulation setup and numerical representation of surrogate PMMA head model showing midsagittal view of surrogate head model (**b**) Experimental set-up of surrogate head model filled with bio gel material

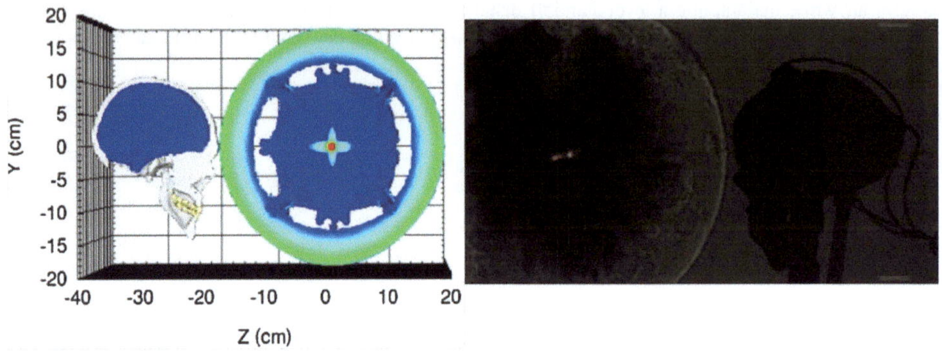

Fig. 30.4 Comparison of simulations (*left*) with experiments (*right*) of RDX initiation and blast wave propagation toward surrogate PMMA head model from180 mm standoff distance at time 140 μs. The explosive reaction products are shown to drive the shock front (Reaction product denoted as *blue* in simulation and *dark brown* in experiment)

expanding detonation product gases were represented by a Jones-Wilkens-Lee (JWL) EOS. A nonlinear tabular EOS model was also used to represent the surrounding air. Adaptive mesh refinement (AMR) capabilities were also used to track and refine the expanding propagating shock front as it moved through the spatial domain.

All the simulations were performed on the Defense Supercomputing Resource HPC computing systems using 64 processor running on 2 nodes in a parallel computation. To capture the details in the head model, the adaptive mesh was refined down to cell sizes on order of 1.5625 mm. A typical simulation required 34 h of CPU time for all calculations running out to 1 msec.

30.1.3 Comparison of Numerical Results with Blast Experiments

To create an accurate representation of the loading imparted to the head surrogate the numerical model was used to track the evolution of the blast waves. The results were compared to shadow graph images of the shock waves captured from the blast experiments. Figure 30.4 shows an instance in time of the shock wave propagation at 140 microseconds after the RDX initiation. Both the simulations and the experiments shows a leading shock front propagated by the reaction products shown in blue for the simulation and black for the experiments. In both the simulation and experiments a secondary initiation was also revealed at the center of the reaction cloud. The qualitative comparison between the simulations and the experiments shows good agreement for the blast wave propagation from a distance of 180 mm from the surrogate head model.

To measure the incident pressure imparted to the head, two pencil gauges shown in Fig. 30.1 were employed. The results from the gauge measurements are shown in Fig. 30.5. The recorded results from gauges c-5 and c-6 produced peak overpressure measurements on the order of 377 and 310 kPa respectively. The calculated results of 300 kPa also compared well with the experimental measurements for the blast evolution shown in Fig. 30.4.

Following the incident pressure wave impact to the skull, Fig. 30.6 shows both a reflected wave from the skull as well as a transmitted wave through the skull into the anterior regions of the surrogate brain material. Early time simulations in Fig. 30.6

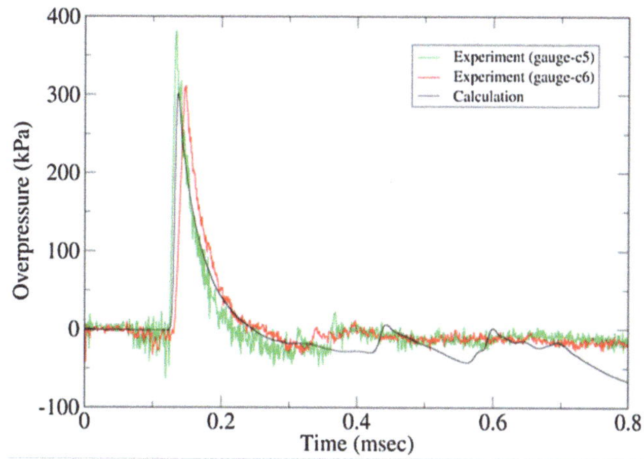

Fig. 30.5 Comparison of calculated overpressure history with pencil gauge data at 180 mm standoff distance from explosive charge

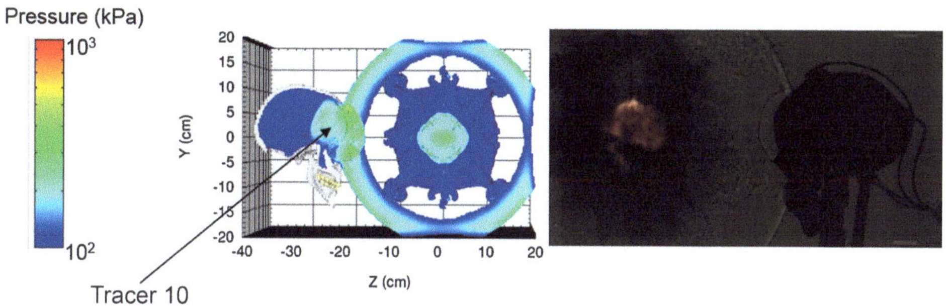

Fig. 30.6 Comparison of simulated blast wave impact on a surrogate head model with experiments at time 0.2 msec for a standoff distance of 180 mm from the explosive charge. The tracer particles shown are used to track the pressure history transition from the air to the skull

shows the intracranial pressure wave expansion from the impact point to the mid-region of the brain. Tracer particles were placed spatially throughout the surrogate brain material to capture the calculated pressure loading. The tracer 10 shown in Fig. 30.6 was positioned to match the location of pressure sensors used in the experiments. Figure 30.7 shows the comparison between the calculated pressure histories from tracer 10 compared with the recorded results from the pressure sensors. From Fig. 30.7, the surrogate brain material experienced both compressive and volumetric tensile loading. The recorded results show peak compressive loading of 95.47 and 57.83 kPa at times 0.237 and time 0.249 msec. Strong volumetric tension of −38.46 kPa were also recorded at times 0.24 msec and 0.263 msec followed by later time volumetric tension −36.27 kPa at time 0.459 msec. The calculated results show reasonable agreement with the experimental data. Early time peak compressive loading of 82.34 kPa was captured at time 0.244 msec followed by a decrease in the compressed loading of 62.2 and 56.07 kPa at times 0.297 msec and 0.472 msec. The calculation also revealed early time volumetric tension of −4.76 kPa at 0.267 msec followed by −31.46 kPa at time 0.499 msec. At later times (after 0.59 msec) the trend in the calculation shows a deviation from the recorded data.

Figure 30.8 below shows more qualitative agreement between the simulation results (left) and the shadow graph images (right) of the shock pressure propagation that impacted the surrogate head at time 0.33 msec. From the shadow graph image (right) the surrogate head experienced both a primary insult from the first shock wave followed by a secondary shock wave. This secondary shock wave was also captured in the simulation (left) but weakened in strength as the wave propagated toward the surrogate head model. Additionally, the simulation also captured the intracranial pressure distribution showing a focal region of elevated pressure at the back of the skull resulting from the internal reflected wave off the back of the skull. The simulated result also showed strong intracranial pressure in the anterior region of the surrogate brain.

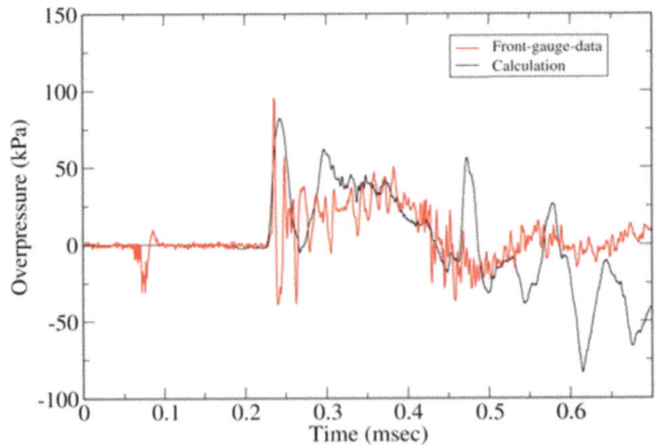

Fig. 30.7 Calculated pressure histories compared with recorded gauge data at location 5 cm in the anterior region of surrogate brain. Results show recorded and calculated peak pressure of 95.47, and 82.34 kPa at times 0.237 and 0.244 msec respectively

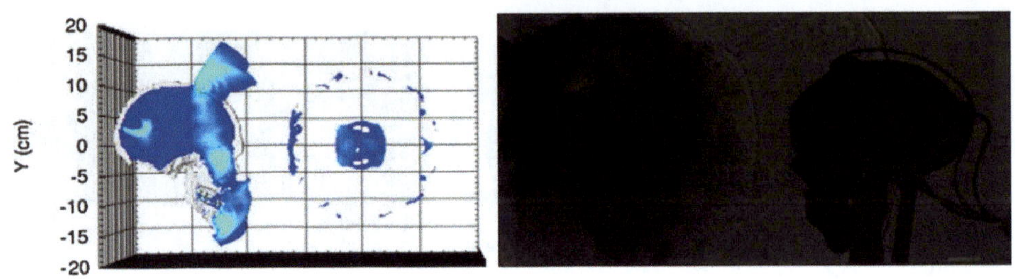

Fig. 30.8 Comparison of simulations (*left*) with experiments (*right*) of RDX blast wave impact to surrogate head model. Simulations show the mid-saggital view of intra-cranial pressure waves reflecting off the inside back skull into incoming pressure waves. Secondary initiation in the experiment produced a second shock wave which impacts the head at a later time after the first wave. Simulation results show break-up of a weak secondary shock wave

30.2 Results and Discussion

In summary, this study represents a concerted effort to address the need for more direct comparison of computationally developed head forms subject to blast impact with real blast experimental data. More specifically, the effects of early time blast impact loading on a surrogate head model was investigated both computationally as well as experimentally. The results revealed sequential shock pressure waves impacting the surrogate head captured via high resolution shadow graph images. This double impact has the potential for elevated damage to the brain beyond the initial damage incurred under a single shock loading. Better experimental diagnosis are needed to capture the influence of the secondary shock wave.

While shadow-graph images provide excellent external flow visualization around the surrogate head there remains a need for internal visualization of the wave mechanics inside the closed skull. The computational models employed in this work was used to fill that void by showing strong intracranial pressure loading in the anterior region of the surrogate brain with elevated focal regions of high pressure distribution in the posterior region. The quantitative comparisons between the calculated results and internal pressure sensors showed reasonable agreement. Both the calculated and experimental results showed compressive loading followed by volumetric tensile loading. The action of these peak positive and negative pressure loading could induce tearing and shearing in brain tissue. No direct comparison of shear stresses were conducted in this study. However, we are building our model to account the biomechanical shear response.

As of the writing of this manuscript additional comparison of the calculated results with experiments were not completed but will be presented in a future paper. However, we can conclude that the primary pressure load imparted to the head surrogate is a dominant injury mechanism. We are developing and validating our numerical models through parallel efforts with physical surrogate head forms subjected to real blast loading.

Reference

1. Moore, D.F., Jaffe, M.S.: Military traumatic brain injury and blast. NeuroRehabilitation. **26**(3), 179–181 (2010)
2. Taylor, P.A., Ford, C.C.: Simulation of blast-induced early-time intracranial wave physics leading to traumatic brain injury. J. Biomech. Eng. **131**(6), 61007 (2009)
3. Warden, D.L., et al.: Case report of a soldier with primary blast brain injury. NeuroImage. **47**, T152–T153 (2009)
4. Blyth, B.J., Bazarian, J.J.: Traumatic alterations in consciousness: traumatic brain injury. Emerg. Med. Clin. North Am. **28**(3), 571–594 (2010)
5. Raul, J.-S., Baumgartner, D., Willinger, R., Ludes, B.: Finite element modelling of human head injuries caused by a fall. Int. J. Legal Med. **120**(4), 212–218 (2006)
6. Willinger, R., Baumgartner, D.: Human head tolerance limits to specific injury mechanisms. Int. J. Crashworthiness. **8**(6), 605–617 (2003)
7. Zhang, L., Yang, K.H., King, A.I.: A proposed injury threshold for mild traumatic brain injury. J. Biomech. Eng. **126**(2), 226 (2004)
8. Chandra, N., et al.: Evolution of blast wave profiles in simulated air blasts: experiment and computational modeling. Shock Waves. **22**(5), 403–415 (2012)
9. Chavko, M., Koller, W.A., Prusaczyk, W.K., McCarron, R.M.: Measurement of blast wave by a miniature fiber optic pressure transducer in the rat brain. J. Neurosci. Methods. **159**(2), 277–281 (2007)
10. Sundaramurthy, A., Alai, A., Ganpule, S., Holmberg, A., Plougonven, E., Chandra, N.: Blast-induced biomechanical loading of the rat: an experimental and anatomically accurate computational blast injury model. J. Neurotrauma. **29**(13), 2352–2364 (2012)
11. Alley, M.D., Schimizze, B.R., Son, S.F.: Experimental modeling of explosive blast-related traumatic brain injuries. NeuroImage. **54**, S45–S54 (2011)
12. Cheng, J., et al.: Development of a rat model for studying blast-induced traumatic brain injury. J. Neurol. Sci. **294**(1–2), 23–28 (2010)
13. Risling, M., et al.: Mechanisms of blast induced brain injuries, experimental studies in rats. NeuroImage. **54**, S89–S97 (2011)
14. Zhu, F., et al.: Using a gel/plastic surrogate to study the biomechanical response of the head under air shock loading: a combined experimental and numerical investigation. Biomech. Model. Mechanobiol. **11**(3–4), 341–353 (2012)
15. Ganpule, S., Alai, A., Plougonven, E., Chandra, N.: Mechanics of blast loading on the head models in the study of traumatic brain injury using experimental and computational approaches. Biomech. Model. Mechanobiol. **12**(3), 511–531 (2013)
16. Hua, Y., Kumar Akula, P., Gu, L., Berg, J., Nelson, C.A.: Experimental and numerical investigation of the mechanism of blast wave transmission through a surrogate head. J. Comput. Nonlinear Dyn. **9**(3), 31010 (2014)
17. Taylor, P.A., Ludwigsen, J.S., Ford, C.C.: Investigation of blast-induced traumatic brain injury. Brain Inj. **28**(7), 879–895 (2014)
18. Grujicic, M., Arakere, G., He, T.: Material-modeling and structural-mechanics aspects of the traumatic brain injury problem. Multidiscip. Model. Mater. Struct. **6**(3), 335–363 (2010)
19. Moore, D.F., Jérusalem, A., Nyein, M., Noels, L., Jaffee, M.S., Radovitzky, R.A.: Computational biology — Modeling of primary blast effects on the central nervous system. NeuroImage. **47**, T10–T20 (2009)
20. Wang, C., Pahk, J.B., Balaban, C.D., Miller, M.C., Wood, A.R., Vipperman, J.S.: Computational study of human head response to primary blast waves of five levels from three directions. PLoS One. **9**(11), e113264 (2014)
21. Nahum, A.M., Smith, R., Ward, C.C.: Intracranial pressure dynamics during head impact (1977)
22. Settles, G.: High-speed imaging of shock waves, explosions and gunshots. Am. Sci. **94**(1), 22 (2006)
23. Cooper, P.W.: Explosives Engineering. Wiley-VCH, New York, NY (1997)
24. Mrozek, R.A., et al.: The relationship between mechanical properties and ballistic penetration depth in a viscoelastic gel. J. Mech. Behav. Biomed. Mater. **44**, 109–120 (2015)
25. Hertel, E.S., et al.: CTH: a software family for multi-dimensional shock physics analysis. In: Brun, R., Dumitrescu, L.Z. (eds.) Shock Waves @ Marseille I, pp. 377–382. Springer, Berlin, Heidelberg (1995)

Dr. Rohan Banton is a Research Engineer at the Army Research Lab (ARL) in Aberdeen Proving Ground, MD. His research focus is in the area of blast Injury research. He received his B.S. degree in Physics from Lincoln University, a M. S. degree in Mechanical Engineering from Drexel University, and his Ph.D. degree in Mechanical Engineering from the University of Maryland Baltimore County (UMBC).

Chapter 31
Pressure Sensing in Clay: A New Metric for Characterizing the Ballistic Backface Deformation Response of Personnel Protection Equipment

M.H. Merrill, C.J. Kindle, and J.P. Thomas

Abstract Clay indentation depth is traditionally used to quantify the performance of personnel protection equipment (PPE) under non-penetrating ballistic impact events. However, plastic deformation of the clay 'witness' media or backing following an impact event does not have a direct correlation to risk of injury. Alternative methods for assessing PPE have been developed including instrumented test dummies or gel-based surrogates, but results from these methods are not backward compatible with decades of existing clay witness data. Adding clay pressure measurements to the current clay-indent methodology will provide time-resolved data related to impact load transmission through the PPE that may better quantify the risks for biological damage. Four different commercial pressure sensors (piezoelectric and piezoresistive types) were selected and tested in air and clay using a shock tube and a step-less pressure generator with various conditioning and data acquisition methods. For both sensor types, sensitivity changed by only a few percent in clay compared to air, and both were able measure the amplitude, rise time, and shape of pressure waves in the clay media. The piezoresistive sensors appear to be better suited to this application due to their smaller size (necessary for helmet testing), greater bandwidth and resonant frequency, and superior mechanical robustness with lower sensitivity to radial loading and mounting torque variations.

Keywords Pressure Sensing • Dynamic Characterization • Ballistic Testing • Soft Armor • Shock Tube

31.1 Introduction

The state of the art in PPE testing lacks scientific links to modes of human injury [1, 2]. Adding clay impact pressure measurements to the current testing methodology would provide a relevant metric (transmitted load), but the viability of this advancement requires clay-pressure sensor systems that are accurate and repeatable. The commercial sensors selected for testing in this study are summarized in Table 31.1; their signal-bandwidth characteristics are shown in Fig. 31.1a. The piezoelectric sensors were 'integrated electronic piezoelectric' (IEPE) devices with on-board amplification to boost the signal levels and avoid cable-length interaction problems. The piezoresistive sensors use a strain-gage bridge on a silicon diaphragm; signal conditioning is simple, but the output is much lower in magnitude than the IEPE voltage signals. Several conditioning/acquisition systems were utilized in this study including the PXIe-4462, 4492, 4331, and 6361 modules from National Instruments, and a 1 MHz-bandwidth IEPE conditioner by PCB (model 438C05). As shown in Fig. 31.1a, the available data acquisition methods fully covered the range of sensor capabilities.

31.2 Shockless Pressure Step Generator

An Aronson Shockless Pressure Step Generator (Modal Shop Model 9907C Series) was used to measure sensor response. The pressure sensors were threaded into a steel plug which was inserted into the system. A large reservoir was then pressurized to the desired step pressure using helium gas. A poppet valve actuated by manual impact rapidly exposed the pressure sensor to the full reservoir pressure. The pressure step rise time for this system is 20–40 microseconds and the maximum pressure is 1000 psi. A typical calibration included 3–5 identical runs at each of 4–5 pressure levels; a straight line fit was used to

M.H. Merrill (✉) • J.P. Thomas
U.S. Naval Research Laboratory, 4555 Overlook Ave., SW, Washington, DC, 20375, USA
e-mail: marriner.merrill@nrl.navy.mil

C.J. Kindle
Aerodyn Engineering, Inc., 1919 S Girls School Rd, Indianapolis, IN 46241, USA

J. Kimberley et al. (eds.), *Dynamic Behavior of Materials, Volume 1*, Conference Proceedings of the Society for Experimental Mechanics Series, DOI 10.1007/978-3-319-62956-8_31

Table 31.1 Pressure sensor specifications

	PCB: 105C02	PCB: 113B21	ME: 8511A	MS: EPIH
Sensor type	Piezoelectric	Piezoelectric	Piezoresistive	Piezoresistive
Max pressure (psi)	200	400	5000	300
Sensitivity (mV/psi)	50	25	0.1	0.3
Resonant frequency (kHz)	>250	>500	>1000	>1700
Length (in)	0.9	1.5	0.9	0.7
Approximate cost (USD$)	1100	585	1410	1130

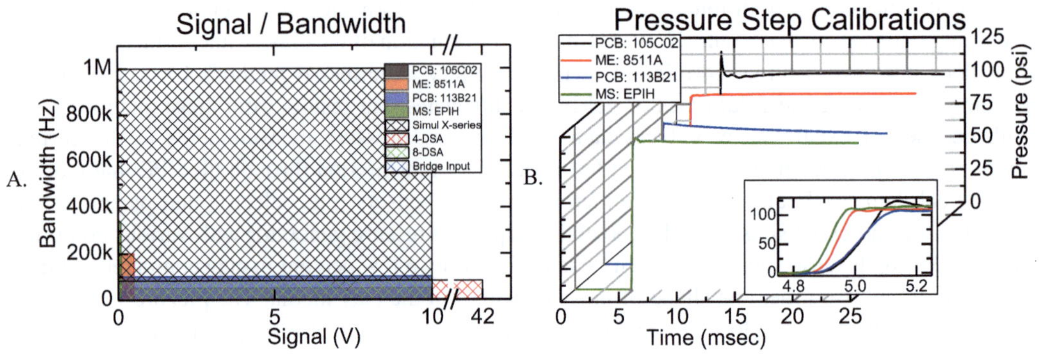

Fig. 31.1 (**a**) The four sensor types, their signal level vs bandwidth (20% of resonance), and data acquisition capabilities (*cross-hatched*). (**b**) Nominal 100 psi pressure step responses; the inset shows the rise times

determine the sensor sensitivity (output voltage per unit pressure). The supplied conditioner and data acquisition system with the Aronson device is limited to piezoelectric sensor types, so we used another system (voltage supply and data acquisition) to measure the response of the piezoresistive sensors. More than 40 sensors in all were tested. Measured sensitivities for the PCB:113B21 sensors and the two piezoresistive sensor models were typically within 1–2% of manufacturer specifications. Calibrations run with a clay covering on the sensor face showed only minor differences. For example, an MS:EPIH sensor showed changes in sensitivity of less than 2.3% (0.2889, 0.2835, 0.2848, and 0.2824 mV/psi) for a series of calibrations with and without clay. The ME:8511A sensors were particularly impressive; they fell within specified calibrations even at 100 psi, which is only 2% of full-scale (5000 psi). In contrast, the PCB:105C02 sensors showed inconsistent behavior with changes in sensitivity as high as 30–40%. This sensor was highly sensitive to radial pressures on the barrel near the sensing face. A polymer adapter plug was used to adapt the sensor to the Aronson plug, and the measured sensitivity would change by as much as 2× as the adapter was tightened beyond finger tight.

The response of the sensors to a step pressure is summarized in Fig. 31.1b for the four sensor types. The PCB: 105C02 sensors (black) had a tendency to overshoot the pressure and also showed some ringing. The other piezoelectric sensor (PCB: 113B21 – blue) showed the expected piezoelectric response with a step followed by a decay in the signal due to the inherent time constant decay. As seen in the inset, the two piezoresistive sensors showed faster rise times. Many of the sensor runs showed slight overshoot in pressure, possibly due to variations associated with poppet actuation.

31.3 Shock Tube Testing

A two-inch diameter shock tube (PCB 901A10) with circumferential ports in a polycarbonate extension (see inset of Fig. 31.2a) was used to test the sensors. Two sets of sensors were examined at different axial locations along the shock tube to measure pressure wave attenuation with distance. Unless otherwise noted, conditioned sensor outputs were zeroed to remove DC bias and low-pass filtered at 20% of the reported resonant frequency. All of the presented data corresponds to a nominal 40 psi (in air) shock caused by bursting 0.01 inch thick Mylar films with nitrogen at a driver pressure of 300 psi. For the clay-pressure measurements, the polycarbonate extension was partially filled (back end) with Roma Plastilina No 1 clay. The first set of sensors were located 1.5 inches in clay from the clay-air interface and the second set were 3.5 inches from the interface. The entire polycarbonate extension system with clay and sensors was thermally conditioned per PPE testing requirements in a 95 °F oven with clay test block verification [2].

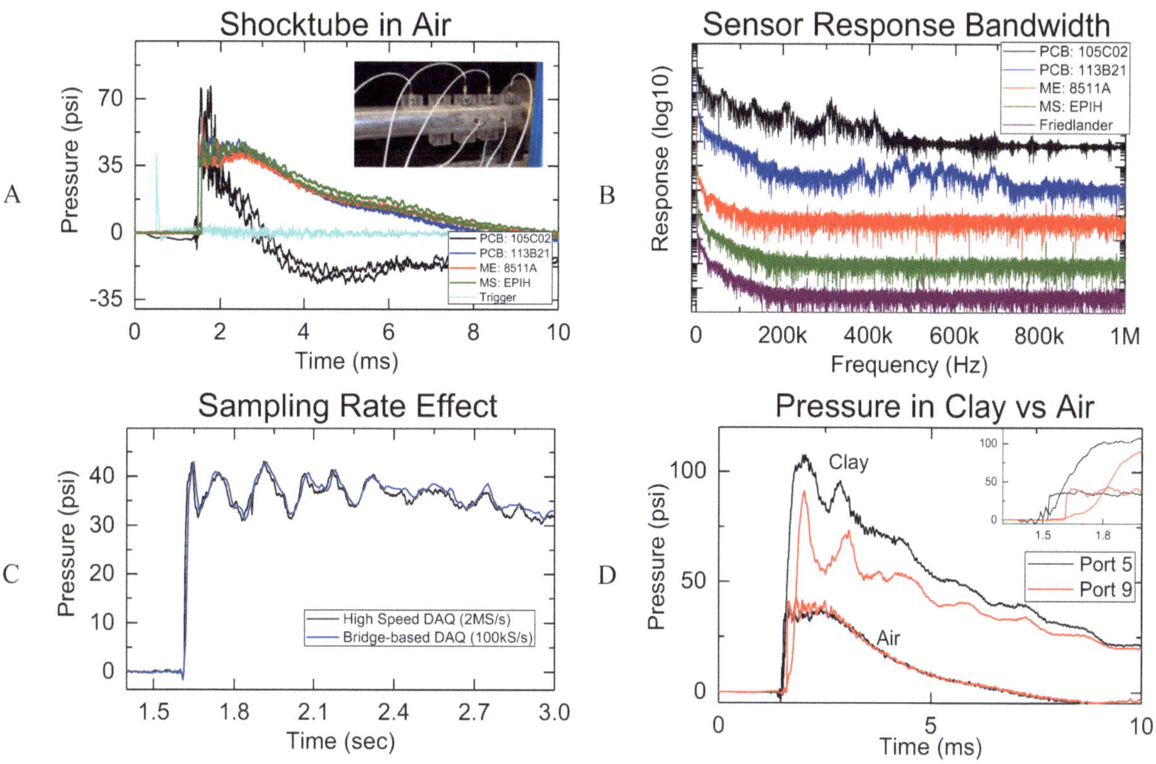

Fig. 31.2 Shock tube results for a nominal 40 psi air shock. (**a**) Typical response in air for all four sensor types; the inset shows the shock tube extension. (**b**) Frequency spectrum for the sensors compared to the ideal Friedlander wave. (**c**) Responses from two (identical) test runs showing the effects of a different data acquisition rate. (**d**) Shock tube response in air and clay for PCB 113B21 sensors. The inset shows the early time response. Port 5 and Port 9 are located 1.5 and 3.5 inches deep in the clay, respectively

Typical shock tube results are shown in Fig. 31.2a with two responses for each sensor type. They all exhibit the same basic pressure-time profile except for the 105C02 sensors (black traces), which show a markedly different peak magnitude and wave shape. In most PPE assessment applications, the primary measurement will be the peak pressure, which is less likely to be affected by test fixture reflections that would alter the wave shape. However, the peak pressure overshoot by the 105C02 sensors, similar to that seen with the Aronson tests, highlights their unsuitability for PPE testing applications. Unfiltered responses are shown in Fig. 31.2b along with the response of an ideal Friedlander wave. The 105C02 shows resonances (the minor peaks) as low as 60 kHz and the 113B21 at roughly 400 kHz. The piezoresistive sensors, by comparison, did not show any sign of resonance up to 1 MHz. In Fig. 31.2c, the responses for two "identical" air shocks with data acquisition rates of 100 kHz and 2 MHz show insignificant differences.

Figure 31.2d shows nominally identically shock tube runs with the PCB:113B21 sensors in air and clay. The peak pressures measured in the clay were more than double those in air, but attenuated with clay transit distance ~7–10%/inch. The peak pressure approximately corresponds to expectation by considering a 1D wave impacting a rigid boundary (transmitted = 2X incident). The pressure wave has a different in shape in clay than in air, and transmission appears to be slower than in air (roughly 833 fps compared to 1960 fps as measured in Fig. 31.2d). However, it is difficult to resolve the signal from internal reflections in the clay as the pressure wave pass back and forth through the 7 inch clay plug in less than 1 ms. Recent improvements in the shock tube extension design will allow for additional measurements including the in-air incident and reflected pressure wave to fully understand the in-clay pressures for shock tube measurements.

31.4 Conclusions & Future Work

In-clay pressure sensing is shown to be a viable mechanism to add quantifiable, time-resolved data to existing clay witness testing of soft armor. Commercial piezoresistive and piezoelectric sensors were evaluated both in clay and air media using an Aronson shockless pressure step generator and a shock tube. All tested sensors, except for the PCB:105C02, showed consistent sensitivities (output voltage per unit pressure) under repeated calibrations with the Aronson system, in air and with clay packing. The shock tube results showed higher bandwidth from the piezoresistive sensors with resonant frequencies greater than 1 MHz. Overall, the ME series of piezoresistive sensors appear to be optimal for our future implementation. Ongoing efforts include additional shock tube testing and ballistic testing of PPE with sensors embedded in clay witness media and silicone gel replacements for the clay. Future efforts will seek to link time-resolved pressure data to specific risk levels for biological damage.

Acknowledgements Funding for this work by Office of Naval Research and Naval Health Research Center under the direction of James Mackiewicz (NHRC) is gratefully acknowledged.

References

1. National Research Council: Review of Department of Defense Test Protocols for Combat Helmets. The National Academies Press, Washington, DC (2014)
2. National Research Council: Testing of Body Armor Materials: Phase III, p. 340. The National Academies Press, Washington, DC (2012)

Dr. M.H. Merrill is a research scientist at the U. S. Naval Research Laboratory focused in testing and processing multiscale and/or multifunctional materials. Projects have ranged from wave propagation in clay to electrospray self assembly. He received his Ph.D. degree from Purdue University in 2009.

Chapter 32
Evaluation of Concrete Penetration Resistance Using Small Caliber Bullets

Stephan Bless, Michael McAleer, and Rey Guzman

Abstract Small caliber armor-piercing bullets were fired at two difference types of concrete, 3.5 ksi or 10 ksi. The projectiles all had the same diameter, 5.6 mm, but had different lengths, and hence different masses. Exit velocity was measured. Therefore the data provided information for penetration resistance as a function of impact velocity, projectile mass, and concrete strength. Data were fit with a Poncelet equation. Superior fits were achieved when both the inertial term and strength terms were included. The empirical Poncelet parameters were compared with those computed from conventional cavity expansion analysis that includes both inertia and compressive strength. The cavity expansion model provided a good fit to residual velocity as a function of impact velocity for the 3.5 ksi targets but not the 10ksi targets.

Keywords Poncelet • Concrete • Bullets • Residual Velocity • Penetration Resistance

32.1 Introduction

The usual framework for analyzing penetration is in terms of penetration resistance (defined as the force on the projectile divided by its cross section area). Penetration resistance is generally considered a property of the target material that is influenced by the penetration rate, the shape of the projectile/target interface, and the strength of the target. The simplest case is a rigid penetrator, since in that case the interface geometry is largely determined by the penetrator nose shape, and the effect of the penetration resistance is only to decelerate the penetrator (and not to erode the penetrator). If target properties do not depend on strain rate, elementary dimensional arguments indicate that for a homogeneous material of unlimited lateral extent, penetration resistance (a stress) does not depend on penetrator size (for geometrically similar penetrators). Most concretes are only weakly rate dependent, and this suggests small scale experiments can be used to predict trends that would occur in large scale tests. Thus, although the particular tests that we have done are for small caliber projectiles, to the extent that strength of concrete is rate dependent, the results for penetration resistance should be expected to apply for larger and longer projectiles as well. (Note that stress being independent of rate does not mean it is independent of penetration velocity, which is approximately equal to diameter times strain rate).

The goal of the experiments reported herein was to evaluate the performance of some exotic projectile materials relative to baseline steel. Only the steel data are discussed in this paper.

32.2 Penetrators

A drawing of the baseline projectile is shown in Fig. 32.1. For shooting, all projectiles were encased in copper jackets to engage 7.62-mm diameter barrel rifling. The copper jacket stripped immediately upon impact on targets.

The baseline projectiles were made from 4340 steel. 4340 steel rods were procured then heat treated to a hardness of 45 HRC and machined to drawing specifications. The mass of the steel baseline projectile was 3.17 g.

In order to vary mass, a shorter length (17.25 mm) projectile was created. This had the same diameter and nose profile as the base projectile. The mass of the shorter projectile was 2.61 g.

S. Bless (✉)
New York University Tandon School of Engineering, 333 Jay St, Brooklyn, NY, 11201, USA
e-mail: sjbiat@me.com

M. McAleer • R. Guzman
Nanohomics Inc., 6201 E. Oltorf St, Austin, TX, 78741, USA

J. Kimberley et al. (eds.), *Dynamic Behavior of Materials, Volume 1*, Conference Proceedings of the Society for Experimental Mechanics Series, DOI 10.1007/978-3-319-62956-8_32

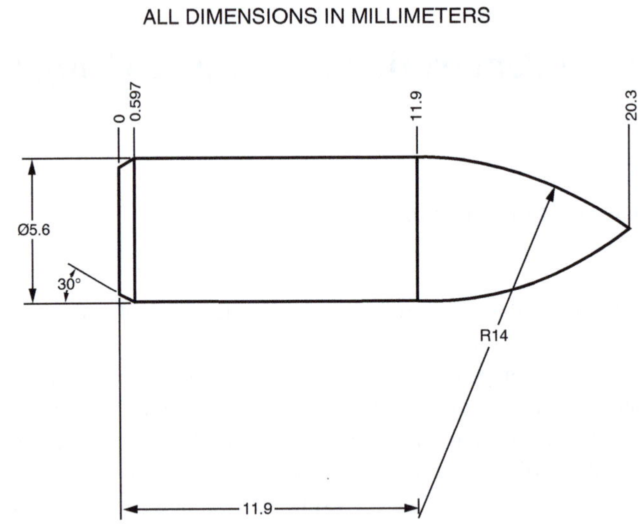

Fig. 32.1 Geometry detail for the fabricated projectile

Fig. 32.2 10 ksi target encased in 6 in pipe

32.3 Targets

There were two target formulations, referred to herein as "strong" and "weak." The strong targets had a nominal strength of 10 ksi (57 MPa), were six-inches in diameter, and encased in steel as shown in Fig. 32.2. The strong concrete meets ASTM C33–03 standard specification for concrete aggregates. The aggregate was sifted through an 18 mesh sieve to ensure a particle size less than 1 mm. The concrete was then poured into 6 inch diameter schedule 40 steel pipe that was cut down to approximately 4 inches long and allowed to cure for 30 days. The steel pipe has an interior diameter of 6.065 inches with a wall thickness of 0.28 inches and a nominal burst pressure of 5 ksi. Compressive tests were performed on three of the concrete cylinders using test method ASTM C39 "Standard Test Method for Compressive Strength of Cylindrical Concrete Specimens". The cylinders had an average compressive strength of 10,880 psi with fracture pattern 3. The fracture pattern was determined using the schematic in ASTM C39. The specific gravity of the strong concrete was 2.12.

The weak targets were nominally compressive strength of 3.5 ksi (23 MPa). They were furnished by the U.S. Army Engineering Research and Development Center (ERDC) in the form of large concrete drums which were cut down to make the 6″×6″×2″ targets blocks. For testing, the targets were stacked in pairs and taped in an attempt to make a 4 inch thick block. The specific gravity of the weak concrete was 1.93.

32.4 Experiment Procedures

Shots were carried out at Chesapeake Testing Services in Belcamp, MD. Velocity was measured from four time-of-arrival detectors in all shots. A high speed camera was used to measure velocity after exit through the target.

Table 32.1 Ballistic data

Shot	Target	Bullet	Mass	V_o	V_R
1–2	Weak	Std	3.18	860	213
1–3	Weak	Std	3.19	951	390
1–4	Weak	Std	3.17	1106	605
2–1	Weak	Std	3.17	917	286
2–2	Weak	Std	3.17	1110	582
2–3	Weak	Std	3.17	1119	625
3–1	Strong	Std	3.17	1070	459
3–2	Weak	Std	3.17	1079	639
3–3	Strong	Std	3.17	1210	544
4–5	Strong	Short	2.60	1021	221
4–7	Strong	Std	3.15	917	174
4–8	Strong	Std	3.15	1045	385
4–9	Strong	Std	3.14	1137	500
4–12	Weak	Short	2.61	1013	347

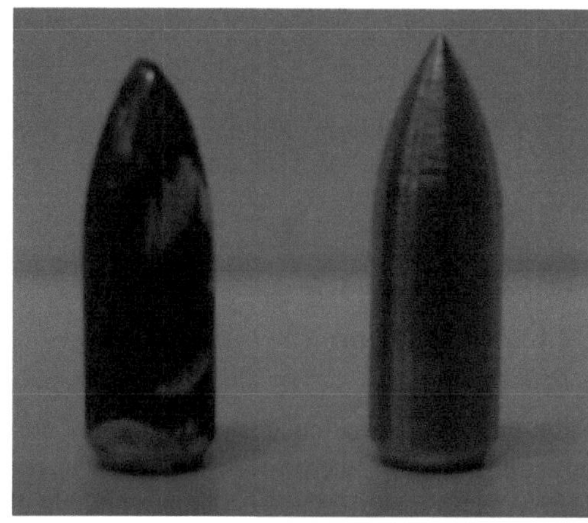

Fig. 32.3 Damage to steel bullet from strong target (Virgin bullet on *right*)

In some shots, striking velocity was also measured with high speed cameras. In one test the target was eliminated and three high speed cameras were used. It was found that the velocities measured by the cameras scattered by 0.6% around the average and the average velocity measured by the cameras agreed within 0.1% with the velocity measured from the time-of-arrival detectors. Applied to data taken with targets, this implies uncertainties in impact velocities are negligible, and uncertainties in exit velocities are typically ± 3 m/s. The targets were held in place by clamps to a steel frame.

In all but the first four shots, pitch and yaw were measured with pre-impact high speed cameras. Angle of attack was always minimal. The exiting projectile was always well ahead of any concrete fragments.

The data are listed in Table 32.1. V_o is striking velocity, V_R is exit velocity. Here lengths are in mm, velocities are in m/s, mass is in grams. Targets are described as "weak" or "strong" as above.

32.5 Residual Velocity Data

It was expected that loss of mass of the projectiles would be minimal, and that was the case. Although in some of the highest velocity shots there was slight nose deformation, since the projectiles were relatively high aspect ratio, the energy absorbed in nose deformation was a negligible fraction of the initial kinetic energy. For example, Figs. 32.3 and 32.4 compares a preshot and postshot steel bullet (Shot 4–5) for a strong target.

Figure 32.4 is a plot of all of the V_o–V_R data. In this plot the blue round points are the baseline steel bullet vs the weak target. The brown round points are the baseline steel bullet vs the stronger target. The lines are spline fits to the data. The offset of these two curves demonstrates the effect of strength of the target on residual velocity.

The blue diamond is the lightweight steel bullet vs the weak target. The brown diamond is the lightweight steel bullet vs the strong target. The offsets show the effect of projectile mass on residual velocity.

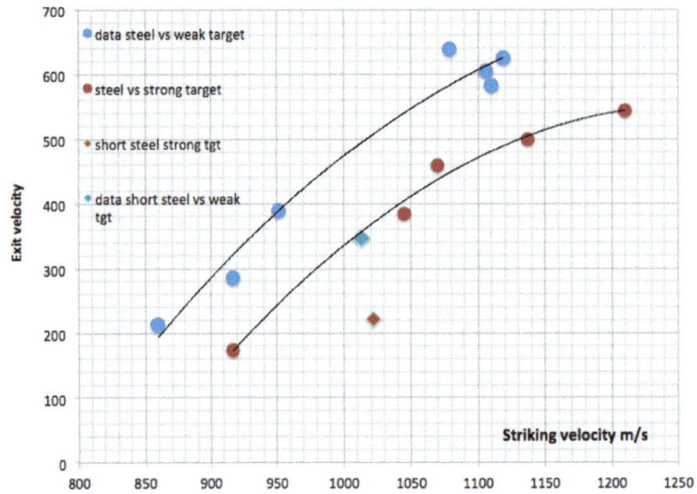

Fig. 32.4 Summary of striking and residual velocity data

The scatter amongst groups of data at similar velocities implies that the variability in any one measurement of exit velocity is ±10–20 m/s. The uncertainty in measuring velocity is much less than this, so the scatter is probably due to target heterogeneity.

32.6 Discussion

Semi-empirical equations for penetration of concrete and other geomaterials usually take for form of a Poncelet equation, in which the force of the projectile is given in eq. (32.1), which is written for the case where the projectile mass is constant. Here A is presented area, ρ is density, C is the Poncelet drag, and R is target strength.

$$F = M\frac{dv}{dt} = -\rho CAv^2 - RA \tag{32.1}$$

Inertia effects give rise to the velocity-dependent term, and target strength effects are manifested in the second term. The equation can be integrated to give the exit velocity if we ignore the influence of the free surfaces (which are small in these targets which are more than 10 diameters thick):

$$v = \left[\left(R/\rho C + v_0^2\right)\exp\left(-2\rho Ax/M\right) - R/\rho C\right]^{1/2} \tag{32.2}$$

Forrestal al et. [1] derived values for C and R from a cavity expansion analysis, and found that the Poncelet parameters are given by

$$\mathrm{C} = \mathrm{N}, \mathrm{R} = \mathrm{Sf} \tag{32.3}$$

Here $N = (8\Psi\text{-}1)/24\Psi^2$; $\Psi = r/2a$, where r is ogive radius giving $\Psi = 2.5$ & $N = 0.127$; f is the unconfined compressive strength and S is an empirical factor $= 82.6\ (f)^{-0.544}$ (for f in MPa). The result is C = 0.127, R = 528PMa for the strong target, and R = 349 MPa for the weak target. Figure 32.5 illustrates Poncelet fits to weak target data. The best fit is given by C = 0.125 R = 380 MPa. This is in excellent agreement with the values predicted by Forrestal et al. However, for the strong data, no Poncelet fit is adequate for describing the data.. This is probably due to erosion of the projectile nose in the strong target.

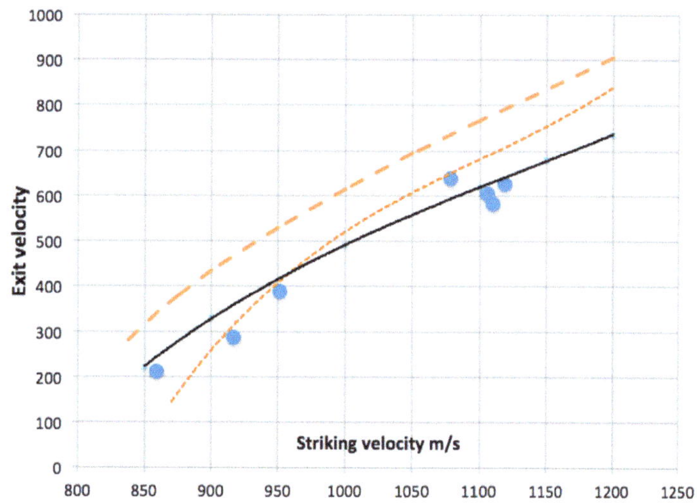

Fig. 32.5 Comparison of data for weak target for a Poncelet fit (*solid line*) and a constant-resistance force (*dotted lines*); the longer dashes use the force-strength correlation from Rosenberg and Kisitoski; the shorter dashes use the best fit value

The Poncelet equation implies that if the mass changes,

$$dV_R/dM = \left[\frac{RAT+AC\rho TV^2}{M^2}\right]\frac{\exp(-AC\rho T/M)}{\sqrt{\left[\left(R/\rho C+V_0^2\right)\exp(-2AC\rho T/M)-R/\rho C\right]}} \quad (32.4)$$
$$= 1.42\mathrm{E}5\ \mathrm{m/s/kg\ when\ V} = 1013\ \mathrm{m/s}.$$

For the difference in mass between the standard and short steel bullets, namely 0.56 g, this formula predicts a change in exit velocity of 79 m/s. The actual change when the standard bullet (interpolated to 1013 m/s using the Poncelet fit) is compared to the light steel bullet at that velocity is 145 m/s. There is relatively little uncertainty in the experimental value for the standard projectile because there are several points at around this same velocity. Thus, somewhat surprisingly, although the Poncelet formulation well describes variations of V_R with V_o, it is not satisfactory for predicting the effect of variations in projectile length.

The extent to which the velocity-dependent term in eq. (32.1) is required has been discussed recently by several authors. Warren et al. [2] analyze accelerometer data for rods penetrating concrete at speeds up to 460 m/s and found that deceleration was effectively constant after an initial transient. Rosenberg and Dekel [3] Rosenberg and Kositski [4] also examine several previous publications and find that penetration depth vs velocity data do not require a velocity-dependent term in eq. (32.1) at speeds up to 1100 m/s.

Thus, for geomaterials in general and concrete in particular, the conditions under which penetration resistance is velocity dependent are not well established. It does seem that from penetration depth data for rigid long rods into strong concrete at velocities up to 1.1 km/s, fitting data does not require a velocity-dependent resisting force. However, final depth of penetration is perhaps not very sensitive to penetration resistance at the highest velocities which only occur near the impact point. It must be that resisting force does indeed depend on velocity for high enough velocity (perhaps where the projectile is no longer rigid), and also for rigid projectiles into enough weak targets (for example, see the data for sand targets in Omidvar et al. [5]).

The constant-force model implies that

$$V_R = \left(V_0^2 - V_L^2\right);\ V_L = \sqrt{2RAT/M} \quad (32.5)$$

where T is target thickness, and V_L is the limiting impact velocity at which the exit velocity is just zero. Empirically Rosenberg and Kositski found R (in GPa) = 0.22 ln f − 0.285, for f in MPa. This gives R = 405 MPa for the weak concrete.

For the weak target, a case in which it seems reasonable to consider the projectile as rigid, Fig. 32.5 compares the Poncelet model (which is nearly the same as the Forrestal et al. prediction) and the constant-force model. The Poncelet-Forrestal prediction gives a satisfactory fit to the data. The constant force predicted by the Rosenberg and Kisitoski model is much less accurate. The best fit to a constant-force model (with R = 480 MPa) is close to the data but not as good as the Poncelet fit.

However, three additional comments should be noted: (a) a Poncelet fit has two parameters, vs one for the constant-force equation, so it is easier to fit a given data set. (b) a best-fit straight line relationship for V_R vs V_S is as good a fit as the Poncelet model, and (c) it cannot be ruled out that at higher velocity some additional energy is being absorbed by a slightly deforming projectile, which in both cases makes it difficult to fit the high velocity data with the same fidelity as the lower speed data.

32.7 Summary

Use of small caliber bullets and finite targets is a convenient means to characterize the penetration resistance of concrete. For targets that are 23 MPa strength and hard steel bullets, rigid body penetration appears to occur and the data are well described by a Poncelet equation using coefficients as predicted from accepted cavity expansion theory. The data are not as well fit by the constant-force model of Rosenberg and Kositski. 57 MPa targets appear to deform projectiles above 1000 m/s and the Poncelet formulation does not adequately describe the experimental results.

Acknowledgements This work was supported in large part by the Defense Threat Reduction Agency under contract HDTRA1-14-C-0025. The authors are grateful for the guidance of the DTRA Project Manager Dr. Richard Lewis.

References

1. Forrestal, M.J., Altman, B.S., Cargile, J.D., Hanchak, S.J.: An empirical equation for penetration depth of ogive-nose projectiles into concrete targets. Int. J. Impact Eng. **15**, 395–406 (1994)
2. Warren, T., Forrestal, M., Randles, P.W.: Evaluation of large amplitude deceleration data from projectile penetration into concrete. Exp. Mech. **54**, 241–253 (2014). doi:10.1007/s11340-013-9767-9)
3. Rosenberg, Z., Dekel, E.: Letter to the editor. Int. J. Impact Eng. **93**, 231–233 (2016)
4. Rosenberg, Z., Kositski, R.: Modeling the penetration of perforation of concrete targets by rigid projectiles. Int. J. Protective Struct. **7**(2), 157–178 (2016)
5. Omidvar, M., Malioche, J.D., Bless, S., Iskander, M.: Phenomenology of rapid projectile penetration into granular soils. Int. J. Impact Eng. **85**, 146–160 (2015)

Dr. Stephan Bless is an expert in impact mechanics and dynamic behavior of materials. He headed the hypervelocity impact team at the University of Texas at Austin for 20 years, and is now Research Faculty in the Department of Civil and Urban Engineering at New York University.

Chapter 33
Dynamic Analysis of a Bi-stable Buckled Structure for Vibration Energy Harvester

Masoud Derakhshani, Thomas Berfield, and Kevin D. Murphy

Abstract Vibration energy harvesting offers a viable alternative to batteries for powering sensors in remote locations. In the past decade, the energy harvesting community has turned to nonlinear structures as an effective means for creating high-performance devices. In particular, researchers have used buckled structures to improve vibration scavenging power production at low frequencies (<100 Hz) and to broaden device operational bandwidths. To achieve these ends, accurate structural models are needed. These models are critical for carrying out a systematic and quantitative device design process. Specifically, the models enable the user to optimize device geometries, arrive at meaningful estimates of power production, and estimate device lifetimes, etc. This work focuses on the dynamic behavior of a bi-stable switching energy harvester made from a buckled beam structure, coupled to two cantilever beams with tip masses via a torsional rod. Results from experimental testing of the energy harvesting structure under different forced vibration conditions are compared with a nonlinear model created of the structure. For the model, linear equations of motion for free vibration of each component have been derived using Hamilton's principle, and shape functions for each individual component are determined by applying boundary conditions for the linear vibration. Nonlinear dynamic behavior effects are integrated through consideration of large deformation of the main beam. The effects of different parameters on the vibrational system, including the geometry of the structure, buckling load and natural frequency of the cantilever arms, have been investigated. These parameters can play an important role in the optimization process of energy harvesters. Finally, parametric results obtained from the presented method are compared with the experimental data in different aspects.

Keywords Vibration Energy Harvesting • Bistability • Nonlinear dynamics • Buckled beam • Coupled components

33.1 Introduction

In recent years, utilizing wasted vibrational energy has become one of the attractive topics in the field of energy harvesting. Plenty of studies have been done so far on bistable oscillators as a promising way to address the issues associated with linear Vibration Energy Harvesters (VEH), including covering a wider bandwidth of frequencies and existence of two stable states, which results in more reliable design for energy harvesting systems. R.L. Harne and K.W. Wang [1] gave a thorough review on the principal concepts of dynamically nonlinear beams as bistable energy harvesters. They also pointed out the challenges with different analysis approaches and popular experimental setups in their study. One of the main concerns in nonlinear energy harvesting systems is proposing a relatively accurate dynamic analysis of the main structure. Theoretical analysis of nonlinear dynamics of a clamped-clamped buckled beam using the Galerkin discretization method was studied by Samir A. Emam and Ali H. Nayfeh [2]. It is shown that by considering more number of modes for the nonlinear beam, its dynamic behavior can be better described for each sequence of motion. Dynamic modeling and active control of a highly nonlinear hinged-hinged beam was studied by B. R. Nana Nbendjo and P. Woafo [3]. The major difference of this model with the classical Duffing oscillator is its ability of describing the dynamical destruction in the system for large deformations in the potential function.

Samuel C. Stanton et al. [4] studied the effects of nonlinearity in vibration energy harvesting on the broadband frequency response both theoretically and experimentally. They found out that bistable behavior can result in getting a broader band frequency response compared to linear energy harvesters. The nonlinear dynamic analysis of piezoelectric beams for vibration energy harvesting has been thoroughly investigated [5–8]. In these studies, it is shown that bistable buckled beams have more reliable performance as energy harvesters compared to unbuckled cases. It is also pointed out that different

M. Derakhshani (✉) • T. Berfield • K.D. Murphy
University of Louisville, 200 Sacket Hall, Louisville, KY, 40208, USA
e-mail: masoud.derakhshani@louisville.edu

J. Kimberley et al. (eds.), *Dynamic Behavior of Materials, Volume 1*, Conference Proceedings of the Society for Experimental Mechanics Series, DOI 10.1007/978-3-319-62956-8_33

boundary conditions can affect the level of bistability and, therefore, the output voltage of the system. Michael I Friswell et al. [9] studied a new configuration of vibration energy harvester, in which a cantilever beam with a tip mass is mounted vertically and excited in the transverse direction at its base. The results show that by increasing the tip mass, the nonlinear behavior mostly dominates the system until it ends up switching to bistable configuration. Asl et al. developed scaled beam models for vibration prediction and dynamic characterization of composite I-beams [10, 11]. They designed scaled beam prototypes to predict the frequency response function and strain field of flexural beams, which can be useful in analyzing the dynamic response of scaled energy harvesters. The obtained results show that the scaled model is able to replicate the dynamic behavior and the strain field of its full-scale beam [12]. Louis Van Blarigan et al. [13] developed a model for describing the dynamic responses of a non-uniform bistable piezoelectric beam used for vibration energy harvesting. The necessity of keeping several modes in addition to the buckled equilibrium mode shape in the analysis in order to properly capture the small strain snap-through effects of the system is found in this study. Effects of piezo-magnetoelastic harvesters are investigated and compared in different cases for achieving the best performance of the energy harvesters at low frequencies by Abdessattar Abdelkefi and Nilma Barsallo [14]. It is shown that the attractive interaction and smaller spacing gap between two magnets is very beneficial to design a low-frequency piezoelectric energy harvester due to its softening behavior. As it is shown in this part, there have been a lot of studies on dynamic behavior of bistable beams as an energy harvesting system; however, in most of the works, a single nonlinear beam is considered in the modeling part. The main drawback of using such structure as an energy harvester is the difficulty of changing states between two stable conditions during dynamic motions. In the next section, the dynamic behavior of a unique structure in which a nonlinear beam is coupled with two cantilever arms via a torsional rod is derived and analyzed. These cantilever arms with tip masses more readily facilitate the switching process between stable states.

33.2 Formulation

The bistable buckled structure shown in Fig. 33.1 consists of a clamped-clamped main buckled beam which is attached to a torsional rod in the middle. The torsional rod couples the main beam to two cantilevers with tip masses at their ends. The bistability in the beam is provided by a central buckling load applied at the ends of the clamped-clamped beam in a way that both ends are movable in the longitudinal direction. Because the effects caused by the PZT layers on the mechanical behavior of the main beam such as mechanical strain and vibration amplitude is not noticeable, they are neglected in the modeling part. Nonlinear Euler-Bernoulli beam theory is considered for modeling the main beam. By substituting nonlinear strain-displacement relations into the energy terms and applying Hamiltonian principle, the equation of motion for each part of the system is obtained as follows:

$$\overline{m}_1\frac{\partial^2 w_1}{\partial t^2}+EI_{z1}\frac{\partial^4 w_1}{\partial x^4}+P\frac{\partial^2 w_1}{\partial x^2}+c\frac{\partial w_1}{\partial t}-\frac{EA_1}{2l_1}\frac{\partial^2 w_1}{\partial x^2}\left[\int_0^{l_1}\left(\frac{\partial w_1}{\partial x}\right)^2 dx\right]=-\overline{m}_1\frac{\partial^2 w_B}{\partial t^2} \tag{33.1}$$

$$\overline{m}_2{r_2}^2\frac{\partial^2\phi}{\partial t^2}+c_2\frac{\partial\phi}{\partial t}-JG\frac{\partial^2\phi}{\partial t^2}=0 \tag{33.2}$$

$$\overline{m}_i\frac{\partial^2 w_i}{\partial t^2}+EI_{zi}\frac{\partial^4 w_i}{\partial x^4}+c_i\frac{\partial w_i}{\partial t}=-\overline{m}_i\frac{\partial^2 w_i}{\partial t^2}\qquad (i=3,4) \tag{33.3}$$

in which $\overline{m}, EI, JG, c, r, w_B$ linear mass density, longitudinal and torsional rigidity, damping coefficient, radius of gyration and base vibration, respectively. It should be noted that the relative displacement with respect to the base is considered for each component and the equations of motion for torsional rod and cantilevers are considered linear in this modeling. In order to find the response of the system to the harmonic motion, the base vibration is considered as a trigonometric function of time multiplying by a constant amplitude, which is moving at a specific external frequency ω. By introducing some new coefficients, the set of dynamic equations for the whole system can be written in the following form:

$$\ddot{w}_1+\overline{k}_1{w_1}''''+\overline{P}{w_1}''+c_1\dot{w}_1-\overline{k}_2{w_1}''\int_0^{l1}{w_1'}^2dx=-\ddot{w}_B \tag{33.4}$$

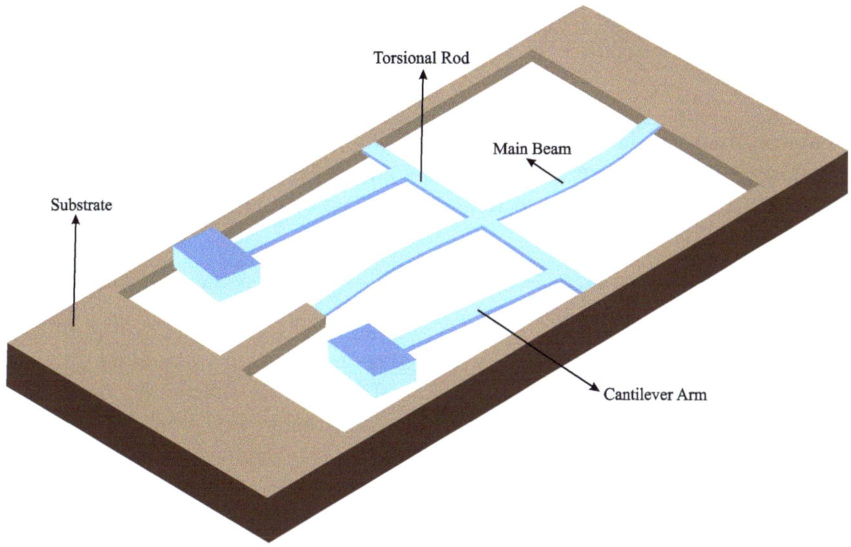

Fig. 33.1 Schematic model of the device

$$\ddot{\phi} + \overline{c}_2\phi - \overline{k}_t\phi'' = 0 \tag{33.5}$$

$$\ddot{w}_i + \overline{k}_i w_i'''' + \overline{c}_i\dot{w}_i = -\ddot{w}_B \qquad (i = 3, 4) \tag{33.6}$$

where

$$\begin{gathered} ()' = \frac{\partial}{\partial x}, \quad (\dot{}) = \frac{\partial}{\partial t} \\ \overline{k}_1 = \frac{EI_{z1}}{\overline{m}_1}, \quad \overline{P} = \frac{P}{\overline{m}_1}, \quad \overline{c}_1 = \frac{c_1}{\overline{m}_1}, \quad \overline{k}_2 = \frac{EA_1}{2\overline{m}_1 l_1}, \quad \overline{c}_2 = \frac{c_2}{\overline{m}_2 {r_2}^2}, \quad \overline{k}_t = \frac{JG}{\overline{m}_2 {r_2}^2}, \quad \overline{k}_3 = \frac{EI_{z3}}{\overline{m}_3}, \quad \overline{c}_3 = \frac{c_3}{\overline{m}_3} \end{gathered} \tag{33.7}$$

The method used to solve the dynamic behavior of the system is based on dividing the solution into spatial and time variables. The shape functions considered for the spatial domain need to be chosen in a way that can satisfy the geometrical boundary conditions. N modes are considered for each component and the presumed solution for each part can be written as a summation of its mode shapes multiplying by time-dependent generalized coordinates:

$$\begin{aligned} w_1 &= \textstyle\sum_{i=1}^{N1} \psi_i^1(x) q_i^1(t); && 0 \le x \le l_1 \\ \phi &= \textstyle\sum_{i=1}^{N2} \psi_i^2(x) q_i^2(t); && 0 \le x \le l_2 \\ w_j &= \textstyle\sum_{i=1}^{Nj} \psi_i^j(x) q_i^j(t); && 0 \le x \le l_j \quad (j = 3, 4) \end{aligned} \tag{33.8}$$

These shape functions can be achieved through solving free linear vibration equation for each part. The nonlinear equations of motion of the system are discretized using Galerkin approach for each component and eventually, a set of nonlinear ordinary differential equations is obtained as follows:

$$\sum_{j=1}^{N_1} m_{ij}^1 \ddot{q}_j^1 + \sum_{j=1}^{N_1} c_{ij}^1 \dot{q}_{ij}^1 + \sum_{j=1}^{N_1} k_{ij}^1 q_{ij}^1 - \sum_{j=1}^{N_1}\sum_{k=1}^{N_1}\sum_{l=1}^{N_1} k_{ijkl}^1 q_j^1 q_k^1 q_l^1 = F_i \quad (i = 1, 2, \ldots., N_1) \tag{33.9}$$

$$\sum_{j=1}^{N_2} m_{ij}^2 \ddot{q}_j^2 + \sum_{j=1}^{N_2} c_{ij}^2 \dot{q}_{ij}^2 + \sum_{j=1}^{N_2} k_{ij}^2 q_{ij}^2 = 0 \quad (i = 1, 2, \ldots, N_2) \tag{33.10}$$

$$\sum_{j=1}^{N_n} m_{ij}^n \ddot{q}_j^n + \sum_{j=1}^{N_n} c_{ij}^n \dot{q}_{ij}^n + \sum_{j=1}^{N_n} k_{ij}^n q_{ij}^n = F_i^n \quad (i = 1, 2, \ldots, N_n) \quad (n = 3, 4) \tag{33.11}$$

The coefficients of mass, stiffness and damping ratio as well as nonlinear stiffness of the main buckled beam can be found as a function of considered mode shapes. The next part is linking individual elements of the system at their connected points. Matching conditions which satisfy geometrical connecting relations can be written as follows:

$$w_1'\left(x = \frac{l_1}{2}\right) = \phi\left(x = \frac{l_2}{2}\right), \quad w_3'(x = 0) = \phi\left(x = \frac{l_2}{8}\right), \quad w_4'(x = 0) = \phi\left(x = \frac{7l_2}{8}\right) \tag{33.12}$$

It needs to be mentioned that each cantilever arm is located a quarter of the way along the gap between connecting point of the main beam with torsional rod and the ending points of the rod. By substituting the presumed solutions into the above equations, we can relate the generalized coordinates of the components. For simplicity, the same number of modes is considered for each part in the calculation process. There are totally three matching conditions and therefore three additional relations between generalized coordinates, which results in the dependency of three of them to the rest of generalized coordinates. By rearranging the obtained relations and eliminating these three dependent coordinates from the whole system, we can write the relationship between the total and independent generalized coordinates through a transformation matrix as follows:

$$\underline{q}_{4N\times 1} = \left\{\underline{q}^{1T}, \underline{q}^{2T}, \underline{q}^{3T}, \underline{q}^{4T}\right\}^T = [B]\,\underline{\overline{q}}_{(4N-3)\times 1} \tag{33.13}$$

where $\underline{q}$ and $\underline{\overline{q}}$ are the vector of total and reduced (independent) generalized coordinate system respectively. By substituting the obtained transformation into the set of above ordinary differential equations and solving them numerically, the independent generalized coordinates and therefore the dynamic solution for the whole system can be obtained.

33.3 Experimental Setup

Figure 33.2 shows the setup of the experiment for capturing the beam displacement results. In order to dynamically analyze the system, a custom built shaker table was used to test the device under harmonic excitation. The shaker system consisted of an anchor platform mounted on a 42 W speaker driven using an audio amplifier. An Arduino Nano (V3.0) was used to read the accelerometer and send its value to a LabVIEW program [15]. The testing device is made of Acrylonitrile-Butadiene-Styrene (ABS) using a GEEETech 3D printer with the dimensions listed in Table 33.1. As it is shown in Fig. 33.2b, the device is fixed on the shaker table at both ends of the main beam and torsional rod and the axial load is exerted by changing

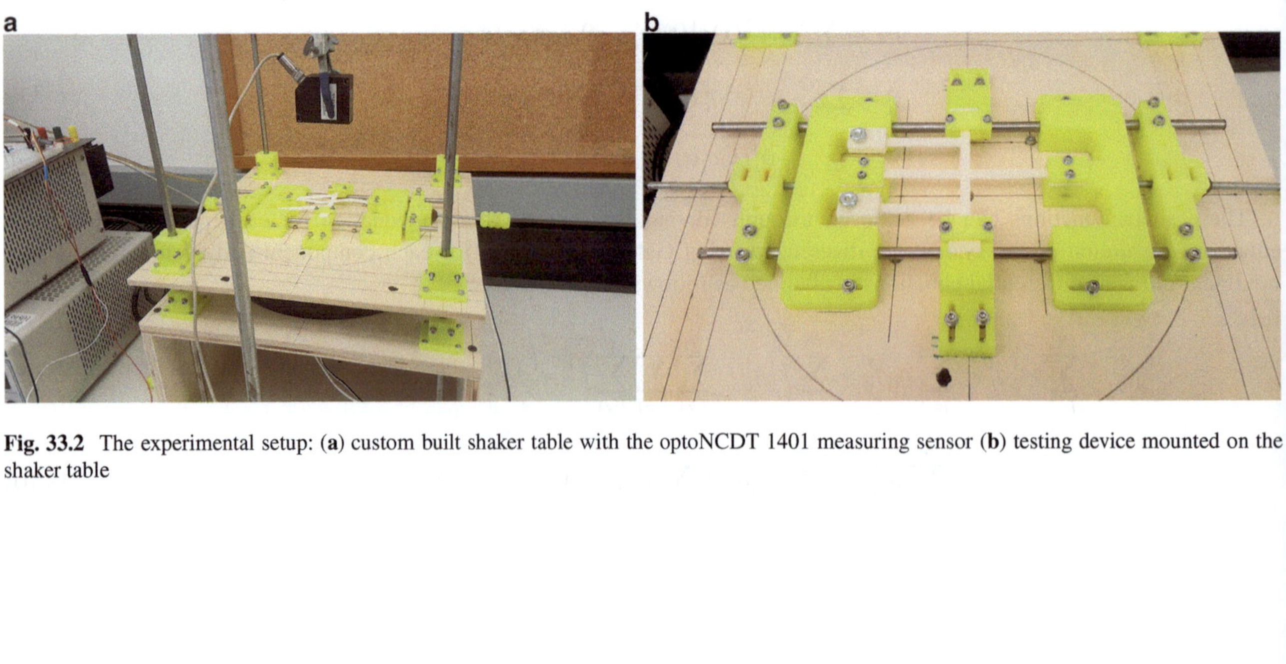

Fig. 33.2 The experimental setup: (**a**) custom built shaker table with the optoNCDT 1401 measuring sensor (**b**) testing device mounted on the shaker table

Table 33.1 Dimensions and physical parameters of the model

Component	Main beam	Torsional rod	Cantilevers
Module of elasticity (GPa)	2.3	2.3	2.3
Density (kg/m^3)	1040	1040	1040
Length (mm)	86	50	63
Thickness (mm)	5	5	5
Height (mm)	1.3	1.8	1.8
Damping coefficient (N.s/m)	0.1	0.1	0.1
Tip mass (gr)	–	–	2.36
Poisson's ratio	0.35	0.35	0.35

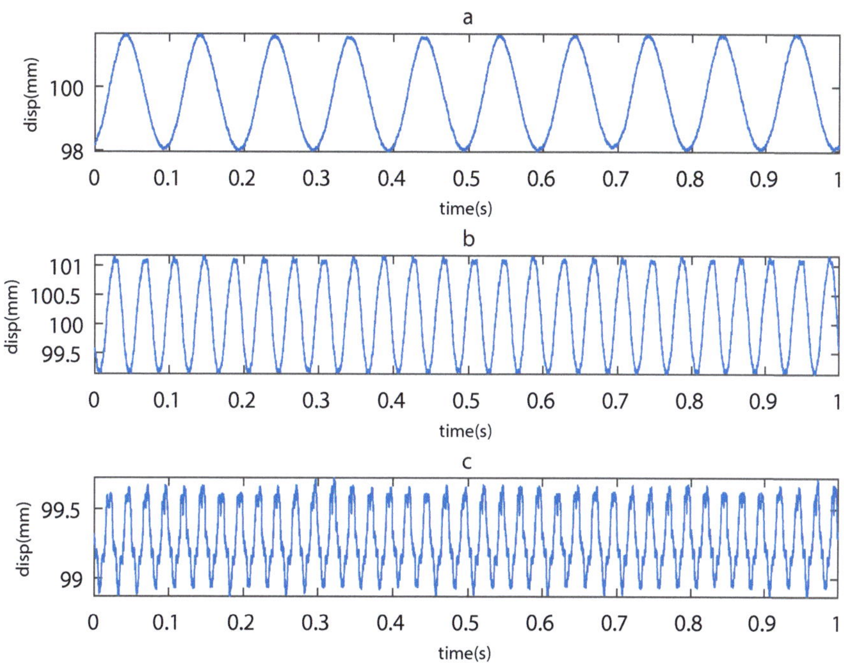

Fig. 33.3 Experimental time response of the midpoint of unbuckled main beam at (**a**) 10 Hz (**b**) 25 Hz (**c**) 40 Hz excitation frequency

the distance between two longitudinal fixed ends. The displacement at the mid-point of the main beam was measured with anoptoNCDT 1401 measuring sensor and the output results of the laser sensor were analyzed using an Infiniium Digital Storage Oscilloscpe.

33.4 Results and Discussion

In this part, the response of the main beam is analyzed by the explained experimental setup and compared with their corresponding modeling results. Figures 33.3, 33.4 and 33.5 show the steady state time-response of the midpoint of the main beam under three buckling conditions and three different excitation frequencies. In this examination, three situations of buckling levels are considered: unbuckled, buckled and highly-buckled. The time response of the beam for each condition is analyzed at three different frequencies: 10 Hz, 25 Hz, and 40 Hz. As it can be seen in Figs. 33.3 and 33.5, the absolute response of the main beam correspondent to the harmonic excitation, which in this case is a sine wave, is almost harmonic without dependency on what frequency the system is vibrating at. This behavior is exactly in agreement with the modeling results, which are shown in the following figures (Figs. 33.6, 33.7 and 33.8). For the modeling part, three numbers of modes are considered for each component in the calculation, which results in 12 total degrees of freedom for the whole system. In Fig. 33.4, the response of the main beam for a buckled condition, in which the buckling load is just higher than the critical buckling level of the device, switching between two stable states of the beam can be observed in steady state motion. This

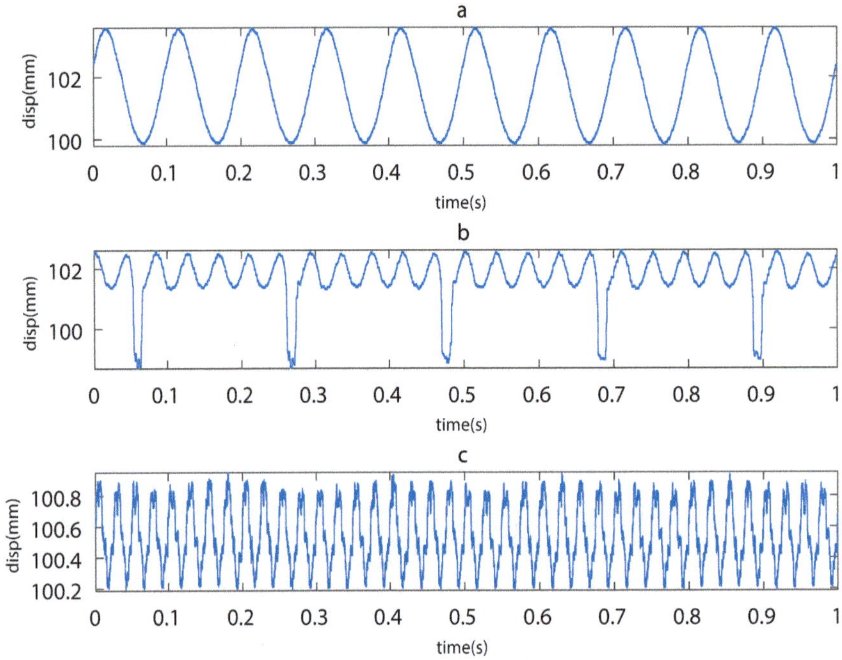

Fig. 33.4 Experimental time response of the midpoint of buckled main beam at (**a**) 10 Hz (**b**) 25 Hz (**c**) 40 Hz excitation frequency

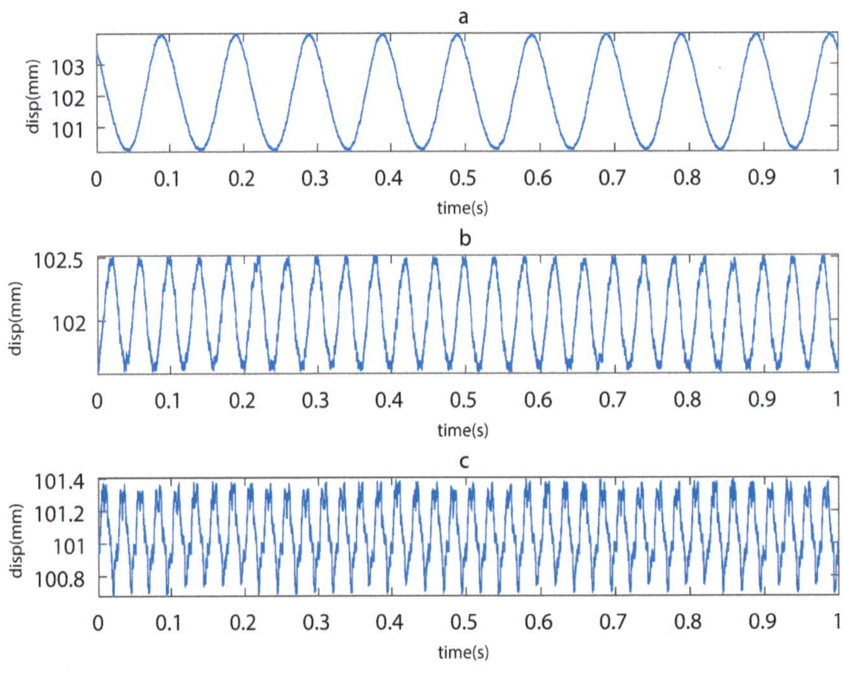

Fig. 33.5 Experimental time response of the midpoint of highly buckled main beam at (**a**) 10 Hz (**b**) 25 Hz (**c**) 40 Hz excitation frequency

behavior is due to the resonance behavior causing the connected cantilever arms vibrating with pretty large amplitude at around 25 Hz exciting frequency (the first natural frequency of the cantilever arms is theoretically calculated at 25.86 Hz). Large motion excited in the connected arms is transferring to the main beam which results in easing the switch between stable states in the system. However, this switching behavior cannot be captured in the modeling (Fig. 33.7). The reason for this inconsistency can be explained due to the mode shapes considered in the modeling formulation, which just satisfy the essential boundary conditions and cannot appropriately describe the force transferring between two connected parts.

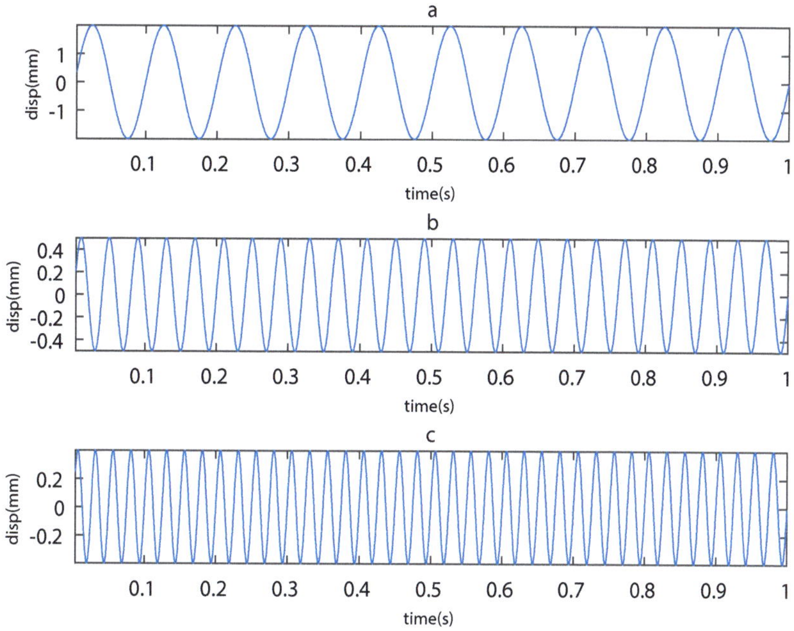

Fig. 33.6 Modeling time response of the midpoint of unbuckled main beam at (**a**) 10 Hz (**b**) 25 Hz (**c**) 40 Hz excitation frequency

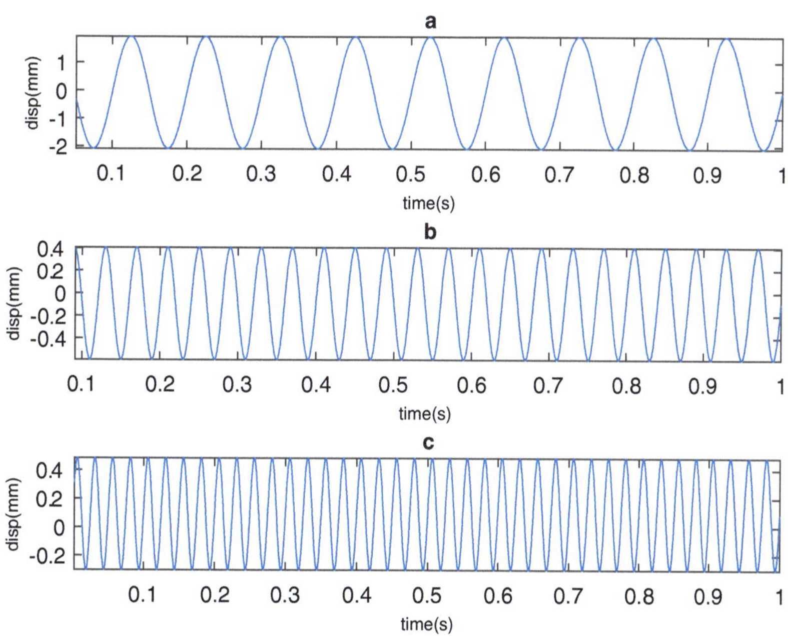

Fig. 33.7 Modeling time response of the midpoint of buckled main beam at (**a**) 10 Hz (**b**) 25 Hz (**c**) 40 Hz excitation frequency

Figure 33.9 shows the amplitude-frequency response at the midpoint of the main beam for both experimental and theoretical results. The amplitude considered in this plot is measured with respect to the base vibration. As it can be noticed from the plot, the relative amplitude of the main beam in the experimental data is relatively small and qualitatively in agreement with the theoretical results except for frequencies close to 25 Hz, at which the cantilevers vibrate at a very large motion due to their resonance behaviors and induce stable-state switching of the main beam. This plot shows that even for higher buckling levels in the system, which can guarantee the existence of two stable states, reaching to the high-energy orbit of the device is possible at relatively low frequencies and poor excitation amplitudes, which is one of the main issues associated with bistable energy harvesters, by appropriately setting the physical parameters of cantilever arms and torsional rod based on the system working conditions.

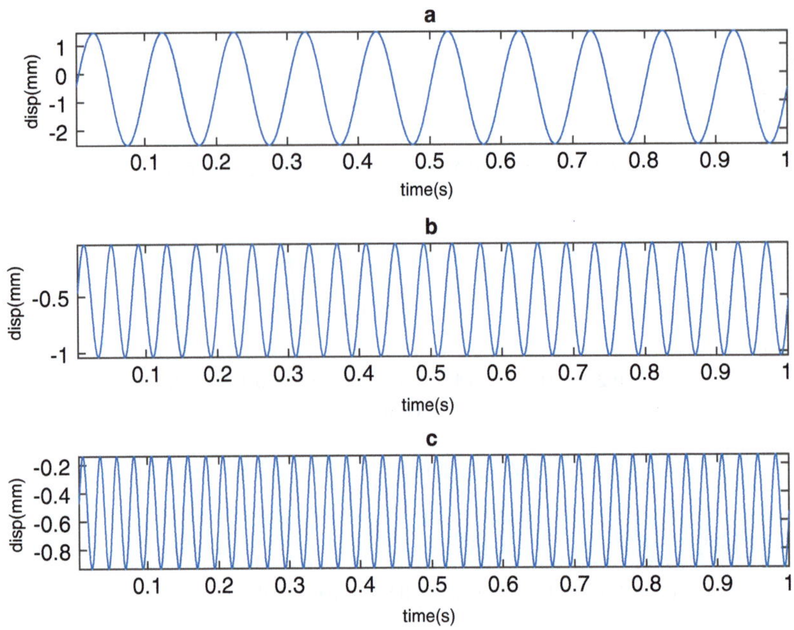

Fig. 33.8 Modeling time response of the midpoint of highly buckled main beam at (**a**) 10 Hz (**b**) 25 Hz (**c**) 40 Hz excitation frequency

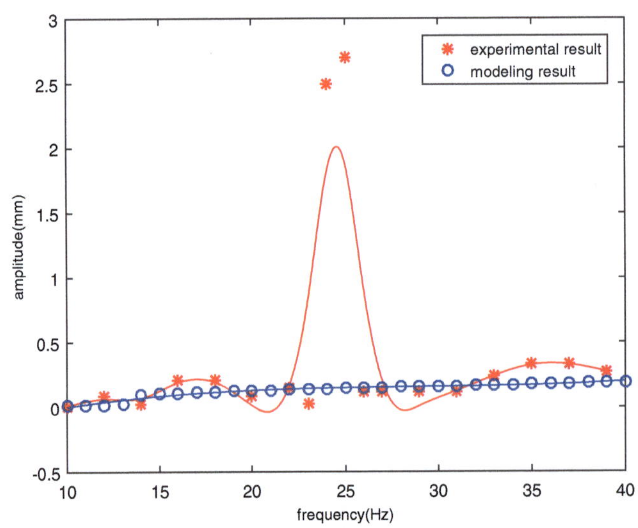

Fig. 33.9 Amplitude-frequency response of the buckled main beam for both modeling and experimental results

In the next plot (Fig. 33.10), the variation of the main beam vibrational amplitude vs. the buckling level is experimentally examined at 25 Hz excitation frequency. The buckling load is swept forward by reducing the distance between two clamped ends of the main beam step by step. The critical load calculated using Euler theory for the first mode of buckling in the system is around 23 N. As it is shown in the figure, the relative vibration amplitude of the main beam significantly increases when buckling happens at $\Delta \simeq 5\,\text{mm}$ due to the softening behavior of the system. Theoretically, this is just a transient state and after the beam is buckled, it will stay at one of its stable positions. But due to the large vibrational energy transferring by the cantilever arms to the main beam, the system can stay in the high-energy loop for a longer time and a wider range of buckling loads, which results in significant strains in the main beam and eventually helps the device harvesting more amounts of energy.

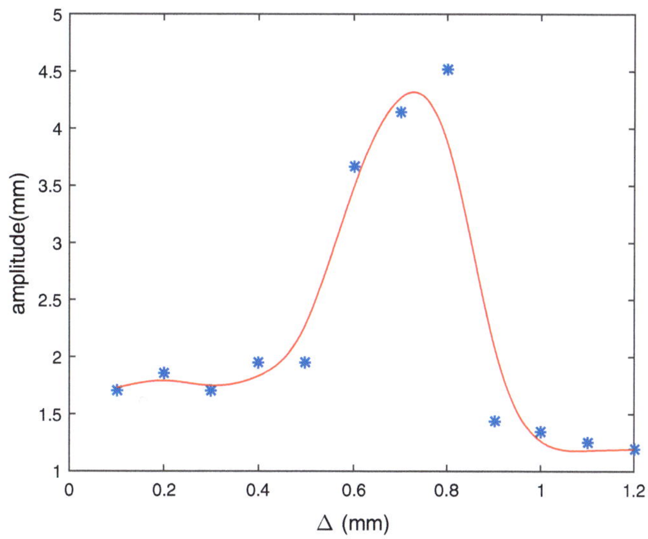

Fig. 33.10 Change of vibration amplitude of the buckled main beam vs. buckling level for 25 Hz excitation frequency

33.5 Summary and Conclusion

The aim of this study is dynamically modeling and experimentally testing a unique bistable energy harvesting system. In the first section, a theoretical model of the system is developed by dividing it into single components and deriving discretized equations of motion for each part using Galerkin approach. To do so, a summation of shape functions which can satisfy the essential boundary conditions along with their corresponding time-dependent generalized coordinates are considered as the general solution. After obtaining the ordinary differential equations for the system, the geometrical matching conditions are written which results in finding transformation matrix relating the general and independent set of generalized coordinates for the whole system. In the next part, the experimental setup and parameters used for testing the device are explained graphically. Eventually, the results obtained from the experimental tests are compared with the theoretical ones in different aspects. It has been experimentally shown that the connected cantilever arms can play an important role in helping the main beam switching between its two stable states. This effect can be significantly highlighted at frequencies close to the first natural frequency of the connected arms and capturing exactly this behavior in the modeling part might require a more hybrid approach. This observation can resolve the main problem of bistable systems used as energy harvesters by optimizing the dynamical characteristics of the connected components, which results in devices with a broader range of buckling loads and need of lower excitation energy.

References

1. Harne, R.L., Wang, K.W.: A review of the recent research on vibration energy harvesting via bistable systems. Smart Mater. Struct. **22**, 023001 (2013)
2. Emam, S.A., Nayfeh, A.H.: On the nonlinear dynamics of a buckled beam subjected to a primary-resonance excitation. Nonlinear Dyn. **35**, 1–17 (2004)
3. Nana Nbendjo, B.R., Woafo, P.: Modeling and optimal active control with delay of the dynamics of a strongly nonlinear beam. J. Adv. Res. Dyn. Control Syst. **1**(1), 57–74 (2009)
4. Stanton, S.C., McGehee, C.C., Mann, B.P.: Nonlinear dynamics for broadband energy harvesting: investigation of a bistable piezoelectric inertial generator. Physica D. **239**, 640–653 (2010)
5. Cottone, F., Gammaitoni, L., Vocca, H., Ferrari, M., Ferrari, V.: Piezoelectric buckled beams for random vibration energy harvesting. Smart Mater. Struct. **21**, 035021 (2012)
6. Vocca, H., Cottone, F., Neri, L., Gammaitoni, L.: A comparison between nonlinear cantilever and buckled beam for energy harvesting. Eur. Phys. J. Spec. Top. **222**, 1699–1705 (2013)
7. Cottone, F., Mattarelli, M., Vocca, H., Gammaitoni, L.: Effect of boundary conditions on piezoelectric buckled beams for vibrational noise harvesting. Eur. Phys. J. Spec. Top. **224**, 2855–2866 (2015)
8. Li, H.T., Qin, W.Y.: Dynamics and coherence resonance of a laminated piezoelectric beam for energy harvesting. Nonlinear Dyn. **81**, 1751–1757 (2015)

9. Friswell, M.I., Ali, S.F., Bilgen, O., Adhikari, S., Lees, A.W., Litak, G.: Non-linear piezoelectric vibration energy harvesting from a vertical cantilever beam with tip mass. J. Intell. Mater. Syst. Struct. **23**(13), 1505–1521 (2012)
10. Asl, M.E., Niezrecki, C., Sherwood, J., Avitabile, P.: Design of Scaled-Down Composite I-Beams for Dynamic Characterization in Subcomponent Testing of a Wind Turbine Blade, vol. 9, pp. 197–209. Shock & Vibration, Aircraft/Aerospace, Energy Harvesting, Acoustics & Optics, Springer International Publishing (2016)
11. Eydani Asl, M., Niezrecki, C., Sherwood, J., Avitabile, P.: Vibration prediction of thin-walled composite I-beams using scaled models. Thin-Walled Struct. **113**, 151–161 (2017)
12. Asl, M., Niezrecki, C., Sherwood, J., Avitabile, P.: Experimental and theoretical similitude analysis for flexural bending of scaled-down laminated I-beams. Compos. Struct. **176**, 812–822 (2017)
13. Blarigan, L.V., Moehlis, J., McMeeking, R.: Low dimensional modeling of a non-uniform, buckled piezoelectric beam for vibrational energy harvesting. Smart Mater. Struct. **24**, 065012 (2015)
14. Abdelkefi, A., Barsallo, N.: Nonlinear analysis and power improvement of broadband low-frequency piezomagnetoelastic energy harvesters. Nonlinear Dyn. **83**, 41–56 (2016)
15. Porter, D.A., Berfield, T.A.: A bi-stable buckled energy harvesting device actuated via torque arms. Smart Mater. Struct. **23**, 075003 (2014)

Masoud Derakhshani started his Ph.D. program at the Mechanical Engineering Department of University of Louisville at 2015. He recieved his B.S. and M.S. degrees in Mechanical Engineering in 2009 and 2011 respectively from University of Mazandaran and Iran University of Science and Technology, Iran.

Chapter 34
Effects of Strain Rate on Mechanical Properties and Fracture Mechanisms in a Dual Phase Steel

Sukanya M. Sharma, Kishlay Mishra, Omar Rodriguez, Wilburn R. Whittington, Paul Allison, Shrikant P. Bhat, Arun M. Gokhale, and Naresh N. Thadhani

Abstract Strain rate sensitivity of sheet steels affects their formability and crashworthiness. This contribution reports strain rate sensitivity and effects of strain rate on fracture micro-mechanisms in a commercial dual phase sheet steel (DP590). Uniaxial tensile tests were performed at strain rates of 10^{-4}/s, 1/s and 3200/s to characterize effects of strain rate on ultimate tensile strength and ductility. Fracture surfaces of the tested specimens were quantitatively characterized using stereological techniques to understand the fracture micro-mechanisms. Obtained data indicates that the basic fracture micro-mechanism remains the same with respect to strain rate but strain partitioning in ferrite and martensite is a strong function of strain rate.

Keywords Dual Phase Steels • Dynamic Deformation • Fracture mechanisms • Stereology • Aspect Ratio

34.1 Introduction

Dual Phase (DP) steels were among the first type of Advanced High Strength steels (AHSS) to be incorporated in weight reduced automotive parts [1–3]. DP steels can be processed to have desirable mechanical properties such as low initial yielding, high initial strain hardening, and a high strength and good ductility which can be controlled by varying the underlying microstructure [1, 2, 4–6]. The nominal microstructure of these steels largely consists of a soft and tough ferrite phase and a stronger and harder martensite phase. Based on the processing employed, other phases such as bainite, or retained austenite may also be present. The chemistry, morphology, and distribution of microstructural phases determine the mechanical properties of these steels [2–5, 7–12].

Numerous experimental and numerical studies conducted on DP steels have concluded that the fracture path and mechanism are sensitive to the size, distribution, and morphologies of the different phases [2, 4, 5, 8, 11]. Ramazani et al. [11] have shown that the failure in DP steels initiates by the cracking of martensite. A significant microstructural attribute which determines the strain at which the cracking begins is martensite bands. The presence of thick bands has been shown to lead to premature fracture [1, 4, 7, 10, 11]. Cingara et al. [7, 8] have shown that the major failure mechanism in DP steels with 20% martensite is decohesion at the ferrite-martensite interface. They also established that the rate of damage accumulation increased substantially at higher strains ultimately leading to fracture. It is important to recognize that the study on distribution of microstructural features made in the previous studies were qualitative and there were no quantitative data available beyond the volume fraction of martensite. Additionally, however, all these studies, have been performed at low strain rates. DP steels are used as structural components of an automobile, thus require demanding characteristics such as high formability during manufacturing and crashworthiness during use. Forming operations subject a material to strain

Sukanya M. Sharma and Kishlay Mishra contributed equally to this work.

S.M. Sharma (✉) • K. Mishra • A.M. Gokhale • N.N. Thadhani
Georgia Institute of Technology, North Avenue, Atlanta, GA 30332, USA
e-mail: msharma47@gatech.edu

O. Rodriguez • P. Allison
University of Alabama, Tuscaloosa, AL 35487, USA

W.R. Whittington
University of Mississippi, Oxford, MS 38677, USA

S.P. Bhat
Global R&D, ArcelorMittal, East Chicago, IN 46312, USA

J. Kimberley et al. (eds.), *Dynamic Behavior of Materials, Volume 1*, Conference Proceedings of the Society for Experimental Mechanics Series, DOI 10.1007/978-3-319-62956-8_34

rates of the order of 10/s to 10^2/s, while crashworthiness testing involves strain rates of the order of 10^2/s to 10^4/s. Hence, to comprehensively determine the mechanical response of Dual Phase steels, a detailed understanding of their behavior as a function of strain rate is required.

DP steels with about 20% martensite are known to have positive strain rate sensitivity and show an increase in flow stress with an increase in strain rate [12–15]. Studies conducted by Dong et al. [14], Wang et al. [15], and Curtze et al. [13], have captured the mechanical behavior of DP steels with different amounts of martensite at intermediate and high strain rates. Curtze et al. [13] have also captured the combined effect of increased strain rate and temperature on various grades of DP steels. They could establish that the steels were more sensitive to temperature fluctuations at dynamic strain rates. Though some of these studies discuss preliminary fractography, there is no detailed study performed in trying to understand the deformation and fracture behavior of DP steels under dynamic strain rates and comparing them to those observed in the quasi-static strain rate regime. Additionally, due to the absence of standard specimen geometry for dynamic tensile testing, authors have used different geometries for conducting dynamic tensile tests on materials. It has been established that the ductility of a material is a geometry dependent property [16–18]. According to a study conducted by Verleysen et al. [17, 18], in which seven specimen geometries were tested at various strain rates, it was observed that at higher values of strain, the strain field is non-uniform in the gage length of the sample. By changing the gauge length, width and radius of the samples, the study concluded that the stress strain behavior of the same material with different geometries should not be compared as the distribution of deformation is strongly affected by specimen geometry. The tests were performed on Transformation Induced Plasticity (TRIP) steels; however, the general results can be extended to all ductile metals.

Above discussion highlights the following; (a) Quantification of microstructures beyond volume fraction of martensite is required to utilize the quantitative details of the morphology and distribution of the different microstructural features of DP steels in predicting fracture mechanisms. (b) Absence of quantitative analysis of the features observed in the fracture surface or the fracture geometry to establish the operative fracture mechanisms at dynamic rates and (c) Although, the effect of geometry on ductility is widely researched, no standard geometry exists for the testing of sheet materials under dynamic rates. Thus, an important objective of this study is to bridge the existing knowledge gaps concerning geometry independent fracture mechanisms of DP steels at higher strain rates via a systematic quantification of fracture surfaces generated at different strain rates to establish the effects of strain rate and microstructure on the fracture mechanisms. To the best of the authors' knowledge, no previous studies have focused on quantifying the fracture profiles or surfaces to determine mechanistic information with a constant geometry at different strain rates. The results of the work described here show that contrary to previous studies, the ductility of the steel increases at higher strain rates. Information regarding the distribution of strain between ferrite and martensite as a function of strain rate is also provided. Additionally, detailed quantitative characterization of fracture surfaces gave an insight on the effect of martensite bands on the fracture mechanism as a function of strain rate. Results pertaining to three strain rates of 10^{-4}/s, 1/s and 3200/s are discussed to cover the quasi-static, transition from quasi-static to intermediate and high strain rates.

In the following section, details of the material used, procedures employed for mechanical testing and quantification of microstructures and fracture surfaces are provided. The obtained results and pertinent discussions are presented in Sect. 34.3.

34.2 Experimental Procedure

The material used in this work was a commercial cold rolled Dual Phase steel, DP590, provided by ArcelorMittal. The sheet steel provided was 1.2 mm thick and was tested in the as received condition. The chemical composition of the steel (form the supplier) is presented in Table 34.1.

Uniaxial tensile tests were carried out in two strain rate regimes: Quasi-static and High strain rate. The Quasi-static tests were performed on a MTS servo-hydraulic machine at strain rates of 10^{-4}/s and 1/s. High strain rate tensile tests were performed using the Split Hopkinson Tension Bar at the University of Alabama at Tuscaloosa. Atleast two tests were performed at each strain rate and the results obtained are averaged over the number of samples. The specimen geometry used was selected due to its compatibility with the high strain rate test equipment. This geometry was adapted and modified slightly from Sun et al.'s [16] study on the dynamic behavior of TRIP steels. Figure 34.1 shows the specimen geometry used in this study. Specimens were machined using water jet machining with the applied tensile load along the transverse

Table 34.1 Chemistry of DP 590 sheet steel used in this study in (%) by weight

C	Mn	Si	Al	P	S
0.089	0.99	0.27	0.04	0.015	0.005

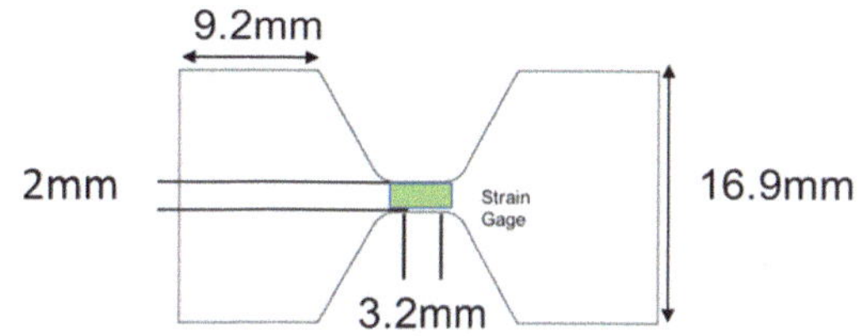

Fig. 34.1 Schematic of the miniature sample geometry adapted from Sun et al.'s work [16]

direction (TD). Tensile specimens showed a traditional necking and fracture behavior in the gage section and the grips underwent complete elastic deformation. The strain in the sample and the total ductility was measured in three different ways at Quasi static strain rates to ensure accuracy: (a) by the displacement reported by the calibrated displacement head of the MTS machine, (b) by measuring the total deformation on the fractures samples directly and (c) by calculating the elongation using Digital Image Co-relation (DIC) on images obtained from the high-speed camera. The ductility obtained by all three methods was very similar. For dynamic testing, strain gages mounted on the bars and high speed camera were used to monitor the strain in the specimen. The total elongation was calculated in a similar manner using the fractured specimens and DIC as in the Quasi-static sate. As the ductility is sensitive to gage length, in both cases the gage length used in DIC was kept the same as the actual gage length of the sample to eliminate any discrepancies in ductility arising from the change in gage length. The ductility results plotted are those obtained from the samples directly to maintain consistency.

Metallographic samples were prepared from sections obtained from three orthogonal planes of the sheet steel, using standard metallographic techniques. Samples were then etched with 2% Nital for 25 s and 10% aqueous sodium metabisulphide for another 20 s, to reveal the ferrite and martensite phases respectively. The Carl-Zeiss optical microscope AxioVert 200M and Hitachi SU-8030 Scanning Electron Microscopy (SEM) were utilized to obtain micrographs for Image based quantitative study of the microstructure.

Image based quantitative fractography and profilometry were performed on the fractured specimens using Axiovision and ImageJ image analyzer software. Systematic random point counting was used to calculate the volume fraction [19–21] and the number density of features were calculated using an unbiased frame [22]. Fracture profile of sections containing the thickness and transverse directions were studied. Specimens were grounded to half the thickness (at the fractured edge) and central one-third of the complete fracture profile was studied to neglect the effects of fast fracture near the edge. The surface roughness was obtained by estimating the true length of the fracture profile in different sections and dividing that by the projected length of the respective section [23]. The aspect ratio of ferrite was estimated by counting the number of intersections of ferrite grain boundaries with horizontal and vertical test lines ($<P_L>$ in the horizontal and vertical directions). The ratio of the two $<P_L>$ gives the aspect ratio of ferrite at varying strain rates. The aspect ratio of martensite was estimated in a similar manner. In the calculation of the aspect ratios, one field of view at a known distance from the fracture surface covered the central 1/3rd of the fracture profile.

Similarly, in image based quantitative fractography, the central 1/3rd region on the fractured surfaces were considered for quantification to neglect the effects of fast fracture near the edge. The area fraction of dimpled regions was calculated using systematic point counting [20, 24, 25] and the dimple size was estimated by considering a region that underwent complete dimple fracture with no facets [26]. The average length of the bands was estimated by measuring each band individually and the number density was estimated using an unbiased frame.

34.3 Results and Discussion

34.3.1 Quantitative Microstructural Characterization

Figure 34.2 shows the micrographs of as received DP 590 sheet steel. Measurements were averaged over orthogonal planes to obtain an unbiased estimate of the microstructural attributes. The calculated values of volume fraction of martensite (V_V (α')), surface area per unit volume between ferrite and martensite and between two ferrite grains (S_V (α-α, α-α')), length density of ferrite-ferrite triple lines (L_V (α-α-α)) and aspect ratios of ferrite and martensite are shown in Table 34.2.

Form the depicted microstructures, it is evident that martensite is the topologically dispersed phase. High surface area per unit volume of martensite implies a uniform distribution of ferrite and martensite in the microstructure. Additionally, the low values of S_V (α-α) and L_V (α-α-α) also indicated that the martensite islands are preferably located along the ferrite grain boundaries and triple points. At a lower magnification, bands of martensite were observed along the rolling direction (RD) and transverse direction (TD) as depicted in Fig. 34.3a, b. The bands observed in the transverse direction were much

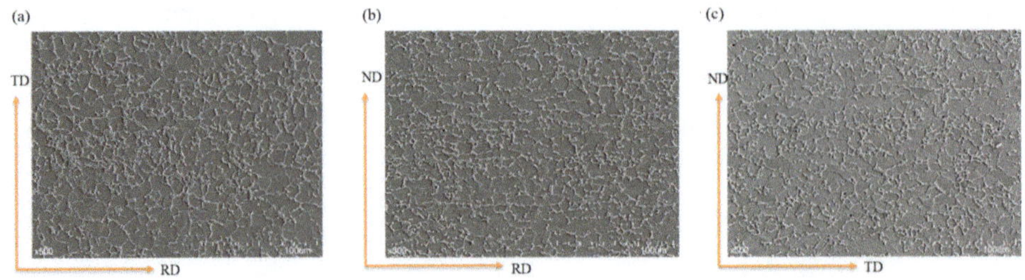

Fig. 34.2 Undeformed microstructures of DP 590 along the (**a**) short transverse direction; (**b**) transverse direction and (**c**) longitudinal direction

Table 34.2 Quantified microstructural parameters of DP 590

V_V (α′) (%)	S_V (α-α′) (/mm)	S_V (α-α) (1/mm)	L_V (α-α-α) (1/mm^2)	Aspect ratio (α)	Aspect ratio (α′)
29 ± 0.05	292 ± 23	80.5 ± 9	16.4 ± 0.56	1.55 ± 0.07	1.47 ± 0.02

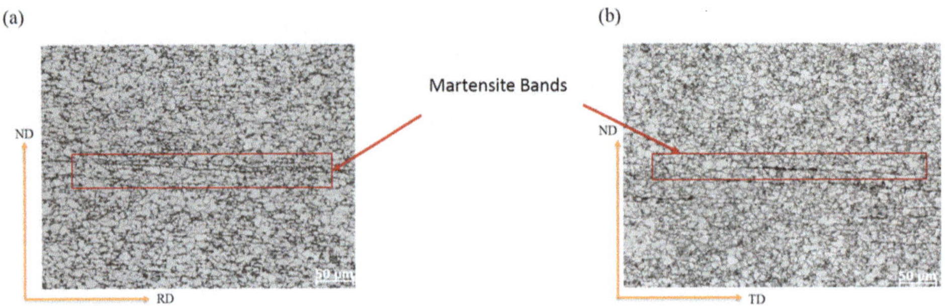

Fig. 34.3 Optical micrographs of (**a**) transverse direction and (**b**) longitudinal directions depicting martensite transformation bands

shorter than those observed in the rolling direction. The average length of these martensitic bands (along RD) was calculated to be approximately equal to 100 μm. These bands are thin compared to the bands observed in some of the previous works [1, 7, 27]. These bands are transformation induced bands in the central region of thickness, formed by linking largely single martensite islands.

34.3.2 Mechanical Properties

In Fig. 34.4a, b the variation of ultimate tensile strength (UTS) and ductility as a function of strain rate are presented. Three strain rates of 10^{-4}/s, 1/s and 3200/s were used in this study. A marked increase in the UTS beyond 1/s strain rate was observed, expected with steels showing positive strain rate sensitivity. The ductility of the samples also shows a pronounced increase with an increase in strain rate.

Various studies in the literature [28–31] have shown that at higher strain rates, as much as 80–90% of the work could be converted to heat and lead to elevated temperatures of as high as 200 °C. These high temperatures in some cases can be sufficient to cause an increased plasticity. Additionally, Hutchinson and Neale [32] have also shown that at dynamic rates the forming abilities of sheet steels can be greater as the deformation in a specimen is highly localized. Further work at intermediate strain rates can shed light on this behavior of DP 590 steels.

34.3.3 Quantitative Profilometry

Figure 34.5 illustrates the unetched fracture profile obtained at the quasi strain rate strain of 10^{-4}/s and the true and projected lengths used for the profile roughness calculations. Surface roughness was calculated to be 1.5 ± 0.2, and was approximately constant for all strain rates.

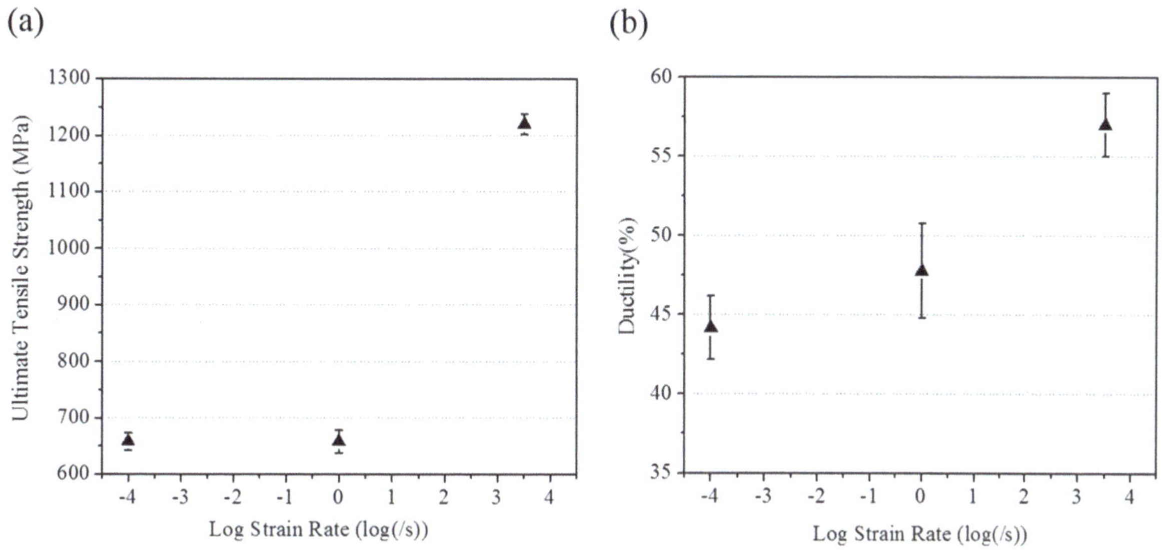

Fig. 34.4 Plots showing the effect of increasing strain rate on (**a**) UTS and (**b**) ductility

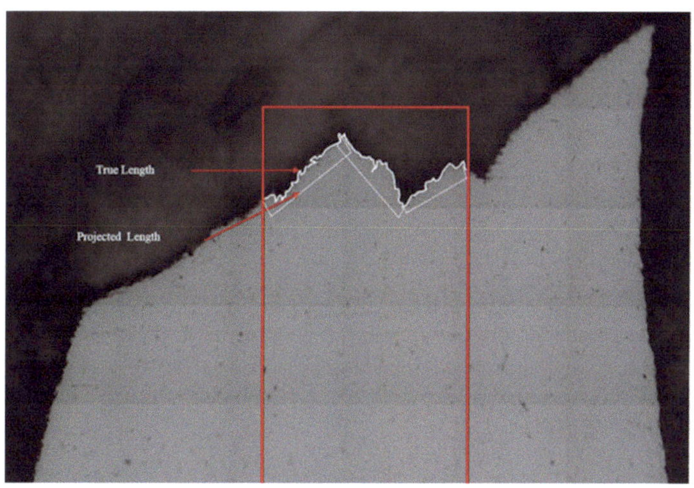

Fig. 34.5 An example of fracture profile roughness measurements

To understand the strain distribution in ferrite and martensite undergoing deformation, and observe their dependence on strain rate, the aspect ratio of ferrite and martensite were plotted near the fracture surface experiencing necking and increased localized deformation, and away from fracture in the gauge length experiencing uniform elongation. Figure 34.6a, b shows the etched fracture profile at the highest strain rates of 3200/s near the fracture surface and in a region experiencing uniform strain. The microstructures show significant deformation near the fracture surface with the presence of some voids. The void density is highest in the central region where martensite islands appear to be connected. Cingara et al. [reference] studied these voids and concluded that they form because of martensite-ferrite interface delamination.

As discussed previously, the aspect ratio of ferrite and martensite in the undeformed as-received microstructures were measured to be 1.55 and 1.47 respectively. The aspect ratios of ferrite and martensite as a function of strain rate provide information on the strain partitioning in each of the phases. From Fig. 34.6c, d it can be observed that at the highest strain rate the aspect ratio of ferrite is the highest and that of martensite is the lowest. The aspect ratio of ferrite atleast doubled at all strain rates and was the maximum at the highest strain rate, where the ratio almost tripled. The aspect ratio of martensite displayed an increase of atleast 20% with the highest ratio at the lowest strain rate, where the ratio increased by 60%. All these results imply that the strain partitioning is strain rate sensitive. At the higher strain rates the deformation is taken up largely by ferrite and martensite tends to approach brittle fracture as it displays less plasticity. Additionally, the strain partitioning is different at the lowest strain rate where martensite takes up higher amounts of strain as depicted by its increased aspect ratio. Further experiments at intermediate strain rates and experiments with a higher volume fraction of martensite can reveal more information on the strain rate dependence of the strain partitioning.

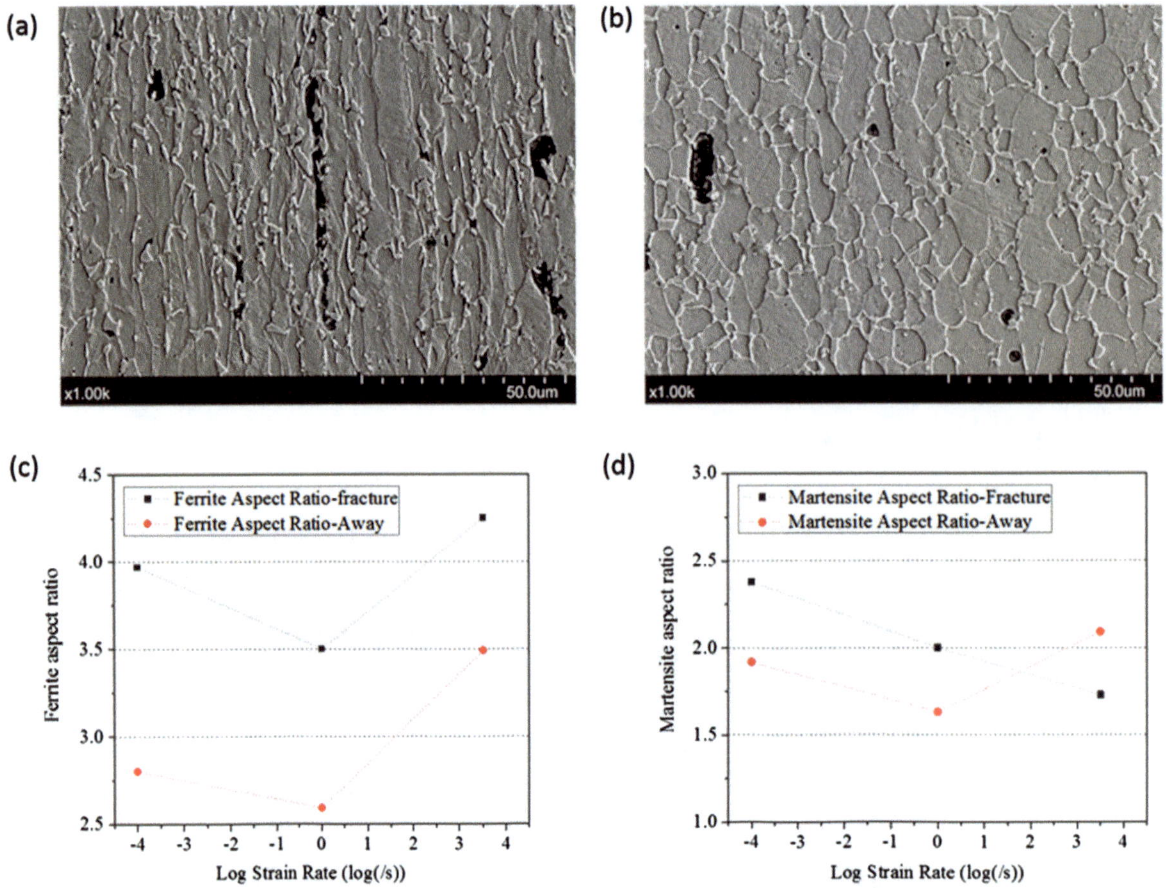

Fig. 34.6 (**a**) and (**b**) An example of deformed microstructure closer to the fracture profile and in a region experiencing uniform strain respectively; (**c**) and (**d**) plots showing variation of aspect ratio as a function of strain rate of ferrite and martensite respect

34.3.4 Quantitative Fractography

Fractured surfaces were quantitatively studied to understand the micro-mechanisms of failure at various strain rates. Figure 34.7 illustrates features observed in fractured surfaces of DP 590 sheet steel upon failure. Qualitative study of the fractured surface revealed predominantly ductile fracture owing to dimpled fracture regions. Some facets signifying brittle fracture were also observed. An interesting observation was the presence of bands that were parallel to the rolling direction on the fracture surface. These bands were observed to be either very small (10–40 μm) or substantially longer (100 μm or more). To quantitatively study the fractured surfaces, area fraction of dimpled and brittle regions, average dimple size, average length of the bands and number density of the bands were calculated.

The area fraction of the ductile region was calculated to be equal to 90 $\pm$ 3%, being approximately constant through all strain rates. Observed bands in the fractured surfaces were categorized into two groups based on their lengths, shorter than 100 μm and greater than or equal to 100 μm. This was done in order to separate any effects of long range order in the microstructure from any local occurrences. An important observation was that the average length of these bands was more or less the same at all strain rates indicating that a common microstructural feature may be responsible for the bands. Figure 34.8 illustrates the number density of long bands on fracture surfaces as a function of strain rate.

The martensite bands observed in the microstructure along the rolling direction were of comparable lengths and hence may be responsible for causing these bands. Additionally, the number density of these bands decreased with an increase in strain rate. These bands may be caused due to the delamination of martensitic bands and have not been previously discussed in the literature. Further quantification regarding the separation of bands in the microstructure and comparing the width of the bands in the fracture surface and microstructure could also lead to more information on whether these bands are due to the delamination of ferrite and martensite.

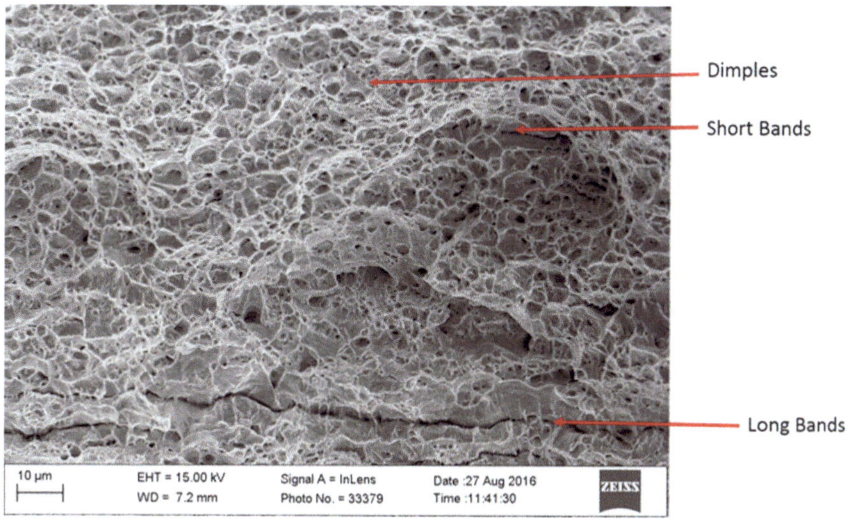

Fig. 34.7 An example of the fracture surface seen with different features observed in the fracture surface

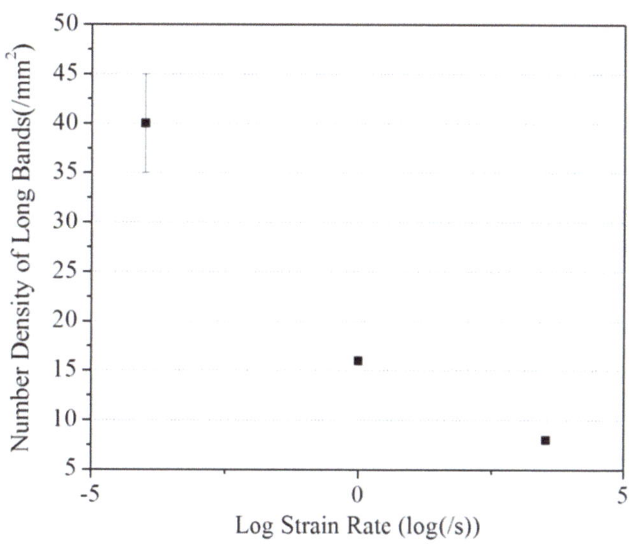

Fig. 34.8 Plot showing the variation of number density of long bands estimated as a function of strain rate

34.4 Conclusions

In this study, we have demonstrated that in the case of DP 590 steel, the strain partitioning in ferrite and martensite are strain rate sensitive. From the calculated aspect ratios of martensite, it was observed that martensite deforms plastically at all strain rates, however, martensite deforms less at the higher strain rate. The flow stress and ductility increase with an increase in the strain rate. Area fraction of dimpled region, number density of dimples and average dimple size in fracture surface does not change with strain rate. Bands greater than 100 μm observed in the fractured surfaces show a correlation with the martensite bands observed in the microstructure. The number density of long bands decreases with increase in strain rate which also corresponds to the increase in ductility.

Acknowledgements This work was funded through an unrestricted research grant from ArcelorMittal Global R&D East Chicago to the Georgia Tech Foundation. The financial support is gratefully acknowledged.

References

1. Davies, R.: Influence of martensite composition and content on the properties of dual phase steels. Metall. Trans. A. **9**(5), 671–679 (1978)
2. Rashid, M.: Dual phase steels. Annu. Rev. Mater. Sci. **11**(1), 245–266 (1981)
3. Tasan, C.C., et al.: An overview of dual-phase steels: advances in microstructure-oriented processing and micromechanically guided design. Annu. Rev. Mater. Res. **45**, 391–431 (2015)
4. Bag, A., Ray, K., Dwarakadasa, E.: Influence of martensite content and morphology on tensile and impact properties of high-martensite dual-phase steels. Metall. Mater. Trans. A. **30**(5), 1193–1202 (1999)
5. Calcagnotto, M., et al.: Orientation gradients and geometrically necessary dislocations in ultrafine grained dual-phase steels studied by 2D and 3D EBSD. Mater. Sci. Eng. A. **527**(10), 2738–2746 (2010)
6. Colla, V., et al.: Strain hardening behavior of dual-phase steels. Metall. Mater. Trans. A. **40**(11), 2557–2567 (2009)
7. Avramovic-Cingara, G., et al.: Effect of martensite distribution on damage behaviour in DP600 dual phase steels. Mater. Sci. Eng. A. **516**(1), 7–16 (2009)
8. Avramovic-Cingara, G., et al.: Void nucleation and growth in dual-phase steel 600 during uniaxial tensile testing. Metall. Mater. Trans. A. **40**(13), 3117 (2009)
9. Granbom, Y.: Effects of process parameters prior to annealing on the formability of two cold rolled dual phase steels. Steel Res. Int. **79**(4), 297–305 (2008)
10. Kim, S., Lee, S.: Effects of martensite morphology and volume fraction on quasi-static and dynamic deformation behavior of dual-phase steels. Metall. Mater. Trans. A. **31**(7), 1753–1760 (2000)
11. Ramazani, A., et al.: Characterization and modelling of failure initiation in DP steel. Comput. Mater. Sci. **75**, 35–44 (2013)
12. Zhao, Z., et al.: Microstructure, mechanical properties and fracture behavior of ultra-high strength dual-phase steel. Mater. Sci. Eng. A. **618**, 182–188 (2014)
13. Curtze, S., et al.: Deformation behavior of TRIP and DP steels in tension at different temperatures over a wide range of strain rates. Mater. Sci. Eng. A. **507**(1), 124–131 (2009)
14. Dong, D., et al.: Microstructure and dynamic tensile behavior of DP600 dual phase steel joint by laser welding. Mater. Sci. Eng. A. **594**, 17–25 (2014)
15. Wang, W., et al.: Experimental study on high strain rate behavior of high strength 600–1000MPa dual phase steels and 1200MPa fully martensitic steels. Mater. Des. **47**, 510–521 (2013)
16. Sun, X., et al.: Effects of sample geometry and loading rate on tensile ductility of TRIP800 steel. Mater. Sci. Eng. A. **541**, 1–7 (2012)
17. Verleysen, P., et al.: Influence of specimen geometry on split Hopkinson tensile bar tests on sheet materials. Exp. Mech. **48**(5), 587 (2008)
18. Verleysen, P., et al.: Effect of strain rate on the forming behaviour of sheet metals. J. Mater. Process. Technol. **211**(8), 1457–1464 (2011)
19. Gokhale, A.M.: Estimation of length density Lv from vertical slices of unknown thickness. J. Microsc. **167**(1), 1–8 (1992)
20. Gokhale, A.M.: Quantitative Analysis of Fracture Surfaces Using Stereological Methods. Georgia Institute of Technology, Atlanta (1994)
21. Karlsson, L., Gokhale, A.: Stereological estimation of mean linear intercept length using the vertical sections and trisector methods. J. Microsc. **186**(2), 143–152 (1997)
22. Gundersen, H.J.G.: Notes on the estimation of the numerical density of arbitrary profiles: the edge effect. J. Microsc. **111**(2), 219–223 (1977)
23. Drury, W., Gokhale, A.M., Antolovich, S.: Effect of crack surface geometry on fatigue crack closure. Metall. Mater. Trans. A. **26**(10), 2651–2663 (1995)
24. Gokhale, A.M.: Quantitative characterization and representation of global microstructural geometry. Metallography and microstructures, ASM Handbook. ASM Int. **2004**, 428–447 (2013)
25. Underwood, E.: In: Pellissier, G., Purdy, S. (eds.) The Mathematical Foundations of Quantitative Stereology, Stereology and Quantitative Metallography, STP36841S, pp. 3–38. ASTM International, West Conshohocken (1972.) https://doi.org/10.1520/STP36841S
26. Jamwal, R.S.: Microstructural Origins of Variability in the Tensile Ductility of Dual Phase Steels. Georgia Institute of Technology, Atlanta (2011)
27. Sarwar, M., Priestner, R.: Influence of ferrite-martensite microstructural morphology on tensile properties of dual-phase steel. J. Mater. Sci. **31**(8), 2091–2095 (1996)
28. Kapoor, R., Nemat-Nasser, S.: Determination of temperature rise during high strain rate deformation. Mech. Mater. **27**(1), 1–12 (1998)
29. Mason, J., Rosakis, A., Ravichandran, G.: On the strain and strain rate dependence of the fraction of plastic work converted to heat: an experimental study using high speed infrared detectors and the Kolsky bar. Mech. Mater. **17**(2–3), 135–145 (1994)
30. Rittel, D.: On the conversion of plastic work to heat during high strain rate deformation of glassy polymers. Mech. Mater. **31**(2), 131–139 (1999)
31. Yi, G., et al.: Response characteristics and adiabatic heating during high strain rate for trip steel and dp steel. J. Iron Steel Res. Int. **22**(1), 48–54 (2015)
32. Hutchinson, J., Neale, K.: Influence of strain-rate sensitivity on necking under uniaxial tension. Acta Metall. **25**(8), 839–846 (1977)

Sukanya M. Sharma joined Georgia Institute of Technology as a PhD student in Materials Science and Engineering in August 2014. Her research is currently on understanding the fracture mechanisms of different kinds of Dual Phase steels as a function of strain rate. Apart from research, Sharma is also actively involved in the Graduate Student Senate.

Chapter 35
Nonlinear and Inertant Acoustic Metamaterials and Their Device Implications

Prateek P. Kulkarni and James M. Manimala

Abstract Acoustic Metamaterials (AM) are a class of artificial structural materials that derive their unique dynamic properties, not just from material constituents but more so from engineered local configurations. Tailoring these local configurations have been shown to impart unusual mechanical wave manipulation capabilities to AM with potentially novel applications in protective structures; acoustic devices for sensing, noise control, and energy harvesting; and MEMS devices. Most AM require the presence of periodic features that locally exhibit dynamic phenomena like resonance or instability within a host material or structure. A key advantage of the AM design approach is the latitude to explore new local configurations to further enrich their dynamic behavior. The present study focusses on nonlinear and inertant AM configurations and their device implications. Using a lumped-parameter effective-mass model, approximate analytical solutions were derived for the amplitude-dependent dispersion curve shifts in nonlinear AM owing to the presence of cubically hardening or softening nonlinearities in local oscillators. Discrete element simulations predict the possibility of realizing passive acoustic control devices such selective filters, amplitude band-pass filters and direction-biased waveguides using nonlinear AM. A numerical routine to generate root profile geometries that enable contact-based hardening response in tip-loaded cantilever beam resonators was developed and implemented. Experiments on a structural waveguide test article verify the existence and extent of bandgaps and also provide an indication of the passive direction-bias phenomenon. Whereas, incorporating inerters, which are mechanical elements that display a force proportional to the relative acceleration across them, could create structural devices that display frequency-dependent negative and even extreme effective-mass and stiffness regimes. Such devices have implications for passive high-pass filters with ultra-low frequency bandgaps that encompass the long wavelength limit and can be realized without the use of any mechanical grounding elements and even for structural networks that can act as a nearly total mechanical wave inhibitor. Further research on the interactive synergies between nonlinear and inertant configurations and practical strategies to scale and fabricate them could have promise for realizing a new class of AM with enriched dynamics.

Keywords Acoustic metamaterials • Cubic nonlinear stiffness • Inerter • Direction bias • Negative effective mass • Negative effective stiffness

35.1 Introduction

Metamaterials are a relatively new class of materials that trace their history back to 1888 when concepts of negative compressibility [1] arose. Metamaterials were first accepted widely in the field of Electromagnetics (EM) following Veselago's postulations of materials with negative EM parameters [2]. The concept of Metamaterials gained more acclaim when Pendry [3] proposed the possibility of making left-handed metamaterials theoretically. Studies for the attainment of the acoustic analogue of EM materials resulted in the investigation of structures containing ordered discontinuities arising from the addition of a resonant substructure to the host material. The most widely studied of this class are Locally Resonant Acoustic Metamaterials (LRAM), one of the earliest among these was reported by Martinez et al. [4] for sound attenuation by sculpture. Many experimental studies on LRAM adopted the model [5] described by Liu et al. [6–8], where sonic crystals were fabricated along the lines of the concept of locally resonant metamaterials leading to frequency-dependent effective-density [5] and modulus [9] which could become negative and give rise to tunable bandgaps [9, 10] and propagation modes even with negative group velocity [11]. Several other studies on LRAM were carried out, employing different mechanisms to achieve local resonance, such as Helmholtz resonator, elastic membranes and sonic crystals [5, 12–14] for instance.

P.P. Kulkarni • J.M. Manimala (✉)
School of Mechanical and Aerospace Engineering, Oklahoma State University, 218 Engineering North, Stillwater, OK, 74078, USA
e-mail: james.manimala@okstate.edu

J. Kimberley et al. (eds.), *Dynamic Behavior of Materials, Volume 1*, Conference Proceedings of the Society for Experimental Mechanics Series, DOI 10.1007/978-3-319-62956-8_35

For the most part, acoustic metamaterials have been studied in several configurations employing local features having different dynamic behavior to derive unique wave phenomena, which have novel applications. Incorporation of instability, nonlinearity [15, 16], and grounded springs have been studied to tailor dynamic characteristics to realize hitherto unobtainable properties.

Owing to the enrichment of dynamic response by the addition of nonlinear local oscillators in an otherwise linear system, there has been an increased interest in the field of acoustic metamaterials that expoit nonlinear response [17, 18]. Nonlinear dispersion relation in a periodic string was experimentally and analytically obtained by Manktelow et al. [19]. Narisetti et al. [20] employed a perturbation approach to derive dispersion relation with first order approximation for one-dimensional nonlinear periodic structures with nonlinearity introduced in different configurations successively, and provided evidence that introduction of cubic nonlinearity caused the propagation to become amplitude-dependent. A few device implications exploiting this behavior were also provided. Further, it was shown from simulations [21] that by exploiting the amplitude-dependent behavior of the AM with cubically nonlinear local attachments, a direction-biased waveguide can be potentially realized. Informed by these simulation, a possible design for such a device was postulated and subjected to experimental testing in this study.

Unidirectional transmission of waves is of significant importance in the area of acoustic switches, diodes or rectifiers to allow wave propagation only in a pre-decided direction. Waves propagating in the opposite direction are attenuated and ideally completely filtered. An attempt to synthesize acoustic diodes were made by employing bistable members with magnetic links [22]. The method proposed in this study is based on frequency modification through nonlinearities. To further enrich the behavior of acoustic metamaterials, inerters have been considered due to their ability to introduce a dynamic mass presence many times more than these static device mass. Inerters, postulated by Smith [23] are a recent addition to the mechanical elements and relate the acceleration difference across its terminals to their restoring force. With the same units as mass, its inertance is denoted by the symbol 'J'. A practically attainable ratio of inertance to device mass as high as 300 has been reported [23] through experimental investigations.

Inerters have been studied as part of vibration isolation systems and their presence has been documented to result in improved performance over traditional vibration isolators and shock control systems. Inerter-based devices have been proposed for vibration suppression, vehicle suspension and steering systems [24–30]. Inerters have also seen cutting edge commercial use, following their adoption under the moniker of 'J-damper' in Formula One cars. Experiments [31] to measure mechanical admittance functions of ball-screw and rack and pinion based inerter designs were conducted to gauge their relative performance. Extending the classical vibration control configuration of the tuned-mass-damper system by incorporating inerters, tuned-mass-damper-inerter systems [32] were shown to outperform them while remaining relatively lightweight. Numerical optimization [33] of vibration suppression devices with inerters predict 10–20% performance increase in a wider frequency band over traditional devices. Although the inerter was explored by many researchers in various domains with and majority researchers focusing on their application in vibration systems, the behavior of inertant systems for wave propagation has not been explored. This study looks at the effect of incorporating inerters in an infinitely long 1-D lattice system representing an acoustic metamaterial in various configurations – introducing the inerter in both host and local attachments.

One limitation present in conventional materials is the generally inverse relationship that exists between density and stiffness. The potential to produce lightweight, stiff structural materials with high damping has driven the investigations into new designs for acoustic metamaterials with improved and passive adaptive wave manipulation properties without loss of load carrying capabilities. This study explores the possibility of enriching dynamic properties of acoustic metamaterials by incorporating nonlinear and inertant local features.

The ability to manipulate cubic nonlinearity in an acoustic metamaterial is of particular interest for applications because it may lead to an improved resolution in acoustic imaging [34] and directional bias in acoustic energy propagation [35]. Tuning the energy dissipation and mechanical properties of such materials [36] to the amplitude of an incoming wave provides a means of creating passive adaptive acoustic devices with unusual wave manipulation capabilities. We discuss some device implications for nonlinear and inertant metamaterials using analytical and numerical studies and verify the possibility of amplitude triggered passive direction bias in propagation through a structural waveguide experiment for a prototype acoustic metamaterial design.

35.2 Nonlinear Acoustic Metamaterials

Consider a discrete 1-D spring mass system with a local mass attachment via a cubically nonlinear spring, whose stiffness is given by Eq. (35.1). A perturbation method based approach is applied to obtain the dispersion relation of the system for wave propagation characteristics. The dispersion relation with first order correction is given in Eq. (35.2). This relation, plotted as

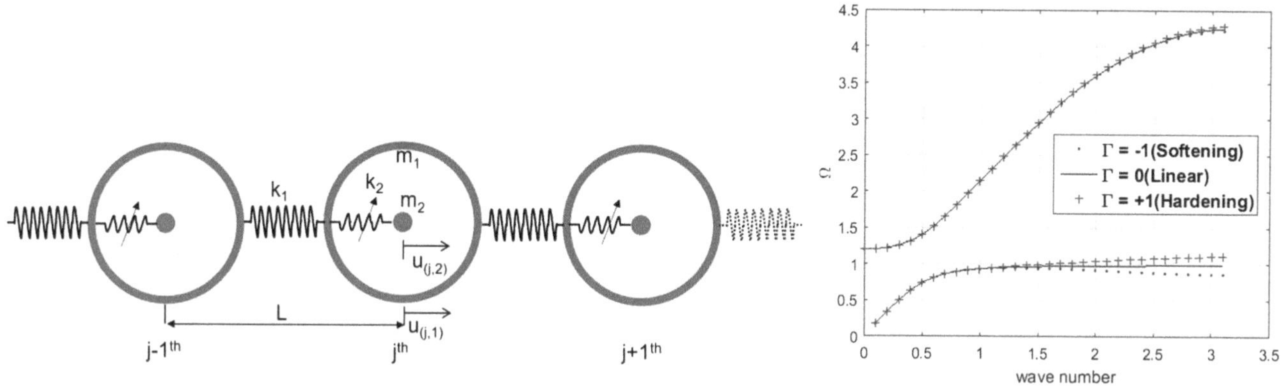

(a) (b)

Fig. 35.1 (**a**) Discrete spring mass lattice representation for an NLAM with nonlinear stiffness in attachment spring and (**b**) Dispersion curves obtained by perturbation method for LRAM type lattice with softening and hardening type nonlinearity in the local attachment. The excitation frequency normalized with local resonance frequency is plotted against the wave number

shown in Fig. 35.1b demonstrates a shift in the lower bound of bandgap, indicating a narrowing of bandgap for a Nonlinear hardening (NLH) type nonlinearity as compared to the LRAM. This modification of dispersion curves provides the potential to tune the AM to have a narrow band or specific frequency filtering characteristics. The terms in the Eqs. (35.1 and 35.2) are consistent with [20].

$$k = k_0 + k_n x^2 \tag{35.1}$$

$$\omega = \omega_0 + \varepsilon \frac{3\overline{\Gamma}|A_1|^2}{8\alpha\overline{\omega}_0}\left(1 - \frac{1}{\left(1 - \overline{\omega}_0^2\right)}\right)^3 \tag{35.2}$$

35.3 Simulations

A nonlinear local attachment similar to that shown in Fig. 35.1a is considered for numerical simulations. The parameters for this system were chosen with manufacturing constraints in view. Nonlinearity is a function of displacement amplitude. Hence, k_n for the system is chosen such that at sufficiently small displacements, it behaves close to an LRAM as nonlinearities are not triggered. Therefore, at low amplitude, the bandgap limits of this NLAM is obtained from Eq. (35.3) as [600–980 Hz]. The procedure to obtain the equation is elaborated in [37].

$$\omega_0 < \omega < \sqrt{1 + \frac{m_2}{m_1}} \tag{35.3}$$

A simulation model similar to LRAM, as shown in Fig. 35.2 with 100 unit cells is set up in this case with the following lattice parameters in compatible units.

$$[m_1, m_2, k_1, k_2, k_n] = [1.67E-3, 2.84E-3, 2.43E6, 4.04E4, 1E11]$$

Firstly, keeping the condition for not triggering of nonlinearity in view, a low amplitude (0.01 mm) single frequency harmonic excitation is applied at 1st mass and resulting displacement of 100th mass is recorded. The output and input displacement plots are shown in Fig. 35.3. The simulations show a marked attenuation at 606 Hz, which lies just within the bandgap while 500 Hz shows propagation as it lies within the predicted propagation zone.

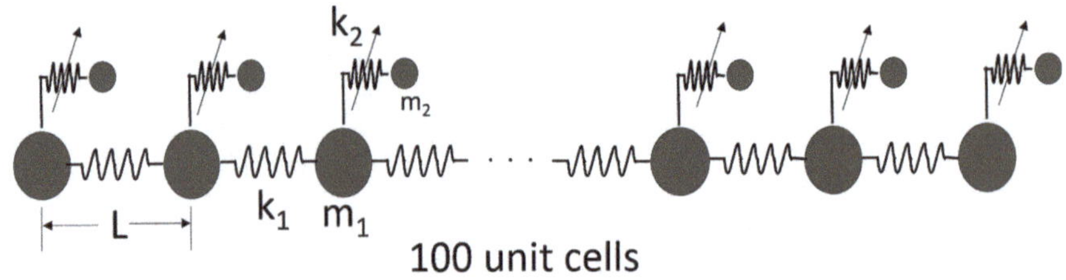

Fig. 35.2 Simulation model for NLAM

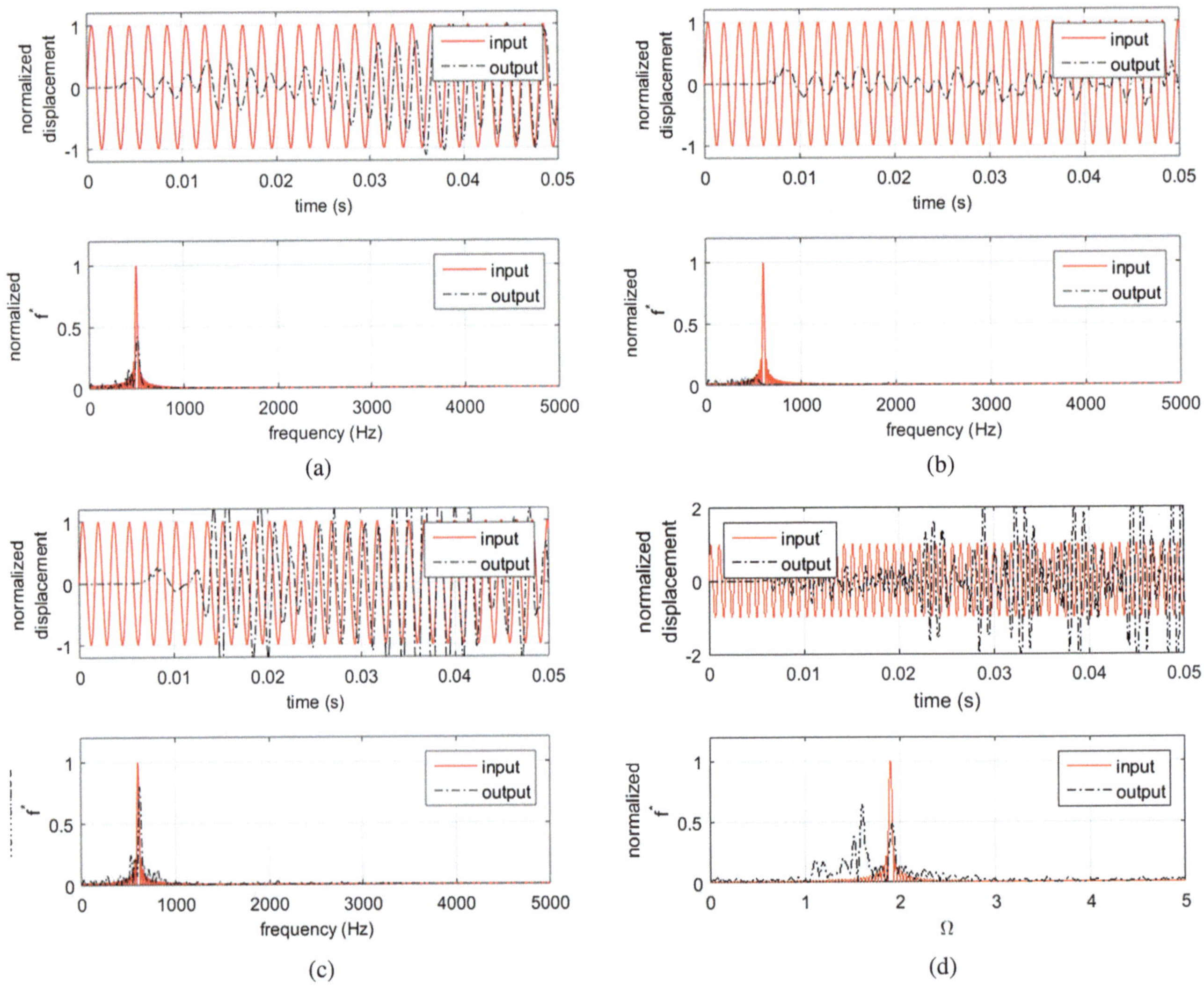

Fig. 35.3 Demonstration of bandgap effect and amplitude dependent propagation. Simulation results showing the displacement history of the queried mass locations at (**a**) 500 Hz, low amplitude, (**b**) 606 Hz, low amplitude, (**c**) 606 Hz, high amplitude and (**d**) 1140 Hz, high amplitude

Next, at the same frequencies, a high amplitude (1 mm) excitation is applied to observe the effect of nonlinearity, which is triggered due to increased amplitude. The plots obtained for this case are shown in Fig. 35.3c and indicate that at high amplitude, the response at 606 Hz switches to propagation. This is a consequence of narrowing of bandgap as predicted and shown in Fig. 35.1b. The reason for such phenomenon is the amplitude-triggered nonlinear hardening, which renders the acoustic mode to shift to higher frequencies for large wavenumbers.

The effect of high amplitude excitation through the NLAM is shown in Fig. 35.3d. The attenuation at low amplitudes is expected as nonlinearities are not triggered and the local attachments still behave as quasi-linear oscillators. But due to nonlinearities being triggered at a higher amplitude (A = 1E-3), we see marked propagation albeit with a modification in the spectrum for the transmitted wave. This switching from attenuation to propagation can be attributed to the shift of the lower limit of the bandgap to higher frequencies for high wavenumbers as shown in Fig. 35.1b. A more detailed analysis of the

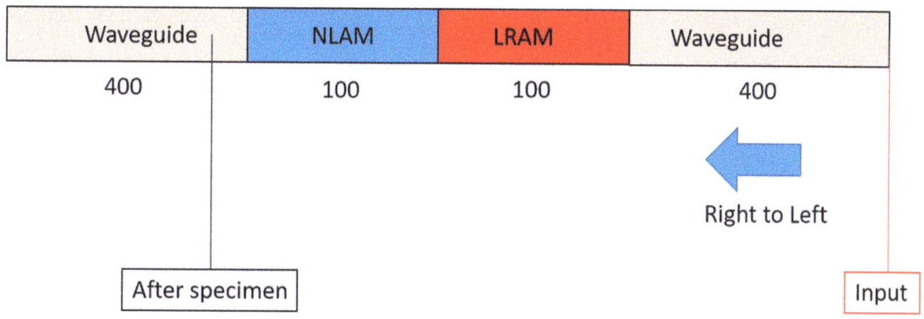

Fig. 35.4 Schematic of the simulation model for the direction biased waveguide

effect of nonlinearity and amplitude on the propagation characteristics of nonlinear AM was undertaken in [21]. Based on the simulation results, an acoustic diode-like device can be constructed, by exploiting the amplitude-dependent propagation and modification of frequency spectrum. Depending on the behavior exhibited by the NLAM chain, a suitably tuned LRAM is selected in order to obtain a direction-bias effect in propagation through a lattice chain.

A simulation model is set up in Abaqus with 1000 unit cells for the direction-bias case to preclude the effects of reflections. This model is a representation of a long waveguide with a test-article inserted at the center, dividing the waveguide into two equal lengths. Each section of the waveguide was assigned with 400 unit cells. The test-article consisted of two parts – NLAM and LRAM, each consisting of 100 mass-in-mass unit cells. The first mass of the waveguide is chosen for application of displacement and this was treated as the 'input'. Displacement response 5 unit cells after the test-article is recorded and is treated as 'output'. Both time history and frequency spectrum of the input and transmitted (output) displacements are recorded for post-processing.

The stiffness curves for the selected NLH are shown in Fig. 35.9a. On setting $k_n = 0$, the linear stiffness of the nonlinear spring is retained and the resonance frequency corresponding to this stiffness is 600 Hz. This could be considered to occur at small displacements as can be deduced from Eq. (35.1) when setting the displacement, x to a very small value. Appreciable nonlinearity of the NLH oscillator is triggered only after a high enough displacement is attained by the attachment mass, thus making the oscillator's response quasi-linear at small amplitude oscillation. It is worthwhile to note here that consistent units (SI) are used throughout the simulations. A schematic of the simulation model is illustrated in Fig. 35.4.

By considering the quasi-linear resonance frequency of the NLAM as the normalizing parameter, the respective band gaps of NLAM and LRAM, represented as lower and upper limit normalized with this local resonance frequency of the NLAM are as follows

$$NLAM : [\Omega_L - \Omega_U] = [1 - 1.6]$$

$$LRAM : [\Omega_L - \Omega_U] = [1.86 - 2.1]$$

The parameters used for the LRAM system are as follows

$$[m_1, m_2, k_1, k_2, k_n] = [1.69E - 3, 0.46E - 3, 2.48E6, 2.27E4, 0]$$

With the selected LRAM parameters, a querying frequency of $\Omega=\omega/\omega_0=1.9$ was applied at two different amplitudes along two directions respectively. In other words, four cases are simulated depending on the amplitude and direction of traverse of the wave as shown in Fig. 35.5, which shows a clear indication of direction-bias effect. At high amplitude, a wave incident first on the NLAM side displays marked propagation. While in the opposite direction the wave is attenuated. Moreover, at low amplitude, the same wave is attenuated irrespective of its direction of traverse. The mechanism of direction-bias is discussed in the following section.

At low amplitude (A = 1E-5) excitation at the queried frequency, a wave moving from left to right arrives at the NLAM first, for which the frequency lies outside the band gap; causing the wave to propagate through it. As the frequency lies within the band gap of LRAM, the wave is attenuated by it and hence the wave undergoes an overall attenuation, across the test article. In the opposite direction, the same wave is attenuated by the LRAM first and thus experiences attenuation across the test article.

At high amplitude (A = 1E-3) of excitation, the wave moving from right to left traverses through the LRAM first whose bandgap results in attenuation of the wave. However, as the same wave traverses along the left-right direction, its frequency is modified by the NLAM and shifted to a lower range, which lies outside the bandgap of the specifically designed LRAM. This results in propagation of the wave in this case, exhibiting bias in propagation depending on direction of traverse.

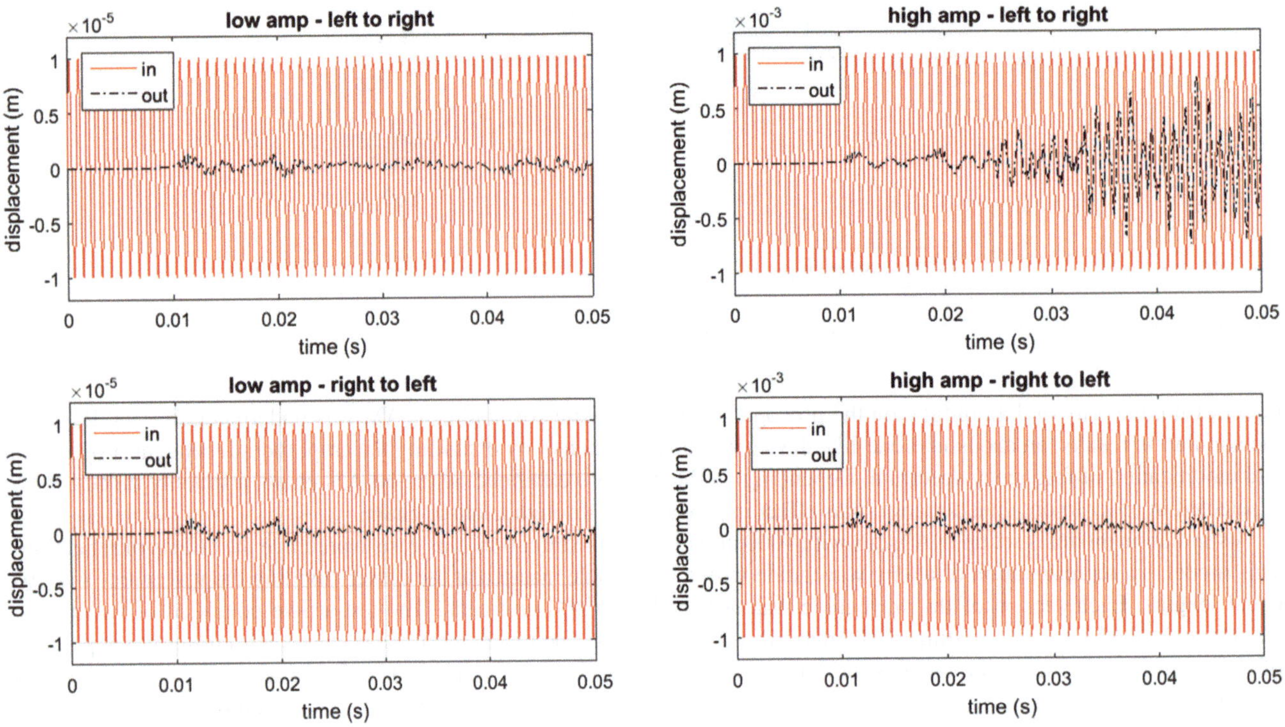

Fig. 35.5 Direction-bias effect obtained for the designed configuration from simulations

35.4 Experiment

The experiment involves measurement of displacement response across the test-article when an excitation is applied to one end. To test the direction bias, two different amplitudes and directions of excitation are required as per the mechanism as detailed above. To enable this type of testing, an experimental waveguide rig was designed. Figure 35.6a shows a schematic of the experimental setup which enables propagation of waves along a waveguide. The polycarbonate waveguide is instrumented with transducers for measurement of response. The far end of the waveguide was tapered to minimize the effect of reflections. This oblique end is embedded in a sand-bed to ensure minimization of reflections from a free boundary. Suspension strings were used to provide periodic supports along the length to the rather slender waveguide assembly. Periodic lateral line-support arrangement is made before and after the test-article to mitigate any mode conversion to transverse mode of the beam.

The components of the system are illustrated in the Fig. 35.6, which shows the (a) shaker assembled with the waveguide with the test-article assembled at the location (e) at the interface of two sections similar to (g). The amplifier is located at (c) which receives the signals from the waveform generator (b). The lateral (f) and vertical suspension (d) supports are employed to act as a guide to the polycarbonate beam against twisting under self-weight. The vertical supports are equipped with special lead screw hooks for a fine leveling of the beam. Leveling of the 35 feet long beam was achieved with the help of laser level markers and bubble indicators. The location of the transverse supports and suspension points enable flexibility to mitigate interruptions from the bending modes of the beam. The instrumentation consists of a Labworks ET-140 shaker, Keysight waveform generator and single axis B&K accelerometers (4507 and 4508). Single frequency excitations are applied using the shaker at the requisite amplitudes and incident and transmitted signals are measured at the aforementioned monitoring locations.

The NI USB 6009 DAQ in conjunction with NI LabVIEW was employed for data acquisition. A sampling rate of 24 kSamples/s collecting 24 kSamples ensured a window of 1 s on the LabVIEW acquisition. Acceleration time histories exported to excel files are processed using Matlab to obtain the frequency spectra. In certain conditions, non-contact measurements are made with a Polytec PDV-100 Laser Vibrometer.

The velocity of wave through the polycarbonate waveguide was obtained experimentally by recording the arrival time of a wave in travelling across two points along the length. The experimentally deduced value of 1400 m/s correlates well with the theoretically calculated value of 1350 m/s Table 35.1, showing a difference of 3.5%. In addition, a wavelength check performed experimentally showed a value of 1.376 m, which is in close agreement with the theoretically established value

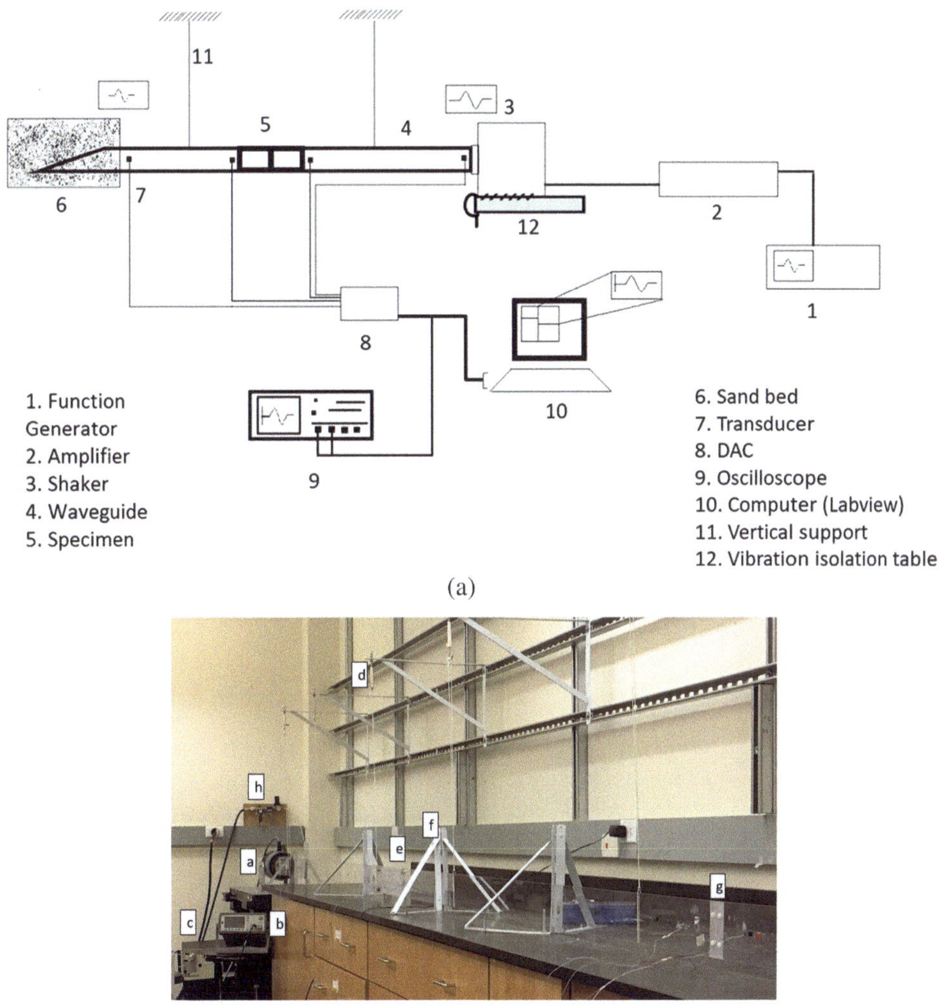

(b)

Fig. 35.6 (**a**) Schematic depiction of the experimental setup and (**b**) assembled experimental setup showing various components of the system

Table 35.1 Mechanical properties of polycarbonate waveguide

Property	Value
Elastic modulus (Pa)	2.2E9
Density (kg/m^3)	1200
Bar velocity (m/s)	1.35E3[a]

[a]note that bar velocity is theoretically calculated using material properties

of 1.35 m at 1 kHz. A sand boundary is enforced on the far side of the waveguide was designed to provide low-reflective boundary condition to minimize any interference with reflections. However, unlike in the case of Smith et al. [38] where the interacting waves were flexural and waveguide was an aluminum beam of cross-section 6.35 mm × 12.7 mm, the case at hand is that of a longitudinal elastic wave traveling in a 6.35 mm × 101.1 mm polycarbonate plate. This reduction in the area of contact could be a cause for the relative reduction in efficiency of this setup as seen in experiments. Given the propagation times and measurement locations, it, however, provides a means to offset the torsional mode otherwise elicited from a free end while providing a notably delayed and diminished reflection owing in part also to the oblique termination.

35.5 Test Article Design

The LRAM part of the test article, as shown in Fig. 35.7 consists of an array of tip-loaded cantilever beams obtained by removing portions of a polycarbonate plate using water-jet cutting. The dimensions were chosen keeping resonant frequency and mass ratios of the representative 1-D system in view. The individual local resonance frequencies are experimentally determined with the help of a laser vibrometer and the spread of obtained frequencies is shown in Fig. 35.7b.

For the NLAM part of the test article, the specified nonlinearity is realized based on varying the effective length of cantilever beams during oscillation [39]. This variation in unsupported length is achieved by creating a supporting profile that mimics the placing of a series of rigid support points along the length of the cantilever along the curve of the profile, as shown in Fig. 35.8. The separation between the dynamic point of contact and the tip mass would become the effective length of the cantilever and as the contact point moves further away from the root (closer to the tip) the effective length decreases and hence results in the desired nonlinear hardening response.

The method of obtaining the stiffness is described in detail by Spreemann et al. [39]. A Matlab code was used to obtain the root profile for a chosen nonlinear stiffness parameter, the algorithm for which is shown in Fig. 35.8a. Root profile obtained for $k_n = 10^{11}$ as obtained from the code is illustrated in the Fig. 35.9b, showing its intersection with the beam deflection curve without considering nonlinear effects. In the case where an integral root profile (as opposed to the present assembly) was planned, it was imperative to ensure that intersection of curves was taken into consideration for the design. This is because if the minimum diameter of cut offered by the tool were more than 1 mm (refer to Fig. 35.9b), the contact would not be initiated and hence the stiffness would be but linear for the amplitude range of interest.

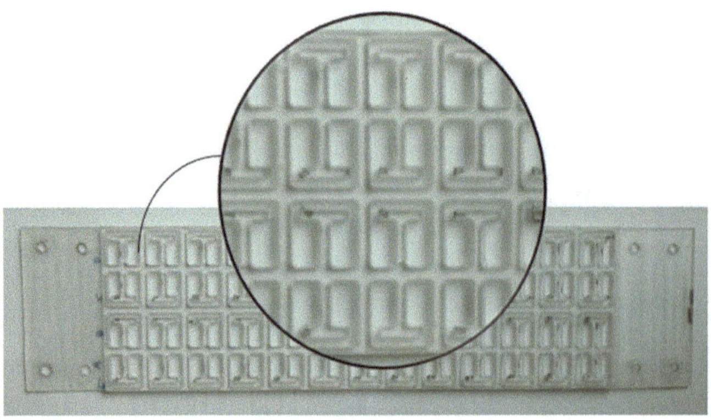

(a)

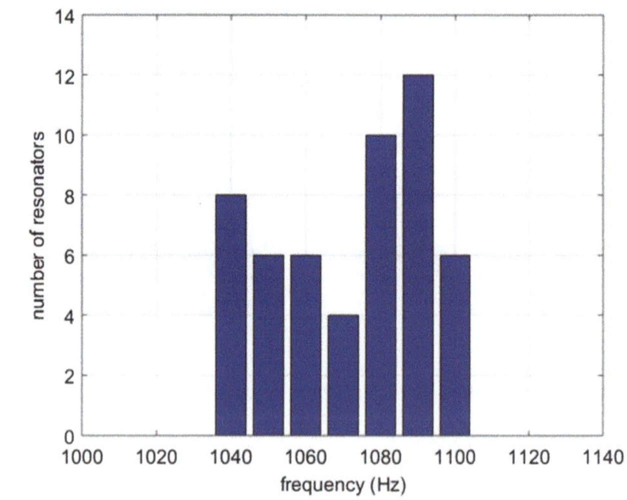

(b)

Fig. 35.7 (**a**) The LRAM test-article and (**b**) experimentally recorded local resonance frequencies

a

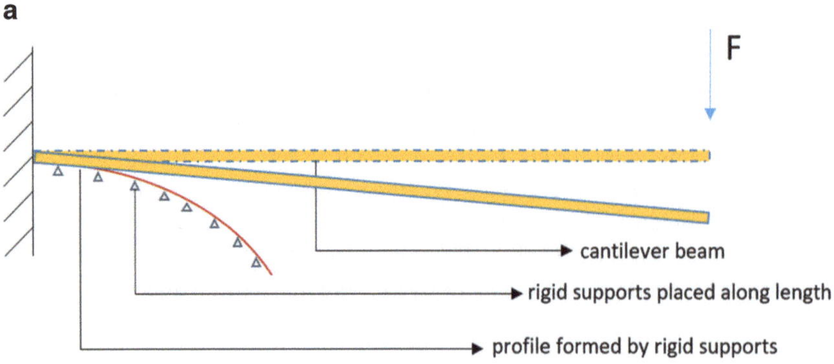

b

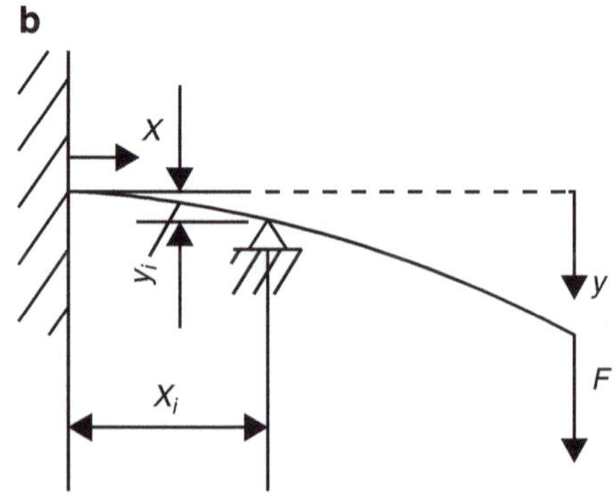

Fig. 35.8 (**a**) Depiction of the mechanism involved in obtaining the nonlinear hardening stiffness through contact with root profile and (**b**) Method of locating coordinates of the support points using beam deflection theory [39]

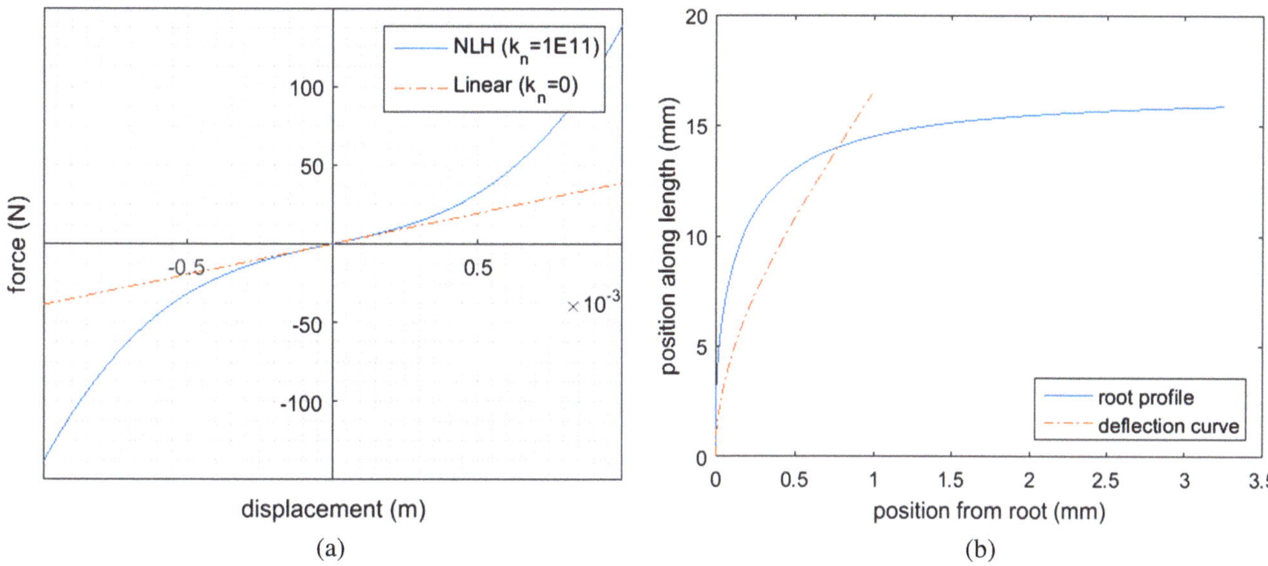

Fig. 35.9 (**a**) Input stiffness curve with Linear and NLH force-displacement data and (**b**) Computed cantilever root profile for obtaining the predefined load-displacement curve

The design demanded the addition of machined steel mass be attached as tip mass, to keep the mass ratio consistent with the defined parameters. The steel mass was machined using CNC milling, while the polycarbonate part was cut with the help of a waterjet. Owing to the limitation imposed by the least diameter of cut achievable by water jet (with abrasive), the root profiles had to be fabricated separately and assembled in the unit cells after the cantilever beam arrays were cut into the polycarbonate plate. These root profiles were printed by using selective laser sintering process and inserted into the machined polycarbonate plate to form the NLAM test-article as shown in Fig. 35.10b. The assembled NLAM test-article and the spread of experimentally obtained local resonance frequency are shown in Fig. 35.11. 3D contact simulations for the cantilever beam with root profile geometries helped verify the designed nonlinear stiffness Fig. 35.12.

35.6 Results

Firstly, the LRAM test-article was assembled with the waveguide and transmissibility obtained at different frequencies of excitation. A similar procedure was followed for the NLAM counterpart as well. An accelerometer placed 2″ away from the test-article along the direction of wave propagation was used to obtain the transmitted wave displacement histories and a point on the shaker head is chosen to represent the incident wave displacement. Single frequency excitation applied for a sufficient amount of time are used to ensure that a steady state is established in the setup, before recording the accelerations. Transmissibility (D*) is calculated as the ratio of transmitted to incident displacement amplitudes at the frequency of excitation. A transmissibility (D*) value less than 1 indicates attenuation while that equal to or above 1 represents propagation.

Next, the D* values obtained by assembling only the NLAM test article in the waveguide were plotted. There is an evident attenuation within the theoretically predicted bandgap (600–980 Hz) at low amplitude, when nonlinearities are not triggered. However, a peak is encountered at 800 Hz, which can be attributed to a spread in the actual local resonances. This is because theoretical predictions assume identical local resonance frequencies for each unit cell, which is not true in the experimental case. The high number of unit cell local resonance frequencies below 600 Hz explains the early initiation of experimental bandgap compared to theoretical predictions.

The direction biased waveguide test-article was put together by assembling the NLAM and LRAM test-articles in series. A procedure similar to that for the LRAM and NLAM test articles was followed to record the displacement responses and to obtain D*. The experimental setup is similar to those modeled in simulations. In order to reverse the direction of wave propagation, the test-article was reversed instead of the shaker location but without loss of symmetry in the waveguide configuration. The configuration where NLAM is to the left as seen in Fig. 35.13 is chosen as a reference. Therefore, the configuration with NLAM to the left is considered to be forward and LRAM to the left is considered reverse. In both the

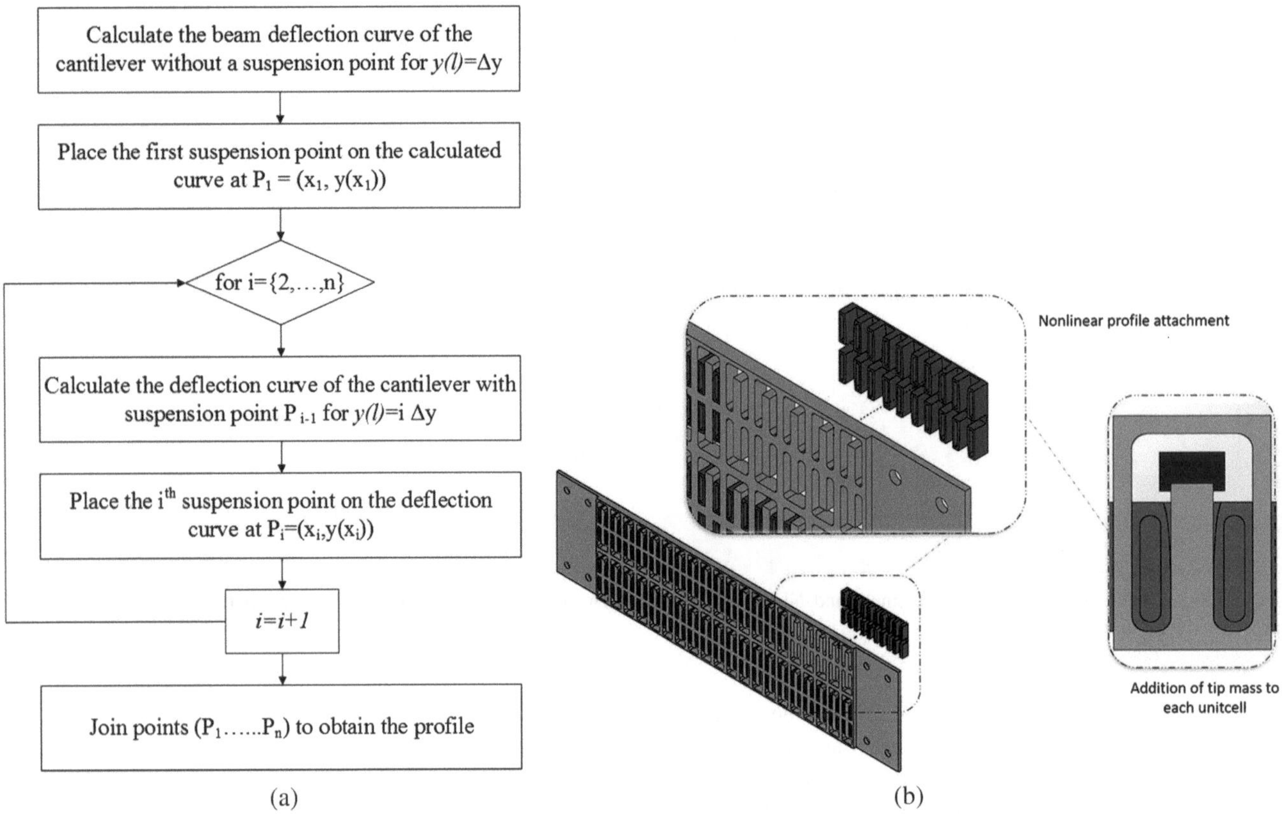

Fig. 35.10 (**a**) Flow chart of the routine used to obtain root profile for pre-defined nonlinearity and (**b**) Assembly procedure to fabricate the NLAM test-article

directions, low and high amplitude excitations were applied at the same excitation frequency. Thus, four cases were obtained for each frequency and the corresponding transmissibility curves are shown in Fig. 35.14. The D* values were obtained from the average of a set of three trials and displayed a maximum deviation of 8% between successive trials.

The design of test article was based on the simulations that display an amplitude-dependent direction bias in wave propagation for a chosen excitation frequency of $\Omega = \omega/\omega_0 = 1.9$ ($\omega = 1140$ Hz). Therefore, a harmonic wave at $\Omega = 1.9$ should experience attenuation along either direction at low amplitude. However, the same wave at higher amplitude propagates while traversing along left to right direction, while at low amplitude, this wave attenuates. The fabricated test article displayed considerable spread in frequency as compared to the design. Therefore, instead of querying at a single frequency, a range of frequencies in the neighborhood of predicted value was utilized. The transmissibility (D*) versus excitation frequency plots obtained from the experiments are shown in Fig. 35.14. In the vicinity of 1080 Hz, an indicative trend can be observed. This region displays direction-bias as predicted by simulations, although at a lower frequency than that was designed for. This provides experimental evidencefor a higher amplitude of propagation relative to the other three cases at the same frequency indicating direction bias in propagation. A closer examination of the time signals and frequency spectra for these cases show a 100% increase in propagated amplitude as may be expected based on the numerical study. The displacement time histories for amplitude and direction variation are shown for comparison in Fig. 35.15.

A strong attenuation arising from the LRAM bandgap is prominent in the ‘right to left’ case. In the neighborhood of the designed excitation frequency of 1140 Hz ($\Omega = 1.9$), there is a minor increase in D* in accordance with the direction-bias mechanism, however a marked direction bias is observed at a slightly lower excitation frequency of 1080 Hz. This deviation from expected result may be attributed to the variation in actual local resonance frequencies for the design value; and the potential difference from the idealized cubic hardening mechanism owing to contact induced effects in the actual test article.

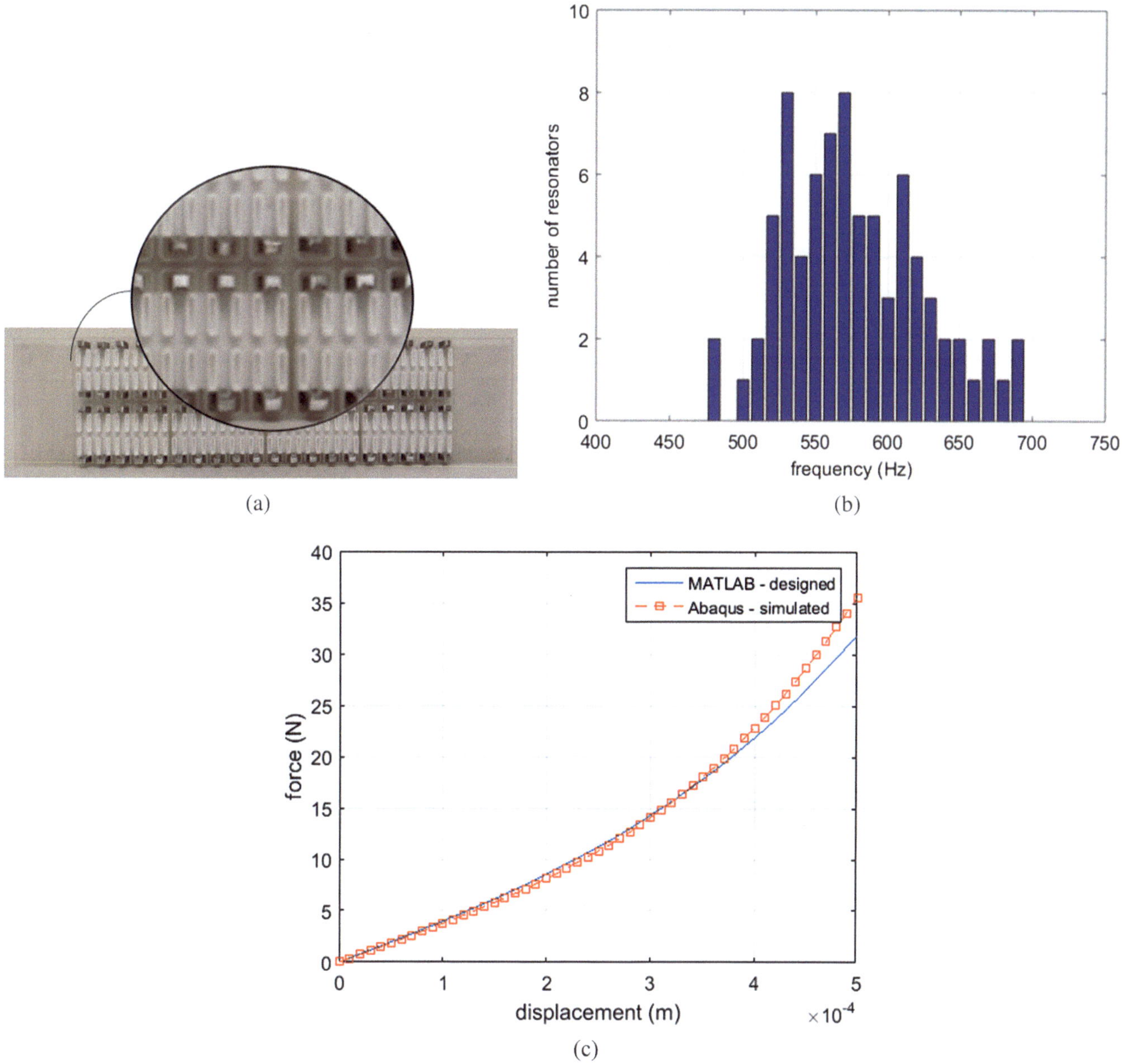

Fig. 35.11 (**a**) Fully assembled NLAM test-article, (**b**) spread of local resonance frequencies and (**c**) numerical verification of nonlinear response of resonators

35.7 Device Implications of Inertant Acoustic Metamaterial

In order to explore the device implications of incorporating inerters, which are mechanical elements that provide a force response proportional to the relative acceleration across them, into acoustic metamaterial configurations, their 1-dimensional discrete element representations were investigated to ascertain their dynamic characteristics. The presence of inerters in either the local attachments or in the host structure were considered. An effective-mass based analytical model [10] was used to obtain the dispersion relations in The following inertant AM configurations were considered: (a) inertance in local resonator attachment – (i) purely inertant, (ii) inertance in parallel, and (iii) Inertance in series; and (b) inertance in the host lattice – (i) inertance in parallel (PL), and (ii) inertance in series (SL). Among these, the device implications for the parallel and series inertant lattice cases shown in Fig. 35.16 are discussed below.

The combination of inerter and stiffness was modeled as an 'effective stiffness' in the configurations with inerter in lattice for ease of parametric interpretation using the effective model. Hence, these configurations are equivalent to an LRAM

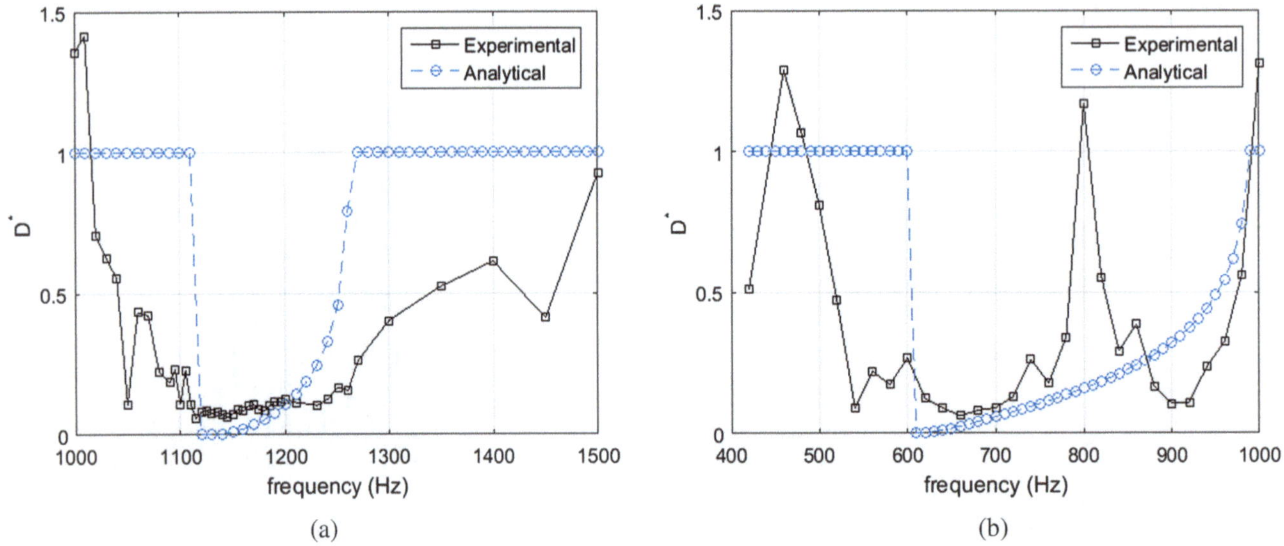

Fig. 35.12 Comparison of experimental and theoretical transmissibility curves for the (**a**) LRAM and (**b**) NLAM test-articles demonstrating the bandgap phenomenon

Fig. 35.13 Assembled DBWG test-article

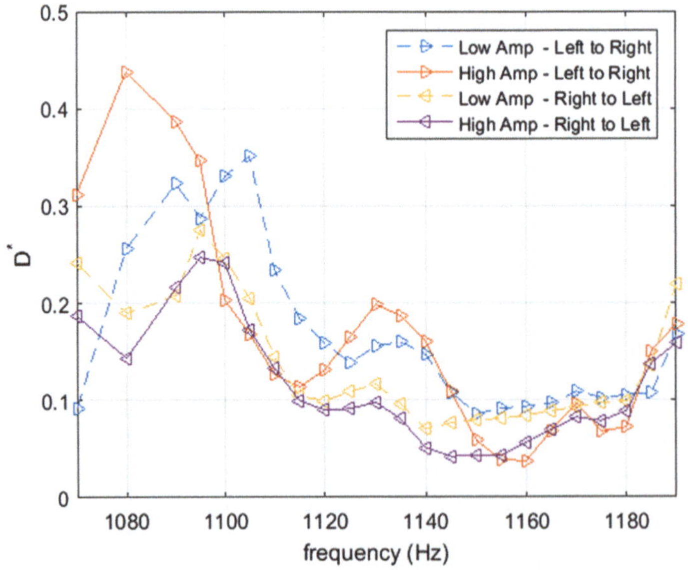

Fig. 35.14 Transmissibility curves for the direction biased waveguide test-article (NLAM and LRAM in series) for the four cases covering both directions and low and high amplitudes of excitation

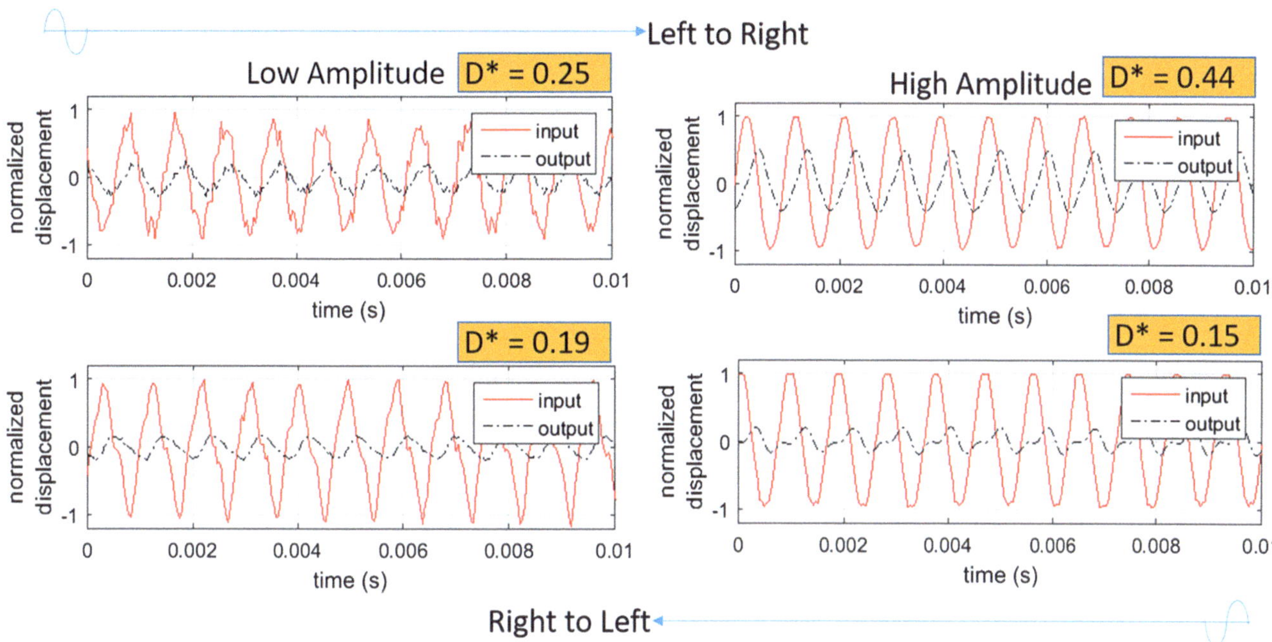

Fig. 35.15 Direction bias effect as observed from displacement time histories at 1080 Hz

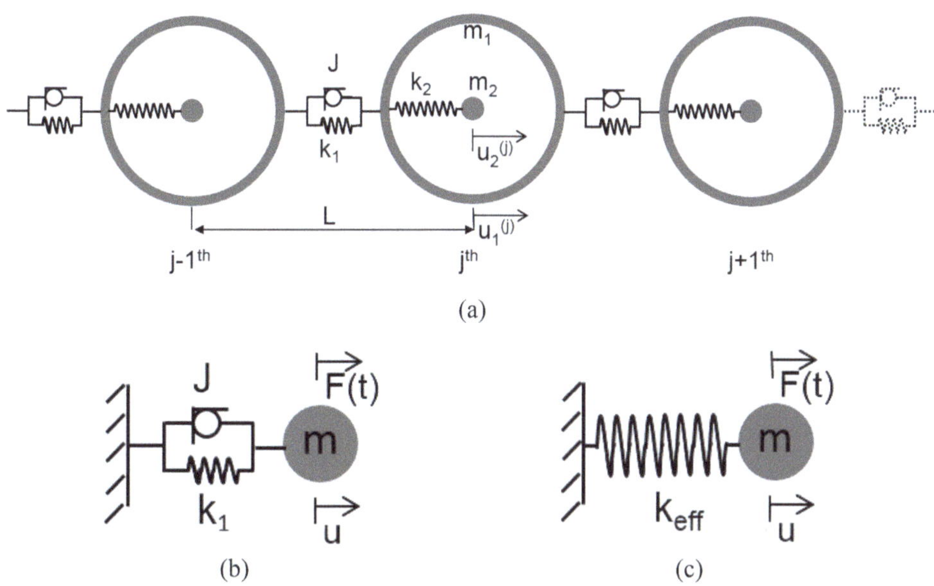

Fig. 35.16 (**a**) Acoustic metamaterial with lattice structure having inerter in parallel with stiffness and (**b**) parallel combination of inerter and spring modeled as (**c**) an effective stiffness

with 'effective mass' and 'effective stiffness'. As with effective mass, the effective stiffness here is found to be frequency dependent and can even become negative in certain frequency ranges. The ratio of the effective mass (m_{eff}) and the effective stiffness (k_{eff}) is normalized to create an 'effective lattice parameter,' $\omega_0^2 m_{eff}^{PL}/k_{eff}^{PL}$which serves as an indicator of propagation characteristics. The general form of the dispersion relation for the effective model, designated for the parallel inertant lattice (PL) case is shown in Eq. (35.4). With lattice parameters $[m_1, m_2, k_1, k_2, L] = [10, 9, 100, 10, 1]$, the behavior of effective lattice parameter for two specific values of J is illustrated in Fig. 35.17. A negative value for this parameter results in a bandgap as seen by the dispersion curves (Fig. 35.17) and relation (Eq. 35.4). Further, a positive value of this parameter due to effective mass and stiffness being simultaneously negative, or in other words, a 'double-negative' region results in propagation with negative group velocity, whereas both effective mass and effective stiffness being simultaneously positive results in a propagation mode with positive group velocity.

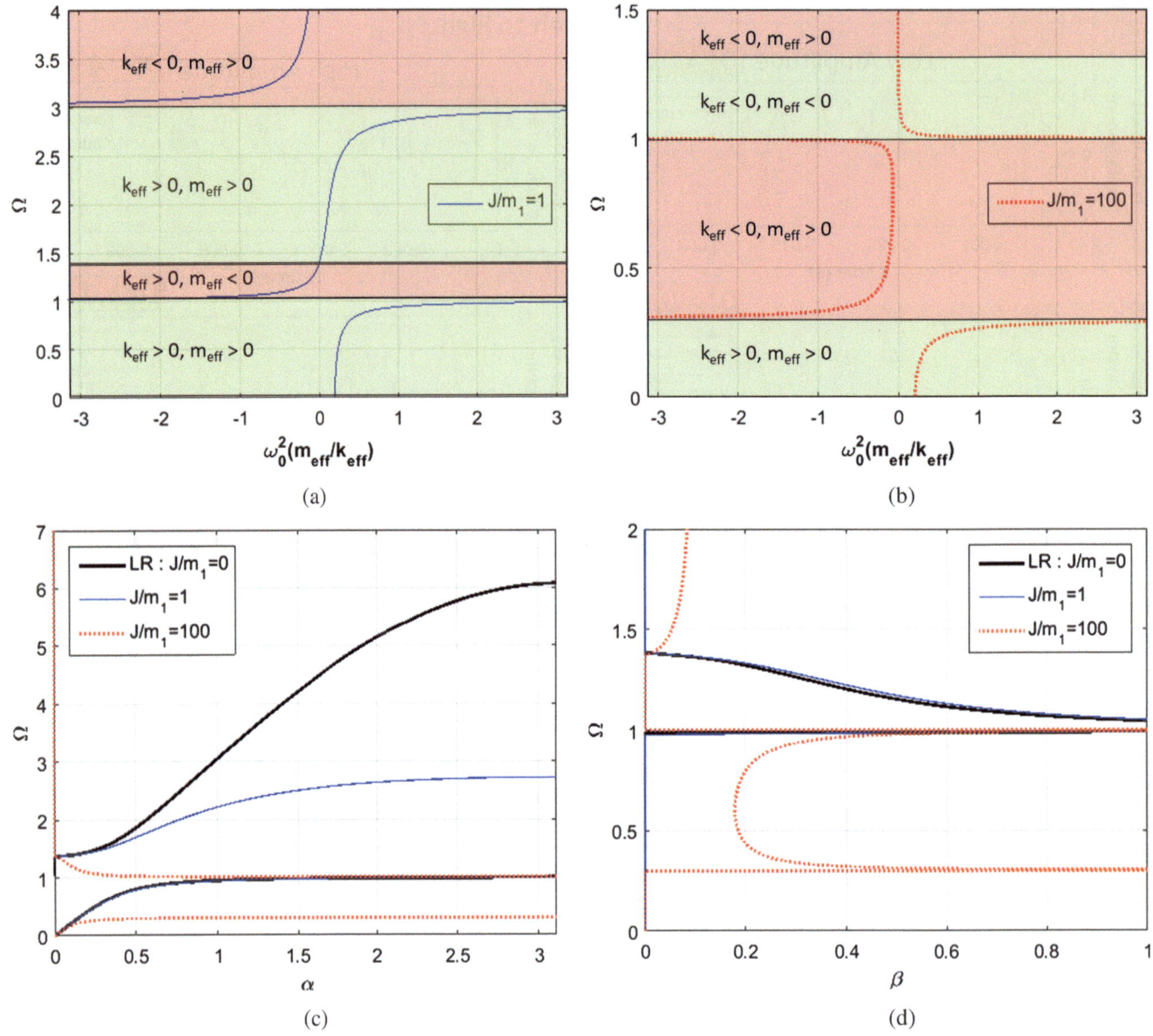

Fig. 35.17 Normalized effective lattice parameter versus normalized excitation frequency for (**a**) J/m$_1$ = 1 and (**b**) J/m$_1$ = 100 along with (**c**) real and (**d**) imaginary parts of the wavenumber versus normalized excitation frequency for these cases for the parallel inertant lattice configuration

$$qL = \alpha + i\beta = \cos^{-1}\left[1 - \frac{m_{eff}^{PL}\Omega^2\omega_0^2}{2k_{eff}^{PL}}\right] \tag{35.4}$$

By enforcing conditions to ensure continuity in the two bandgap regions, two special cases emerge. For J values of $J_{LB}^{PL} = 90$ and $J_{UB}^{PL} = 47.4$, the AM has a continuous higher frequency bandgap as seen in Fig. 35.18 and hence acts as a low pass filter for mechanical waves.

With the same lattice parameters, the series inertant configuration is studied and is found to have two special cases for obtaining continuous bandgaps. The values of inertance (J) are the same as those obtained for the 'PL' case. The dispersion curves for these two special cases (Fig. 35.19) reveals an interesting feature of this class of AM, in that they could display ultra-low frequency bandgaps even encompassing the long wavelength limit and exhibit the potential to act as high-pass filters for mechanical waves.

Further, it is possible to postulate a structural waveguide that acts as an almost total mechanical wave inhibitor by cascading a tuned parallel inertant lattice (PL) and a series inertant lattice (SL) AM configurations in sequence. When a

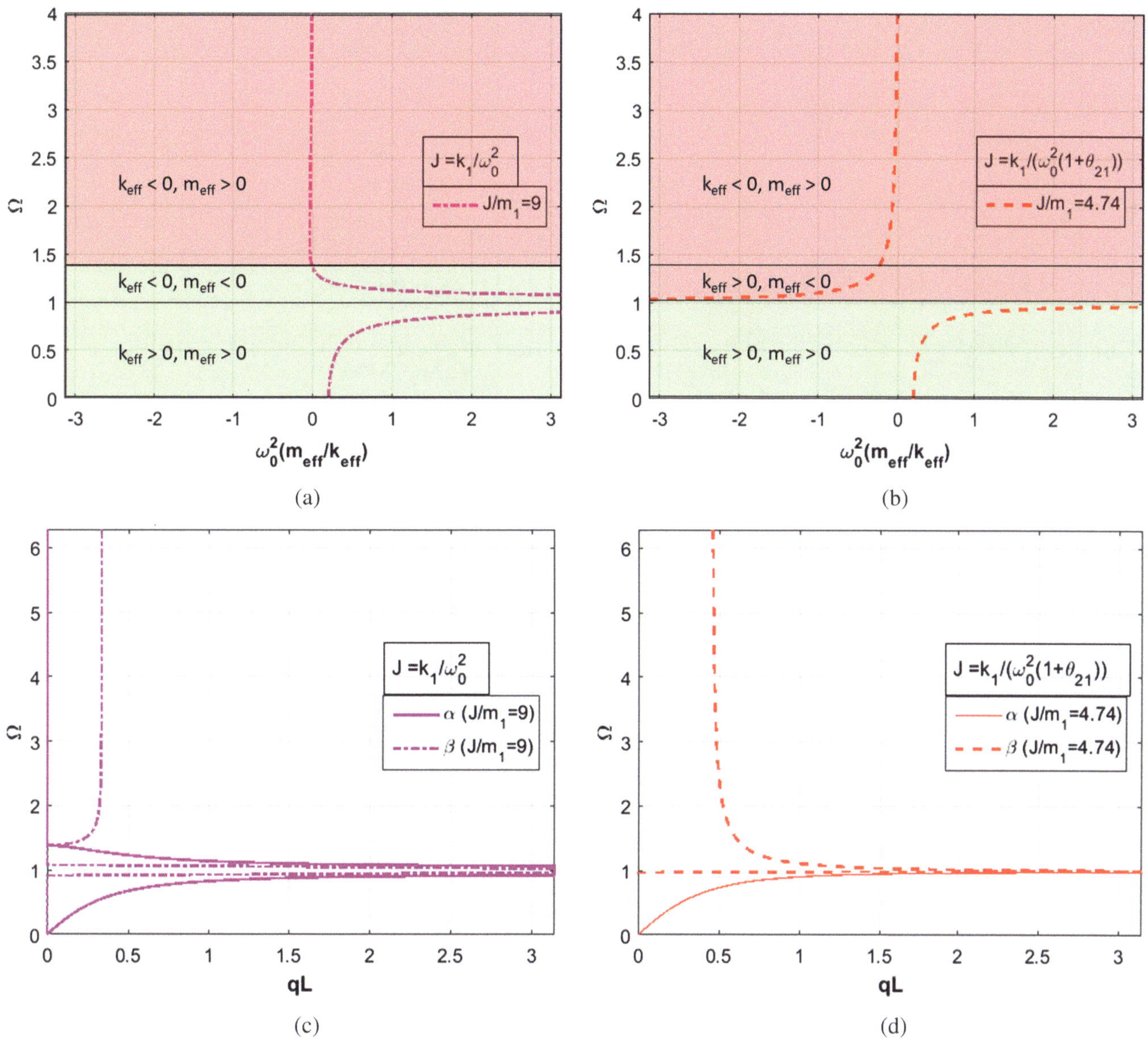

Fig. 35.18 Normalized effective lattice parameter versus normalized excitation frequency for (**a**) $J/m_1 = 9$ and (**b**) $J/m_1 = 4.74$ along with (**c**) real and (**d**) imaginary parts of the wavenumber versus normalized excitation frequency for these cases for the parallel inertant lattice configuration

PL configuration with $J/m_1 = 4.74$ and an SL configuration with $J/m_1 = 9$ that are connected sequentially, the PL behaves as a low-pass filter with $\Omega = 1$ as the upper bound and the SL as a high-pass filter with lower bound of $\Omega = \sqrt{1 + \theta_{21}} = 1.378$. This results in an overlap in their bandgap frequency ranges. As can be seen from the plots in Fig. 35.20 for the attenuation factor, (β) for PL and SL, the overall stop band for this device covers all excitation frequencies leading to it act as a complete longitudinal elastic wave inhibitor. Choosing $J/m_1 = 4.74$ for both PL and SL results in the low and high frequency bandgaps becoming continuous at $\Omega = 1$ where β is also unbounded for the two configurations. By exploring new combinations of inertant metamaterial configurations and tuning the inertance in the various sub-structures, the degree of attenuation over the entire frequency range may be optimized. Also, more interesting characteristics for specialized applications may be realized. The ability to realize such inertant acoustic metamaterials with omnipresent bandgap may result in designing materials with the ability to absorb mechanical vibrations and shock.

The dispersion characteristic for acoustic metamaterial configurations with inerters in the local attachment or in the host lattice shows the possibility of enriching the tunable bandgap behavior obtained in LRAMs especially for lower frequencies without the addition massive elements. Thus inertant acoustic metamaterials could have the potential to overcome some of the limitations of mass ratio encountered in LRAMs. This could have implications for realizing devices such as tunable

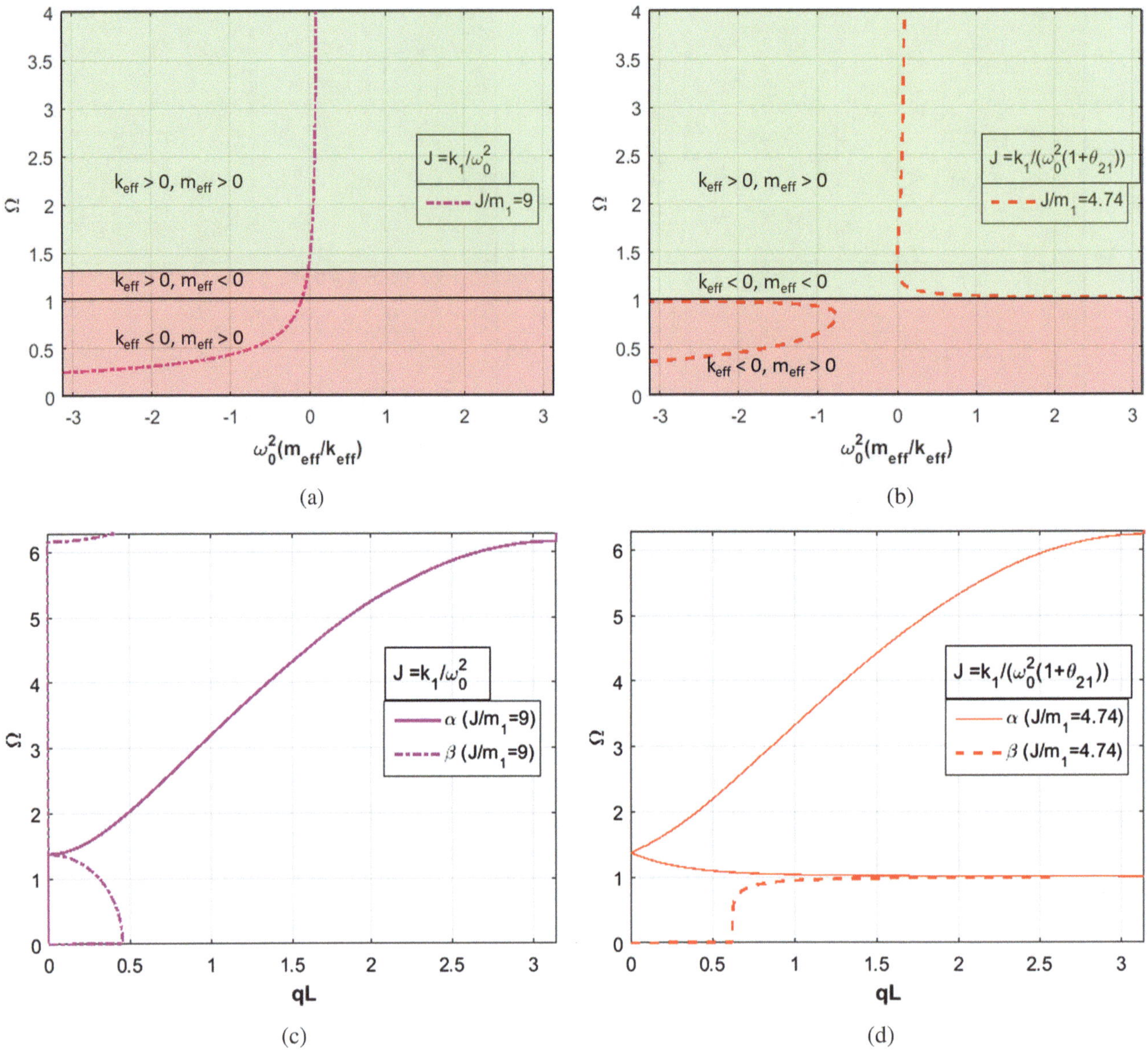

Fig. 35.19 Normalized effective lattice parameter versus normalized excitation frequency for (**a**) $J/m_1 = 9$ and (b)$J/m_1 = 4.74$ along with (**c**) real and (**d**) imaginary parts of the wavenumber versus normalized excitation frequency for these cases for the series inertant lattice configuration

narrow-banded low-frequency acoustic filters and high-pass filters with ultra-low frequency bandgaps that extend up to the long wavelength limit as well as nearly total mechanical wave inhibitors. Low-pass filters with the possibility of propagation modes having negative group velocity may be possible too.

35.8 Conclusions

Analytical, numerical and experimental investigation of nonlinear and inertant acoustic metamaterial configurations were conducted. Using a perturbation approach, approximate dispersion curve shifts owing to the presence of cubically nonlinear hardening and softening stiffnesses in local oscillator attachments were obtained. The predicted amplitude-dependent shifts in bandgap frequency range were verified using discrete element model simulations. Based on the observed shifting of propagated spectra to lower frequencies for the nonlinear hardening case, a waveguide device consisting of a tuned combination of a locally resonant acoustic metamaterial (LRAM) and a nonlinear acoustic metamaterial (NLAM) that

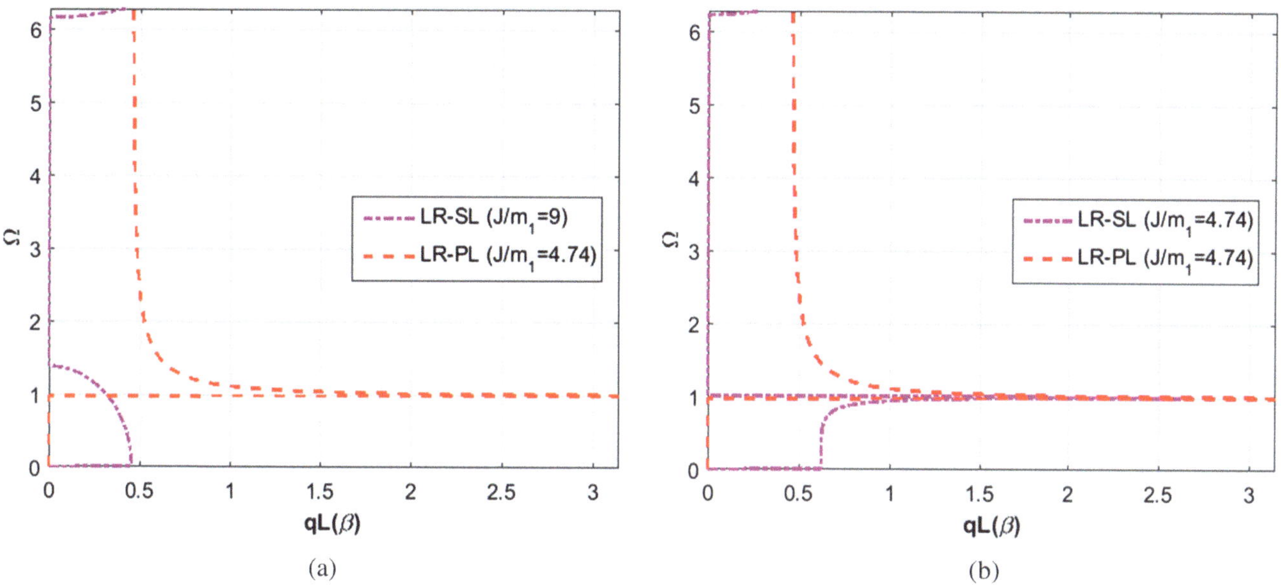

Fig. 35.20 Attenuation factors (β) for the longitudinal wave inhibitor with combined parallel and series inertant lattice configurations having (**a**) $J/m_1 = 4.74$ for PL and $J/m_1 = 9$ for SL and (**b**) $J/m_1 = 4.74$ for both PL and SL. Nonzero values for β indicate the occurrence of a bandgap frequency region

displays direction biased propagation for the same excitation frequency was postulated and verified using simulations. Further, experimental investigations conducted using a prototype direction biased waveguide test article verified the existence and extent of bandgaps and also provide an indication of the passive direction-bias phenomenon. The nonlinear hardening behavior for the local oscillators in the test article was achieved using contact with root profile constrains for the tip-loaded cantilever beams. In addition, the device implications for inertant acoustic metamaterial configurations were investigated using an effective mass model for their discrete element representations. Inerters, which are mechanical elements that provide a force response proportional to the relative acceleration across them, deliver a dynamic mass presence much higher than their static device mass. Structural acoustic devices that display frequency-dependent negative and even extreme effective-mass and stiffness regimes are shown to be possible using inertant acoustic metamaterials. Passive high-pass filters with ultra-low frequency bandgaps that extend up to the long wavelength limit realized without the use of any mechanical grounding elements and even structural networks that can act as a nearly complete mechanical wave inhibitor are potential device implications. It is anticipated that further research on interactive synergies between nonlinear and inertant configurations and practical strategies to scale and fabricate them could have promise for realizing a new class of AM with enriched dynamics.

References

1. Thomson, W.: XLVI. On the reflexion and refraction of light. Philos. Mag. Series 5. **26**, 414–425 (1888)
2. Viktor, G.V.: The electrodynamics of substances with simultaneously negative values of ε and μ. Physics-Uspekhi. **10**, 509 (1968)
3. Pendry, J.B.: Negative refraction makes a perfect lens. Phys. Rev. Lett. **85**, 3966–3969 (2000)
4. Martinez-Sala, R., Sancho, J., Sanchez, J.V., Gomez, V., Llinares, J., Meseguer, F.: Sound attenuation by sculpture. Nature. **378**, 241–241 (1995)
5. Liu, Z., Zhang, X., Mao, Y., Zhu, Y.Y., Yang, Z., Chan, C.T., Sheng, P.: Locally resonant sonic materials. Science. **289**, 1734–1736 (2000)
6. Goffaux, C., Sánchez-Dehesa, J., Yeyati, A.L., Lambin, P., Khelif, A., Vasseur, J.O., Djafari-Rouhani, B.: Evidence of Fano-like interference phenomena in locally resonant materials. Phys. Rev. Lett. **88**, 225502 (2002)
7. Sainidou, R., Djafari-Rouhani, B., Pennec, Y., Vasseur, J.O.: Locally resonant phononic crystals made of hollow spheres or cylinders. Phys. Rev. B. **73**, 024302 (2006)
8. Wang, G., Wen, X., Wen, J., Shao, L., Liu, Y.: Two-dimensional locally resonant phononic crystals with binary structures. Phys. Rev. Lett. **93**, 154302 (2004)
9. Huang, H.H., Sun, C.T.: Anomalous wave propagation in a one-dimensional acoustic metamaterial having simultaneously negative mass density and Young's modulus. J. Acoust. Soc. Am. **132**, 2887–2895 (2012)
10. Huang, H.H., Sun, C.T., Huang, G.L.: On the negative effective mass density in acoustic metamaterials. Int. J. Eng. Sci. **47**, 610–617 (2009)

11. Huang, H.H., Sun, C.T.: Theoretical investigation of the behavior of an acoustic metamaterial with extreme Young's modulus. J. Mech. Phys. Solids. **59**, 2070–2081 (2011)
12. Fang, N., Xi, D., Xu, J., Ambati, M., Srituravanich, W., Sun, C., Zhang, X.: Ultrasonic metamaterials with negative modulus. Nat. Mater. **5**, 452–456 (2006)
13. Lee, S.H., Park, C.M., Seo, Y.M., Wang, Z.G., Kim, C.K.: Acoustic metamaterial with negative density. Phys. Lett. A. **373**, 4464–4469 (2009)
14. Lee, S.H., Park, C.M., Seo, Y.M., Wang, Z.G., Kim, C.K.: Composite acoustic medium with simultaneously negative density and modulus. Phys. Rev. Lett. **104**, 054301 (2010)
15. Lazarov, B.S., Jensen, J.S.: Low-frequency band gaps in chains with attached non-linear oscillators. Int. J. Non Linear Mech. **42**, 1186–1193 (2007)
16. Manimala, J.M., Sun, C.T.: Numerical investigation of amplitude-dependent dynamic response in acoustic metamaterials with nonlinear oscillators. J. Acoust. Soc. Am. **139**(6), 3365 (2016)
17. Boechler, N., Theocharis, G., Daraio, C.: Bifurcation-based acoustic switching and rectification. Nat. Mater. **10**, 665–668 (2011)
18. Bertoldi, K., Boyce, M.C.: Mechanically triggered transformations of phononic band gaps in periodic elastomeric structures. Phys. Rev. B. **77**, 052105 (2008)
19. Manktelow, K.L., Leamy, M.J., Ruzzene, M.: Analysis and experimental estimation of nonlinear dispersion in a periodic string. J. Vib. Acoust. **136**, 031016 (2014)
20. Narisetti, R.K., Leamy, M.J., Ruzzene, M.: A perturbation approach for predicting wave propagation in one-dimensional nonlinear periodic structures. J. Vib. Acoust. **132**, 031001 (2010)
21. Manimala, J.M.: Dynamic Behavior of Acoustic Metamaterials and Metaconfigured Structures with Local Oscillators (2014). https://search.proquest.com/openview/13d9c9f5209af1b55697f6f2aa9e02b5/1.pdf?pq-origsite=gscholar&cbl=18750&diss=y
22. Nadkarni, N., Arrieta, A.F., Chong, C., Kochmann, D.M., Daraio, C.: Unidirectional transition waves in bistable lattices. Phys. Rev. Lett. **116**, 244501 (2016)
23. Smith, M.C.: Synthesis of mechanical networks: the inerter. IEEE Trans. Autom. Control. **47**, 1648–1662 (2002)
24. Lazar, I.F., Neild, S.A., Wagg, D.J.: Using an inerter-based device for structural vibration suppression. Earthq. Eng. Struct. Dyn. **43**, 1129–1147 (2014)
25. Hu, Y., Chen, M.Z.Q., Shu, Z., Huang, L.: Analysis and optimisation for inerter-based isolators via fixed-point theory and algebraic solution. J. Sound Vib. **346**, 17–36 (2015)
26. Wang, F.-C., Liao, M.-K., Liao, B.-H., Su, W.-J., Chan, H.-A.: The performance improvements of train suspension systems with mechanical networks employing inerters. Veh. Syst. Dyn. **47**, 805–830 (2009)
27. Smith, M.C., Wang, F.-C.: Performance benefits in passive vehicle suspensions employing inerters. Veh. Syst. Dyn. **42**, 235–257 (2004)
28. Chen, M.Z.Q., Yinlong, H., Baozhu, D.: Suspension performance with one damper and one inerter. In Control and Decision Conference (CCDC), 2012 24th Chinese, Taiyuan, pp. 3534–3539 (2012)
29. Liu, X., Huang, X., Hua, H.: On the characteristics of a quasi-zero stiffness isolator using Euler buckled beam as negative stiffness corrector. J. Sound Vib. **332**, 3359–3376 (2013)
30. Chen, M.Z.Q., Hu, Y., Li, C., Chen, G.: Performance benefits of using inerter in semiactive suspensions. IEEE Trans. Control Syst. Technol. **23**, 1571–1577 (2015)
31. Papageorgiou, C., Smith, M.C.: Laboratory experimental testing of inerters, in Decision and Control, 2005 and 2005 European Control Conference. CDC-ECC '05. 44th IEEE Conference on, Seville, pp. 3351–3356 (2005)
32. Marian, L., Giaralis, A.: Optimal design of a novel tuned mass-damper–inerter (TMDI) passive vibration control configuration for stochastically support-excited structural systems. Probab. Eng. Mech. **38**, 156–164 (2014)
33. Hu, Y., Chen, M.Z.Q.: Performance evaluation for inerter-based dynamic vibration absorbers. Int. J. Mech. Sci. **99**, 297–307 (2015)
34. Rudenko, O.V.: Giant nonlinearities in structurally inhomogeneous media and the fundamentals of nonlinear acoustic diagnostic techniques. Physics-Uspekhi. **49**, 69–87 (2006)
35. Guo, X., Lin, Z., Tu, J., Liang, B., Cheng, J., Zhang, D.: Modeling and optimization of an acoustic diode based on micro-bubble nonlinearity. J. Acoust. Soc. Am. **133**, 1119–1125 (2013)
36. Lakes, R.S., Drugan, W.J.: Dramatically stiffer elastic composite materials due to a negative stiffness phase? J. Mech. Phys. Solids. **50**, 979–1009 (2002)
37. Kulkarni, P.P., Manimala, J.M.: Longitudinal elastic wave propagation characteristics of inertant acoustic metamaterials. J. Appl. Phys. **119**, 245101 (2016)
38. Smith, T. L., Rao, K., Dyer, I.: Attenuation of plate flexural waves by a layer of dynamic absorbers. Noise Control Eng. J. **26**(2), 56–60(5) (1986). http://www.ingentaconnect.com/content/ince/ncej/1986/00000026/00000002/art00001
39. Spreemann, D., Folkmer, B., Manoli, Y.: Realization of nonlinear hardening springs with predefined characteristic for vibration transducers based on beam structures, MikroSystemTechnik Kongress, Darmstadt (2011)

James M. Manimala is assistant professor in the School of MAE at Oklahoma State University. He earned his BTech in Mechanical Engineering in 2004 from the University of Kerala, India and his MS and PhD in Aero. & Astro. Engineering from Purdue University, Indiana in 2010 and 2014 respectively. Recently he received the 2016 DARPA Young Faculty Award.

Chapter 36
Evaluation of Stress Equilibrium in Dynamic Tests on Agglomerated Cork

Marco Sasso, Fabrizio Sarasini, Gianluca Chiappini, Edoardo Mancini, and Jacopo Tirillò

Abstract In this work, experimental compression tests have been performed on rectangular specimens cut from a cork slab. The tests have been performed both using a quasi-static testing machine and a Split Hopkinson Bar. The dynamic tests revealed a high sensitivity of the material to the strain rate, with stress strain curves significantly higher than in quasi-static condition. Given the low density of the cork, the specimen material may suffer from non-equilibrium; for this reason, a high speed camera was used to frame acquire pictures of the sample during the deformation, which have been used to perform DIC analyses. Indeed, inhomogeneous strain distributions were found, especially due to the low density of samples tested at the higher speed. Moreover, the low impedance of the tested material also determines difficulties in signal synchronization and, hence, in the correct calculation of the stress in the sample. Data from DIC analyses were then used to calculate the true strain in different portions of the specimen and to evaluate the stress due to inertia effect.

Keywords Agglomerate cork • Hopkinson bar • Dynamic tests • DIC • High speed imaging • Inertia stress

36.1 Introduction

The Cork Oak is a three whose bark is extremely thick, useful for multiple application and when cut grow again. Cork exhibits particular physical, chemical and mechanical behavior. Indeed, it is chemically stable and fire resistant, impervious to liquid and gases, its mechanical behavior allows its use in sealing, packaging, insulation, vibration control and thanks to its physical low density it is often used for applications like sound damping, weight reduction etc.

The main characteristics of cork porosity can be found in [1]; the particle density for finely grounded material was found to range from 1400 to 1800 kg/m^3. From the mechanical point of view, density and cell morphology are the key parameters governing the material stiffness; Gibson et al. [2], considering an hexagonal cell morphology, provided formulas for estimating the elastic modulus of cork for the axial, tangential and radial directions, with respect to vertical axis of the tree. They also showed that the mechanical properties at large strain varies significantly on the different loading axis. Fortes [3] studied the Poisson's ratio, showing that it can be as high as 0.05–0.1 at small strain and tends to vanish at higher strain.

The Hopkinson bar is one the most famous apparatus for dynamic test [4]. The study of the behavior at high strain rate of soft material, like foams, cork and rubbers, gained increasing importance in recent year, mainly because of their good performance for crashworthiness and energy absorption; strain rate sensitivity of metallic foams [5, 6], as well as rubber have been widely investigated [7, 8].

In low density materials, inertia effect may become important when the impact speed is high; and this phenomenon has investigated by high speed imaging on aluminium [9] and polymeric foams [10].

The dynamic mechanical properties of cork up to large deformation is not extensively investigated, to the author knowledge only in Gameiro et al. [11], it has been performed high strain rate tests on radial, non-radial and agglomerate cork. About of agglomerate cork they highlighted that its mechanical behavior, in the dynamic range considered (200 s^{-1}

M. Sasso (✉) • G. Chiappini
Università Politecnica delle Marche, via Brecce Bianche, 60100, Ancona, Italy
e-mail: m.sasso@univpm.it

F. Sarasini • J. Tirillò
La Sapienza-Università di Roma, via Eudossiana 18, 00184, Roma, Italy

E. Mancini
Università degli Studi eCampus, via Isimbardi, 22060, Novedrate, CO, Italy

J. Kimberley et al. (eds.), *Dynamic Behavior of Materials, Volume 1*, Conference Proceedings of the Society for Experimental Mechanics Series, DOI 10.1007/978-3-319-62956-8_36

and 600 s^{-1}), was not influenced by variation of strain rate. Instead, the dynamic plateau stress was higher than the static one, so they conclude that agglomerates is dependent from the strain rate applied. However, the important inertia effect has not been investigated.

In this work, experimental compression tests on rectangular specimens cut from one agglomerate cork slab have been carried out. The tests have been performed both using a quasi-static testing machine and a direct Split Hopkinson Tension-Compression Bar (SHTCB). Images captured by a digital camera and a fast cam have been used for a Digital Image Correlation (DIC) to extract strain and acceleration distributions on the specimen surface and evaluate the stress due to inertia effect.

36.2 Experimental Procedure

Samples, whose shape is shown in Fig. 36.1, have been cut from a slab of agglomerate cork, that for sake of simplicity hereinafter is referred to as "*Cork*".

The specimens have an approximately square cross section of 12 × 12 mm, which is close to the largest square that can be inscribed into the 18 mm diameter Hopkinson bar. In order to vary the strain rate different length, 12 and 30 mm, have been used. The detailed samples dimension, density and dynamic conditions are reported in Table 36.1.

These samples have been tested in quasi-static and dynamic conditions by a standard electromechanical machine (model Zwick® Z050) and by an Direct Tension-Compression Hopkinson Bar developed in house [12]. An optical acquisition system described in [13] has been used. Due to the low mechanical impedance of cork a polymeric (PET) version of the bar was adopted, whose details on the viscoelastic properties and wave propagation analysis are discussed in [14].

During the quasi-static tests, pictures of one lateral surface of the specimens have been acquired by a digital camera (model Pixelink® B371F), collecting about 500 images, while the force was measured by the 5kN load cell of the machine.

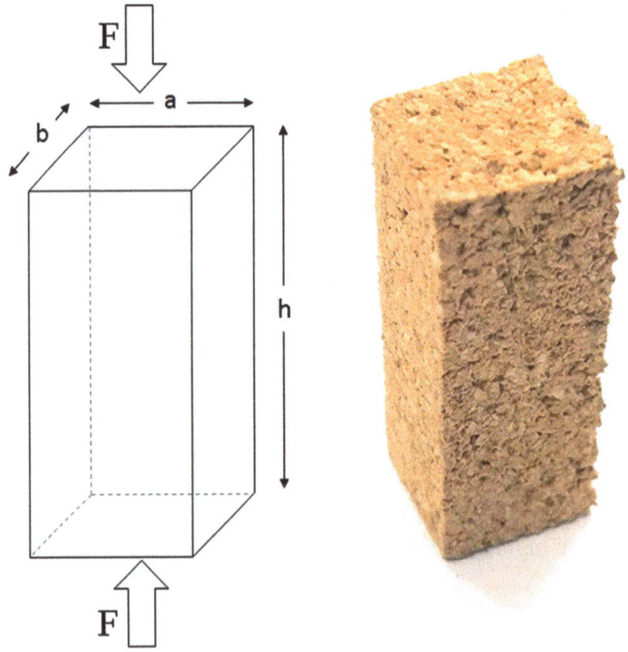

Fig. 36.1 Cork sample

Table 36.1 Sample data for quasi-static and dynamic tests

Sample #ID	a [mm]	b [mm]	h [mm]	Density [Kg/m^3]	Speed [m/s]	Avg rate [1/s]
QS_1	10.9	12.1	30.1	201.0	0.03	0.001
Dyn_1	11.9	11.9	30.3	213.9	3.0	90
Dyn_2	12.10	12.2	30.3	206.9	12.0	400

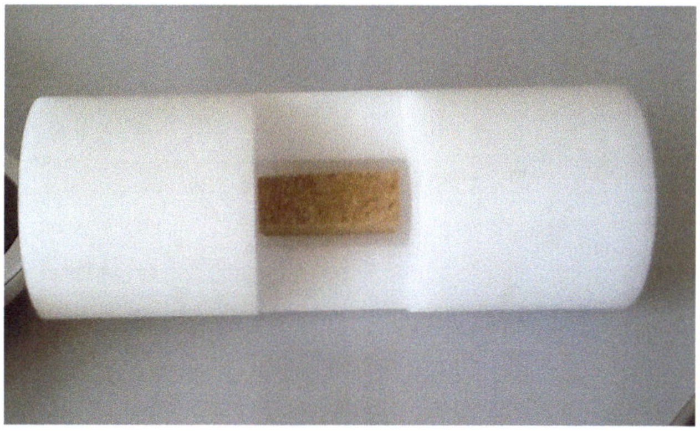

Fig. 36.2 Windowed support at Bar/Specimen interface for the optical access

The adopted Hopkinson Bar, the compression wave is generated by pre-tensioned part of the input bar until a sacrificial disc (here made of glass fiber reinforced polyamide) fails in shear, determining the fast release of the pulling force. The generated compression wave, travels along the input bar and reaches the specimen, which is sandwiched between the input and transmitted bar; then, the wave is partially transmitted to the output bar, and partially reflected backwards in the input bar. These strain waves are measured by strain gauges placed at half length of the input bar and at the beginning of the transmitted bar. The main data of the bars are: diameter 18 mm, pretension bar length: 3 m, input bar length: 7.5 m, transmitted bar length: 4.0 m. The input pulse duration is 4 ms. The optical access to the specimen is guaranteed by the support shown in Fig. 36.2.

During the dynamic tests, one lateral surface of the samples has been framed by a high speed camera (model Photron® SA4) with 192×128 pixel resolution at 100 kpfs.

The signals acquired by the strain gauges are post-processed in the frequency domain; the forces $\tilde{F}(\omega)$ at the bars-specimen interfaces are computed as:

$$\begin{cases} \tilde{F}_I(\omega) = \dfrac{\rho_b A_b \omega^2}{\gamma^2(\omega)} \left[\tilde{\varepsilon}_I(\omega) + \tilde{\varepsilon}_R(\omega)\right] \\ \tilde{F}_T(\omega) = \dfrac{\rho_b A_b \omega^2}{\gamma^2(\omega)} \tilde{\varepsilon}_T(\omega) \end{cases} \tag{36.1}$$

where A_b is the cross section area of the bars, and the subscripts *I* and *T* refer to the input and transmitted bar extremity, respectively. The velocity at the bars-specimen interfaces are computed as:

$$\begin{cases} \tilde{V}_I(\omega) = \dfrac{i\omega}{\gamma(\omega)} \left[\tilde{\epsilon}_I(\omega) - \tilde{\epsilon}_R(\omega)\right] \\ \tilde{V}_T(\omega) = \dfrac{i\omega}{\gamma(\omega)} \tilde{\epsilon}_T(\omega) \end{cases} \tag{36.2}$$

In Eqs. (36.1 and 36.2), $\gamma(\omega)$is the wave propagation coefficient given by $\sqrt{-\rho_b\omega^2/\tilde{E}_b(i\omega)}$.

The pictures acquired during the quasi-static and dynamic tests have been post-processed by an in-house developed DIC software. The surface heterogeneity has been used itself as a speckle pattern, without necessity of any paint spray. The displacements of all nodes of the grid are obtained by minimizing the correlation error computed all over the current frame with respect to the reference one [15–17].

Two typical pictures recorded by the two types of camera with the subsets discretization (60x60 and 12x12 pixels for the quasi-static and dynamic tests) used for the image analysis are shown in Fig. 36.3. In Fig. 36.4 the calculated axial strain, at an intermediate instant of test, is reported.

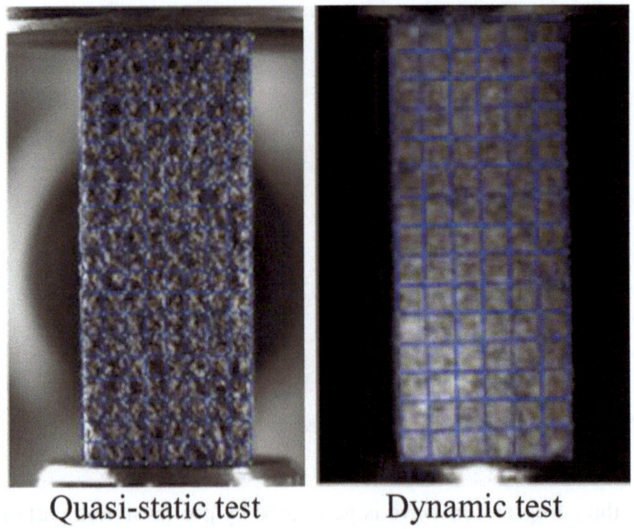

Fig. 36.3 Samples with grid for quasi-static and dynamic test

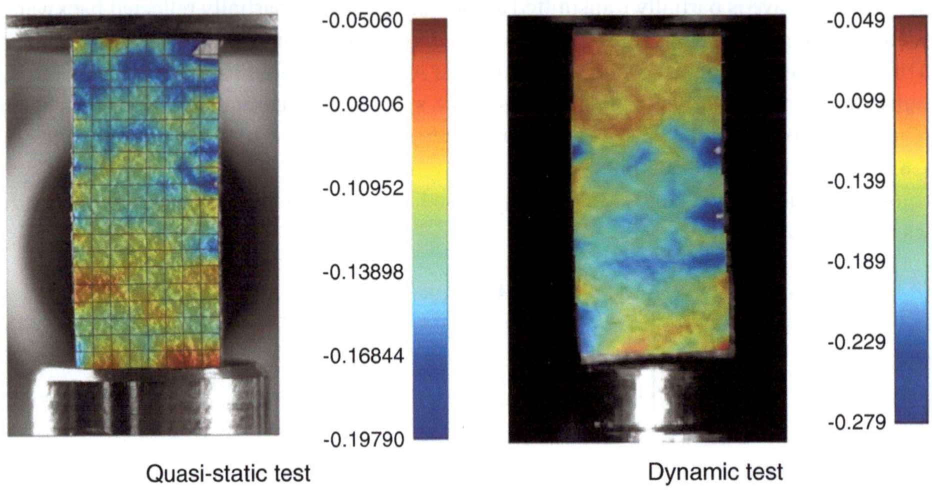

Fig. 36.4 Axial strain distribution maps

36.3 Inertia Stress Evaluation

The DIC analysis provided the strain and displacements distribution maps of each test. An example of temporal histories of displacements and strain of a dynamic test (#ID Dyn_2) is shown in Fig. 36.5a and b. Each curve in different colour refers to the average result of different rows of the grid nodes.

The acceleration is obtained from double temporal differentiation of the displacement history of each node of the subsets grid, and can be put into relation with the inertia stress. As depicted in Fig. 36.6a for a generic time instant t, a slice of thickness Δy experiences a stress difference which equals its inertia stress $\Delta\sigma_i$, according to the relation:

$$\sigma(y+\Delta y,t)=\sigma(y,t)+\Delta\sigma_i(y,t)=\sigma(y,t)-\rho(y,t)\,a(y,t)\,\Delta y \tag{36.3}$$

where ρ is the density and a is the axial acceleration, that cannot be considered constant here. Accordingly to [10], assuming zero Poisson's ratio and integrating (36.3) from a generic y^* coordinate to $y = L$ (the specimen interface with the output bar), the temporal and spatial distribution of the inertia stresses within the specimen are calculated by Eq. (36.4).

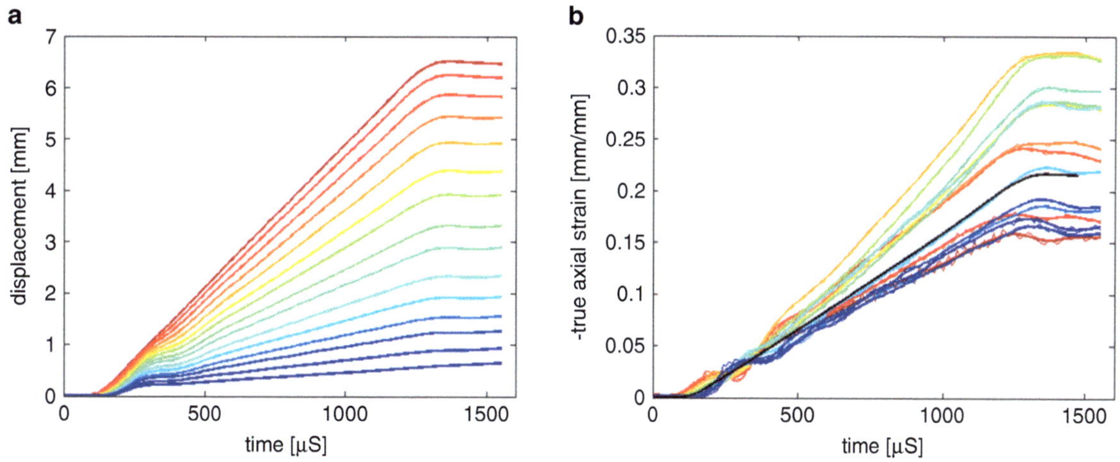

Fig. 36.5 Temporal histories of displacements and strain in a dynamic test

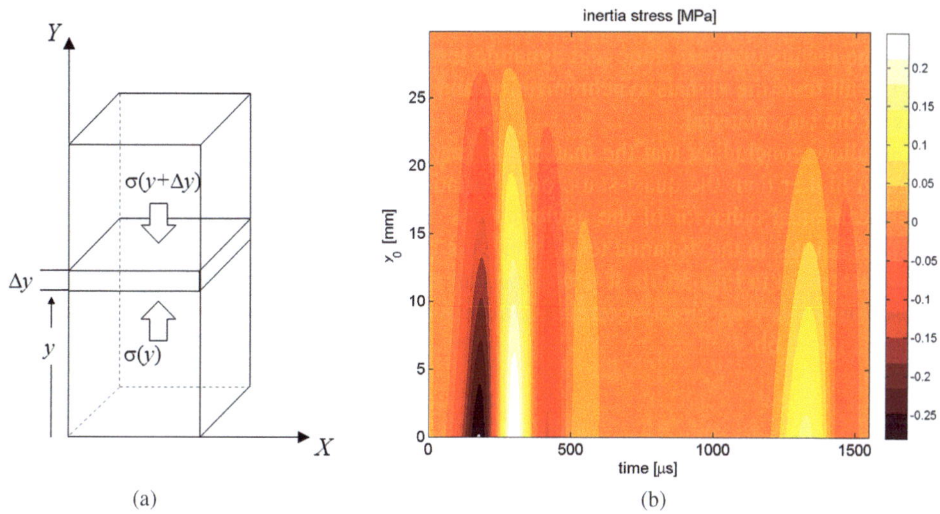

Fig. 36.6 (**a**) Scheme for inertia stress evaluation, (**b**) inertia stress distribution

$$\sigma\left(y^{*}\right)=\sigma\left(L,t\right)-\rho_{0}\int_{y=y_{*}}^{y=L}a\left(y,t\right)\exp\left[-\epsilon_{y}\left(y,t\right)\right]dy \tag{36.4}$$

When the integral in Eq. (36.4) is small, the specimen is in dynamic equilibrium and the usual formulas used in the SHPB analyses provide accurate stress measure, i.e., the stress can be computed by the only transmitted force. Conversely, when the integral is large, the stress in each transverse section of the specimen should be corrected accounting for inertia effect using Eq. (36.3).

36.4 Results

An example of the spatial-temporal evolution of inertia stress, computed applying Eq. (36.4) to the experimental results of specimen ID# Dyn_2, is given in Fig. 36.6b. It seen that the maximum value of inertia stress is about 0.25 MPa. Even if this value is apparently small, actually it is definitely not negligible considering that the cork has a low strength; moreover, the inertia stress peak takes place in the first part of the test (i.e. after 300 μs), when the axial strain and stress are still very small.

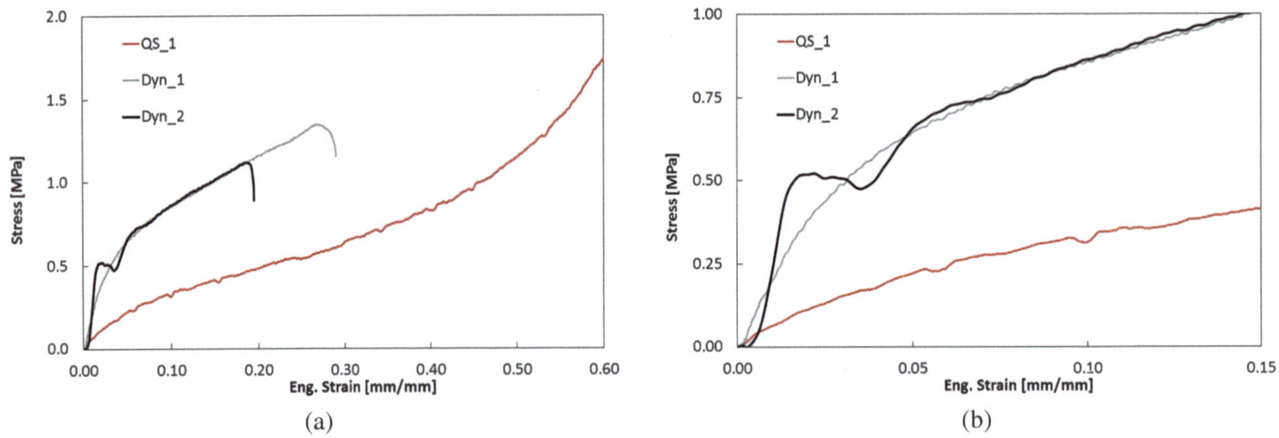

Fig. 36.7 Dynamic Eng stress–strain curves; comparison with the quasi-static response. (**a**) global behavior, (**b**) enlargement of the small strain region

Figure 36.7 shows the results of quasi-static and dynamic tests in term of engineering stress-strain curves for the tested agglomerate corks. For all tests the signals synchronization and treatment has been accomplished taking into account the viscoelastic behavior of the bars material.

These observations allow concluding that the material is very sensitive to load application rate, since the curves of the dynamic tests are much higher than the quasi-static one. Variation of strain rate, for the dynamic range considered, have an influence on the mechanical behavior of the agglomerates. However, no significant influence is observed within the investigated range of strain rates in the dynamic tests (from 90 s^{-1} to 400 s^{-1}).

Considering the enlargement in Fig. 36.7b, it is observed an anomalous fluctuation of the stress curve related to the test ID# Dyn_2. This oscillations is then clearly correlated, both for intensity and instant of occurrence, to the inertia stress represented in previous Fig. 36.6b.

36.5 Conclusions

In this work, quasi static and compression tests have been carried out on parallelepiped samples made of agglomerated cork. The global response highlighted a relevant difference in material strength when passing from slow to fast loading rate, even if no significant difference in strain rate sensitivity is observed between the dynamic tests. The tests have been accompanied by DIC analysis of the specimen surfaces, which showed that the deformation fields are not homogeneous. Moreover, from temporal differentiation of the reconstructed displacements, it was possible to evaluate the acceleration and, hence, the inertia stress distribution within the samples; it was found that the amplitude of the inertia stress is comparable to the fluctuations observed in the stress-strain curve related to the fastest test.

References

1. Ordovas, J., Ortega, M.: Characteristics of internal porosity of cork container media. Hortscience. **31**(7), 1177–1179 (1996)
2. Gibson, L., Easterling, K., Ashby, M.: The structure and mechanics of cork. Proc. R. Soc. Lond A. **377**, 99–117 (1981)
3. Fortes, M.: The Poisson effect in cork. Mater. Sci. Eng A. **122**(2), 227–232 (1989)
4. Song, B., Chen, W.: Split Hopkinson Kolsky Bar: Design, Testing and Applications. Springer, New York (2010)
5. Deshpande, V., Fleck, N.: High strain rate compressive behaviour of aluminium alloy foams. Int. J. Impact. Eng. **24**, 277–298 (2000)
6. Campana, F., Mancini, E., Pilone, D., Sasso, M.: Strain rate and density-dependent strength of AlSi7 alloy foams. Mater. Sci. Eng. A. **651**, 657–667 (2016)
7. Song, B., Chen, W.: One-dimensional dynamic compressive behavior of EPDM rubber. Trans. ASME J. Eng. Mater. Technol. **125**, 294–301 (2003)
8. Antonelli, M.G., Lonzi, B., Mancini, E., Martarelli, M., Sasso, M.: Dynamic testing and constitutive modelling of NBR rubbers. Conf. Proc. Soc. Exp. Mech. Ser. **2**, 187–193 (2017). doi:10.1007/978-3-319-41543-7_24

9. Wang, P., Xu, S., Li, Z., Yang, J., Zhang, C.: Experimental investigation on the strain-rate effect and inertia effect of closed-cell aluminum foam subjected to dynamic loading. Mater. Sci. Eng. A. **620**, 253–261 (2015)
10. Koohbor, B., Kidane, A., Lu, W.-Y., Sutton, M.: Investigation of the dynamic stress–strain response of compressible polymeric foam using a non-parametric analysis. Int. J. Impact Eng. **91**, 170–182 (2016)
11. Gameiro, C., Cirne, J., Gary, G.: Experimental study of the quasi-static and dynamic behaviour of cork under compressive loading. J. Mater. Sci. **42**, 4316–4324 (2007)
12. Mancini, E., Sasso, M., Rossi, M., Chiappini, G., Newaz, G., Amodio, D.: Design of an innovative system for wave generation in direct tension–compression split Hopkinson bar. J. Dyn. Behav. Mater. **1**(2), 201–213 (2015)
13. Sasso, M., Fardmoshiri, M., Mancini, E., Rossi, M., Cortese, L.: High speed imaging for material parameters calibration at high strain rate. Eur. Phys. J. Spec. Top. **224**, 1–14 (2015)
14. Sasso, M., Antonelli, M., Mancini, E., Radoni, M., Amodio, D.: Experimental and numerical characterization of a polymeric Hopkinson bar by DTMA. Int. J. Impact Eng. **103**, 50–63 (2017)
15. Gao, Z., Wang, Y., Gioia, G., Cuitino, A.: A Global Approach for Digital Speckle Correlation. In Proc. of SEM Annual Conference and Exposition, Milwaukee (WI) (2002)
16. Amodio, D., Broggiato, G., Campana, F., Newaz, G.: Digital speckle correlation for strain measurement by image analysis. Exp. Mech. **43**(4), 396–402 (2003)
17. Sasso, M., Chiappini, G., Rossi, M., Cortese, L., Mancini, E.: Visco-hyper-pseudo-elastic characterization of a Fluoro-silicone rubber. Exp. Mech. **54**(3), 315–328 (2014)

Marco Sasso Ph.D. in Mechanical Engineering in 2004.
Assistant Professor in 2007.
Associate Professor since 2016.
Research topics:

- Optical methods.
- Dynamic behavior of materials.

Chapter 37
High Strain Rate Induced Phase Transition of Polymer

G. Montella and C. M. Roland

Abstract In this study we report on the uniaxial compression stress-strain behavior of a representative polymer at a strain rate over which the material undergoes a viscoelastic phase transition. This means that the frequency of its segmental dynamics correspond to the loading frequency. This resonance condition results in large energy absorption, accounting for much of the enhanced impact resistance of polymer coated structures. The transition in the stress-strain behavior as strain rate increase is hard to characterize due to experimental limitations in accessing the full range of strain rates; herein a polymer with a higher Tg is used to overcome this issue. The temperature history of the specimen has been monitored during the experiment duration in order to quantify the temperature rise due to adiabatic conditions during loading. That is, we want to address the question of where does the energy go when a polymer-coated hard substrate is impact loaded.

Keywords High strain rate tests • Glass transition • Polymers • Energy dissipation • Temperature measurements

37.1 Experimental Procedure

1,2 polybutadiene (PBD) disks with average diameter of 8 mm and thickness of 1 mm were punched out of a plate. A T-type (copper-constantan) thermocouple was placed between two PBD disks, which were then pressed together. Following this procedure the thermocouple junction falls at the center of a 2 mm thick and 8 mm diameter specimen. The thermocouple wire was 0.075 mm in diameter and its output was recorded using an oscilloscope. According to Rabin and Rittel [1] the thermal diffusivity of the thermocouple is required to be at least one order of magnitude higher than that of the measured body in order to obtain useful results, this assumption is particularly valid for polymers. Here the absence of time lag in the thermocouple signal confirms the validity of the temperature measurements.

The specimens described above were loaded by means of a standard Kolsky apparatus [2] where bars are made of 6061-T6 aluminum with a diameter of 15.9 mm and a specific acoustic impedance measured to be to 16.9 $\pm$ 1 MRayl at 1 MHz. The incident and transmission bars had a common length of 1830 mm; the striker bar was 304 mm long. An annealed copper disk was employed to shape the incident pulse and allow a more gradual rise in the applied stress. The sample geometry is chosen to minimize inertial effects and friction between the sample and bars. Silicone lubricant was applied to the faces to ensure slippage and no bulging of the sample was observed during high speed camera recording of the test. Comparison of the forces applied on both sides of the specimen showed in each case that the assumption of homogeneous deformations was reasonable. In addition to the above setup a controlled temperature chamber was built and fitted to the Kolsky bar to allow to test the material at different temperature (Fig. 37.1).

G. Montella (✉) • C.M. Roland
Naval Research Laboratory, Chemistry Division, 4555 Overlook Ave., SW, Washington, DC 20375, USA
e-mail: giuseppe.montella.ctr.it@nrl.navy.mil

J. Kimberley et al. (eds.), *Dynamic Behavior of Materials, Volume 1*, Conference Proceedings of the Society for Experimental Mechanics Series, DOI 10.1007/978-3-319-62956-8_37

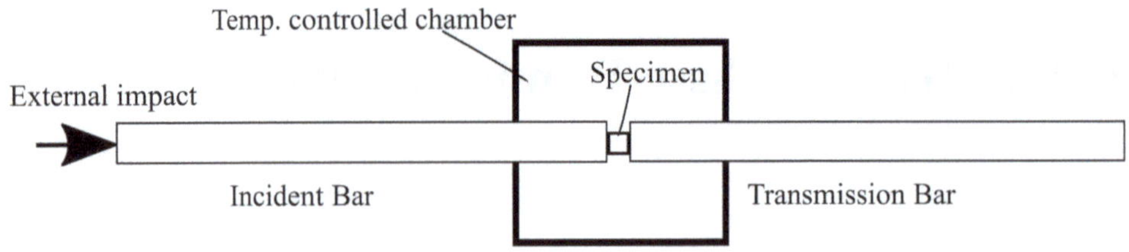

Fig. 37.1 Kolsky bar setup with a temperature controlled chamber

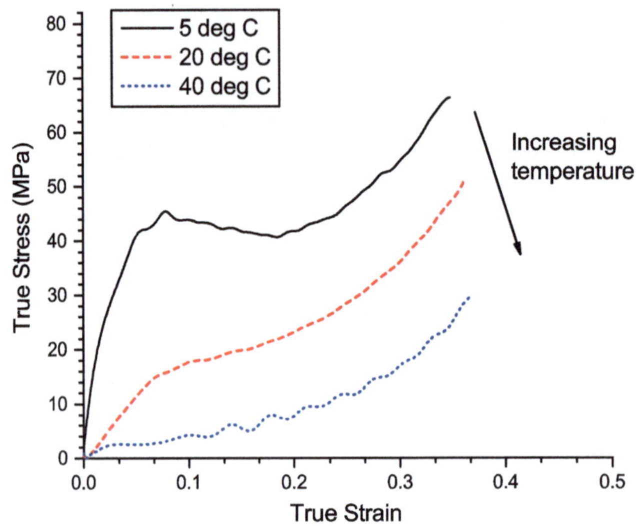

Fig. 37.2 Stress vs Strain curves for 1, 2 PBD at different temperature at 3500 s^{-1}

37.2 Results and Discussion

Performing measurements of polymer behavior at high rates can be challenging, and the very properties which make them useful contribute to this challenge. Over a given range of temperatures and rates, a single polymer may exhibit a wide range of mechanical responses, from rubbery to ductile plastic to brittle.

The shape of the stress–strain curve for the PBD at different temperatures is shown in Fig. 37.2, it exhibits an "N" shape with a temperature dependent yield stress, post yielding strain softening followed by post yield strain hardening. The yield strength increases with decreasing temperature.

Due to the short time duration of the tests (ca. 200 μs) the heat loss due to convection is negligible and adiabatic conditions can be considered [3]. The adiabatic temperature changes for the measurements are shown in Fig. 37.3. The temperature increase at 5 °C was ca. 7 °C, which is about five times larger than the temperature rise (≅1.5 °C) at the same strain rate at 40 °C. This is mainly connected to the greater dissipation due to the dramatically higher yield stress at 5 °C (45 MPa) compared to the yield stress (2.5 MPa) at 40 °C.

Interestingly to see that the largest temperature increase (16 °C) was recorded during the tests at 20 °C. This is due to the fact that the polymer used in this study has a Tg of about 1.5 °C, due to its high vinyl content, for this reason the segmental dynamics at 20 °C is in the same order magnitude of the strain rate during the tests. As expected, in this case, other dissipation mechanisms are activated [4]. It is well known that the glass transition zone is associated with the largest energy dissipation achievable with a rubbery polymer. Further studies on this matter will help understand the mechanism governing ballistic impact where glass transition phenomena may be the reason for enhanced impact resistance of polymer coated structures.

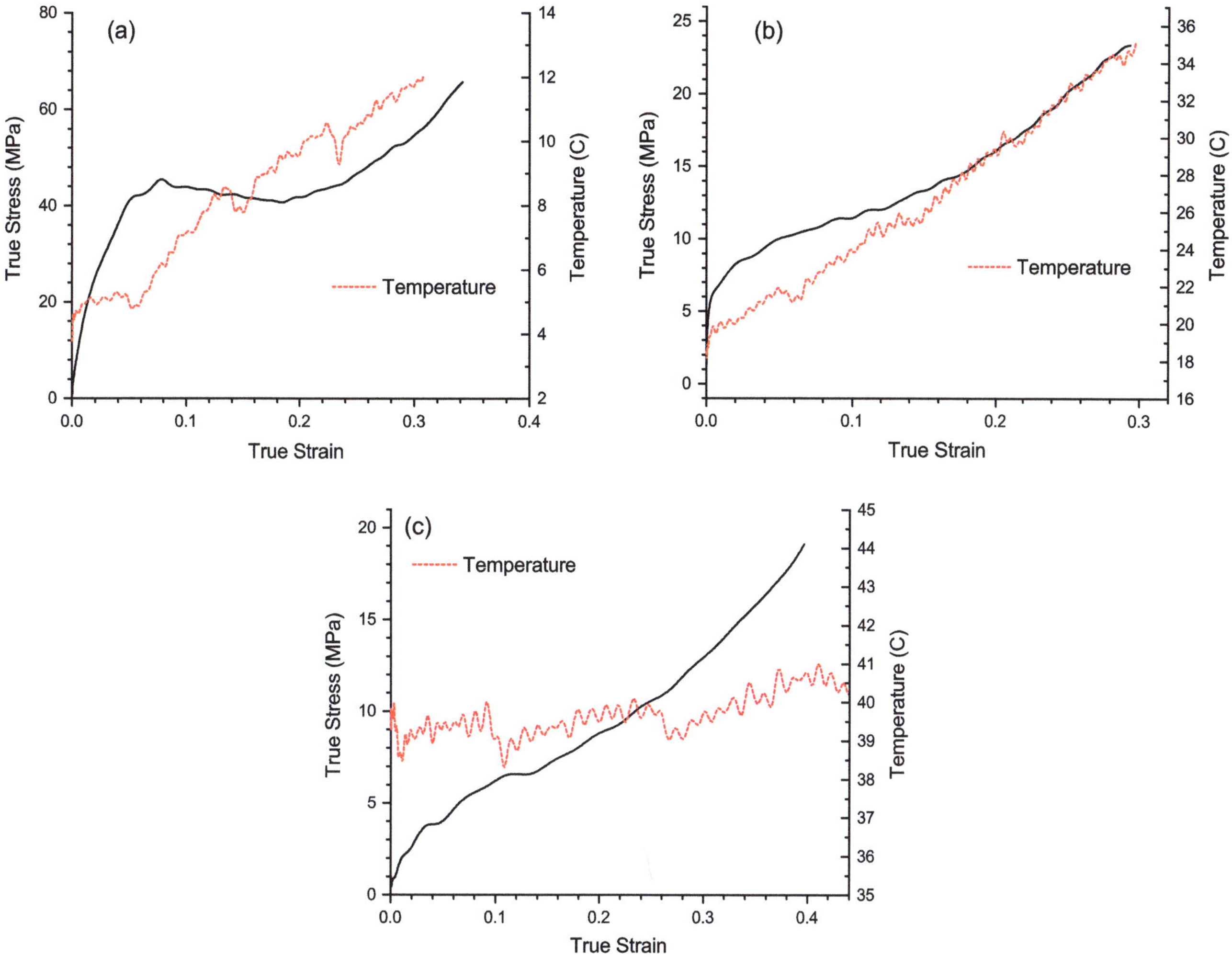

Fig. 37.3 True stress vs True strain and temperature time history during compression at 3500 s^{-1}. (**a**) 5°C, (**b**) 20°C, (**c**) 40°C

References

1. Rabin, Y., Rittel, D.: A model for the time response of solid-embedded thermocouples. Exp. Mech. **39**(2), 132–136 (1999)
2. Kolsky, H.: An investigation of the mechanical properties of materials at very high rates of strain. Proc. Phys. Soc. Sect. B. **62**, 676 (1949)
3. Mott, P.H., Giller, C.B., Fragiadakis, D., Rosenberg, D.A., Roland, C.M.: Deformation of polyurea: where does the energy go? Polymer. **105**, 227–233 (2016)
4. Garg, M., Mulliken, A.D., Boyce, M.: Temperature rise in polymeric materials during high rate deformation. J. Appl. Mech. **75**, 011009, 8 p (2008)

G. Montella I am a postdoc at the Naval Research Laboratory in the polymer physics department. I received my PhD in mechanical engineering from the University of Naples in Italy in 2015.

MIX
Papier aus verantwortungsvollen Quellen
Paper from responsible sources
FSC® C105338

If you have any concerns about our products,
you can contact us on
ProductSafety@springernature.com

In case Publisher is established outside the EU,
the EU authorized representative is:
Springer Nature Customer Service Center GmbH
Europaplatz 3, 69115 Heidelberg, Germany

Printed by Libri Plureos GmbH
in Hamburg, Germany